Recent Developments in Automatic Control Systems

RIVER PUBLISHERS SERIES IN AUTOMATION, CONTROL AND ROBOTICS

Series Editors:

ISHWAR K. SETHI
Oakland University, USA

TAREK SOBH
University of Bridgeport, USA

FENG QIAO
Shenyang JianZhu University, China

The "River Publishers Series in Automation, Control and Robotics" is a series of comprehensive academic and professional books which focus on the theory and applications of automation, control and robotics. The series focuses on topics ranging from the theory and use of control systems, automation engineering, robotics and intelligent machines.

Books published in the series include research monographs, edited volumes, handbooks and textbooks. The books provide professionals, researchers, educators, and advanced students in the field with an invaluable insight into the latest research and developments.

Topics covered in the series include, but are by no means restricted to the following:

- Robots and Intelligent Machines
- Robotics
- Control Systems
- Control Theory
- Automation Engineering

For a list of other books in this series, visit www.riverpublishers.com

Recent Developments in Automatic Control Systems

Editors

Yuriy P. Kondratenko

Petro Mohyla Black Sea National University of the Ministry of Education and Science of Ukraine, Ukraine

Vsevolod M. Kuntsevich

Space Research Institute of the National Academy of Sciences of Ukraine and State Space Agency of Ukraine, Ukraine

Arkadii A. Chikrii

Glushkov Institute of Cybernetics of the National Academy of Sciences of Ukraine, Ukraine

Vyacheslav F. Gubarev

Space Research Institute of the National Academy of Sciences of Ukraine and State Space Agency of Ukraine, Ukraine

River Publishers

Routledge
Taylor & Francis Group

LONDON AND NEW YORK

Published 2022 by River Publishers
River Publishers
Alsbjergvej 10, 9260 Gistrup, Denmark
www.riverpublishers.com

Distributed exclusively by Routledge
4 Park Square, Milton Park, Abingdon, Oxon OX14 4RN
605 Third Avenue, New York, NY 10017, USA

Recent Developments in Automatic Control Systems / by Yuriy P. Kondratenko, Vsevolod M. Kuntsevich, Arkadii A. Chikrii.

Routledge is an imprint of the Taylor & Francis Group, an informa business

ISBN 978-87-7022-674-5 (print)
ISBN 978-10-0079-560-8 (online)
ISBN 978-10-0333-922-9 (ebook master)

While every effort is made to provide dependable information, the publisher, authors, and editors cannot be held responsible for any errors or omissions.

Contents

Preface

This book provides an overview of the recent developments in modern control systems including new theoretical findings and successful examples of practical implementation of the control theory in different areas of industrial and special applications.

The monograph provides an overview of the recent developments in modern control systems including new theoretical findings and successful examples of practical implementation of the control theory in different areas of industrial and special applications.

This monograph consists of research-oriented chapters presented by invited well-known scientists from Poland, Norway, Ukraine, United States of America and Uzbekistan.

The monograph is divided into three main parts: (a) Advances in Theoretical Research of Control Systems; (b) Advances in Control Systems Application; and (c) Recent Developments in Collaborative Automation.

The *first part*, "Advances in Theoretical Research on Automatic Control", includes seven contributions:

The chapter "*Control of Moving Object Groups in Conflict Situation*", by Chikrii A., presents an overview of the methods for analysis of conflict situations involving groups of controlled objects on each of the counteracting sides. The principle of interval decomposition, proposed in the paper, incorporates solving typical problems of the target distribution and the problems of group and successive pursuit. To attack the latter, the method of resolving functions and Krasovskii's extremal aiming rule are used. The method of resolving functions makes it feasible to describe the situation of encirclement of an evader by a group of pursuers, as well as the pursuit under state constraints.

In "*Applications of Variational Analysis to Controlled Sweeping Processes*", B. Mordyukhovich discusses recent developments and applications of variational analysis and generalized differentiation to rather new classes of optimal control problems governed by discontinuous differential inclusions that arise in the so-called sweeping processes and the like. Such problems are particularly important for numerous applications to automatics control systems, engineering design, hysteresis, etc. To study highly challenging control problems of this type, the author develops appropriate constructions of the method of discrete approximations, which allows the investigation of various theoretical and numerical aspects of optimal control for discontinuous differential systems.

L. Zhiteckii and K. Solovchuk in the chapter *"Robust and Robustly-Adaptive Control of Some Noninvertible Memoryless Systems"* discuss the robust control of some uncertain multivariable memoryless (static) discrete-time systems in the presence of arbitrary unmeasurable bounded disturbances, whose bounds may be unknown, in general. The non-adaptive robust control approach applicable to the case when the lengths of these intervals are relatively short is proposed. The robustly-adaptive control utilizing a modification of the inverse-model concept is designed to cope with large parametric uncertainty. The asymptotic properties of both non-adaptive and adaptive feedback control systems are derived. To support the theoretic study, numerical examples, and simulation results are given.

The chapter *"Nonlinear Integral Inequalities and Differential Games of Avoiding Encounter "*, by L.P. Yugay, presents a survey of the application of integral inequalities in various theoretical branches of mathematics and their applied fields. New integral inequalities of the Hölder type are introduced and corresponding estimates are proved. Also, the possibilities of applications of proven results in nonlinear differential games of avoiding encounters are indicated.

In *"Principle of time stretching for motion control in condition of conflict"*, G.Ts. Chikrii discusses the linear differential game of approaching a cylindrical terminal set. The research is done in the frames of the Method of Resolving Functions. The gist of the method consists of constructing some scalar functions associated with the parameters of a conflict-controlled process and characterizing the gain of the first player at each moment. If the total gain achieves the predetermined value then this means that the game terminates in a certain guaranteed time. In so doing, the pursuer's control, realizing the game goal, is constructed based on the Filippov-Castaing theorem on measurable choice.

L.V. Baranovska in the chapter *"Method of Upper and Lower Resolving Functions for Pursuit Differential-Difference Games with Pure Delay"* considers a modification of the method of resolving functions for differential-difference games with pure delay is considered. The lower and upper resolving functions. For the local game of approach, the scheme of a method is constructed, and sufficient conditions for the completion of the game are formulated.

The chapter "*Adaptive method for variational inequality over the set of solutions of equilibrium problem*", by Vedel Y.I., Denisov S.V. and Semenov V.V., considers a two-level problem: a variational inequality over the set of solutions of the equilibrium problem. An example of such a problem is the search for the normal Nash equilibrium. To solve this problem, the authors propose an iterative algorithm that combines the ideas of a two-stage proximal method, adaptability, and iterative regularization. In contrast to the previously applied rules for choosing the step size, the proposed algorithm does not calculate the values of the bifunction at additional points; it does not require knowledge of information about the Lipschitz constants of the bifunction, the Lipschitz constant and strong monotonicity constant of the operator. For monotone bifunctions of Lipschitz type and strongly monotone Lipschitz continuous operators, a theorem on the strong convergence of the algorithm is proved. It is shown that the proposed algorithm applies to monotone two-level variational inequalities in Hilbert spaces.

The *second part*, "Advances in Control Systems Applications", includes five contributions:

Gubarev V., Salnikov N. and Melnychuk S. in the chapter "*Identification of complex systems in the class of linear regression models*" discuss a general approach to identification within the regression model class, which searches for approximate solutions that are consistent in accuracy with errors in the data. Since identification problems in complex systems are often incorrectly posed, the proposed method includes regularization that provides practically suitable solutions.

In "*Fuzzy Systems Design: Optimal Selection of Linguistic Terms Number*", Kozlov O., Kondratenko Y., Skakodub O., and Gomolka Z. focus on the development and research of the advanced information technology for fuzzy systems design and structural optimization with optimal selection of linguistic terms number. The application of the proposed information technology allows for increasing the fuzzy system's performance and accuracy, reducing the computational costs spent on the rule base composing and parameters optimization, as well as simplifying its further hardware and software implementation. The effectiveness study of the developed information technology is carried out on a specific example, in particular, when designing and optimizing the fuzzy control system for the quadrotor unmanned aerial vehicle. The obtained results of the conducted research confirm the high effectiveness of the proposed information technology and the feasibility of its use at creating various types of fuzzy control and decision-making systems.

The chapter *"Analysis of the Dynamics and Controllability of Autonomous Mobile Robot with a Manipulator"*, by Ashchepkova N., Zbrutsky A. and Koshevoy N., presents the results of the study of the dynamics and control of an autonomous mobile robot with a manipulator. An all-wheel-drive model with a four-wheeled chassis layout and equipped with a four-link manipulator is considered. The design of the manipulator consists of a docking ring rotating around a vertical axis and three-rod links of the arm, connected by rotary kinematic pairs of the fifth class. The mathematical model of controlled movement of an autonomous mobile robot with a manipulator is proposed. With the relative motion of the manipulator, the tensor of inertia of an autonomous mobile robot in the coordinate system associated with the chassis is non-diagonality and non-stationarity. Simplifications of the mathematical model for several driving modes are substantiated. An algorithm for integrating the equations of dynamics is developed, taking into account certain features. Based on the results of mathematical modeling, an algorithm for adaptive control of an autonomous mobile robot with a manipulator is proposed.

Kondratenko Y., Roshanineshat A., and Simon D. in the chapter *"Safe Navigation of an Autonomous Robot in Dynamic and Unknown Environments"* focus on the problem of fuzzy control system design for mobile robots operating in dynamic environments with obstacles. The decision-making method, based on the transformation of radar-sensor information to fuzzy set "Obstacles" and the robot's heading course to fuzzy set "Direction to Target Point" is considered. Different scenarios of the working environment (with static and dynamic obstacles) and the adjusted speed of the autonomous robot as well as peculiarities of "robot-obstacle" interaction are discussed in detail. The simulation results confirm the efficiency and safety of the robot's autonomous navigation in dynamic, a priori unknown environments.

The chapter *"Algorithmic Procedures Synthesis of Robust-Optimal Control for Moving Objects"*, by Timchenko V.L. and Lebedev D.O., considers the control of transient processes of a wide class of moving objects described by ordinary nonlinear differential equations with robust-optimal systems and variable structure. The solution to this problem is suggested using general algorithmic procedure for constructing optimal trajectories, determining switching moments, and synthesizing control functions for multidimensional systems. The control of the mismatch between the trajectory of a physical object and the optimal model calculated trajectory allows to take into account the values of the optimal control and to form a robust subsystem, which provides invariance to incomplete information about the moving object. The simulation of a quadrotor UAV stabilization and a sea vessel maneuvering under conditions of external disturbances and parametric noise demonstrates the required control accuracy in the vicinity of the formed optimal trajectory of controlled coordinates. The creation of algorithms and circuit solutions for the synthesis of robust-optimal variable-structure systems for multidimensional nonlinear moving objects has practical significance as a basis for formation of automated procedures for the synthesis of robust-optimal systems and development of software tools for various types of moving object control systems.

The **third part**, "Recent Developments in Collaborative Automation", consists of five contributions:

In "*Modeling of Cyber–Physical Systems*", Holovatenko I. and Pysarenko A. present the current research in the field of constructing cyber-physical systems. For validation of the designed system, authors came up with a live scenario from the logistical systems: the autonomous object delivers goods from one warehouse to another taking into account obstacles along the way. The system should proactively react and recalculate the optimal path. This will lead to the timely delivery of goods to the destination.

Atamanyuk I., Kondratenko Y. and Solesvik M. in the chapter "*Reliability Control of the Technical Systems based on Canonical Decomposition of Random Sequences*" consider a method for determining the suitability of a technical object for further operation based on the information about the current state and the history of its functioning during an arbitrary period of time. The check is carried out based on the analysis of the state of the object under study at future points in time and the value of the posterior probability of no-failure operation. The method for predicting the individual reliability of objects is based on the canonical decomposition of a random sequence describing the change in the value of the controlled parameter over time. As an estimate of the future state, the conditional mathematical expectation is used; to estimate the probability of a no-failure operation, statistical modeling of a random sequence beyond the observation area is performed. The method can also be used to control the reliability of the objects that are characterized by many parameters.

The chapter "*Petunin Ellipsoids in Automatic Control Systems Design*", by Klyushin D.A., Lyashko S.I., and Tymoshenko A.A., proposes a new way of describing the optimization domain with linear constraints in the design space using Petunin ellipsoids. The statistical properties of these ellipsoids are given, making them an effective tool for describing data in the design space. Authors consider a new distribution-free approach to computing ellipsoidal conformal prediction set based on Hill's assumption and the Petunin ellipsoids. The resulting prediction set is a Petunin ellipsoid with a precise confidence level depending on the number of points only.

In *"On Real-Time Calculation of the Rejected Takeoff Distance"*, Vyshenskyy V.I., Belousov A.A., and Kuleshyn V.V. focus on a rejected takeoff in the situation when it is decided to cancel the takeoff of an airplane. The goal of the authors' investigation is to obtain real-time estimations of the distance to a full stop. The proposed real-time algorithm for the calculation of the distance of the interrupted flight uses the step-wise interpolation of the thrust. Using this interpolation, the Riccati equation of the aircraft motion has explicit solutions. Besides using known time intervals of interpolation, the time to reach a given speed can be explicitly calculated. To check and study the proposed algorithm for estimating the rejected takeoff distance in real-time, a computer simulation system was created.

Romanenko V., Miliavskyi Y., and Kantsedal G. in the chapter *"Automated Control Problem for Dynamic Processes Applied to Cryptocurrency in Financial Markets"* develop a cognitive map (CM) of the use of cryptocurrency in the financial market and describe the dynamic model of CM impulse processes as a difference equation system (Roberts equations). The authors chose an external control vector for the CM impulse process provided through CM nodes varying. A closed-loop CM impulse process control system was implemented. This system includes a multidimensional discrete controller, based on automated control theory methods, that generates a selected control vector and directly affects respective CM nodes by varying their coordinates. The authors solved three problems of a discrete controller design for automated dynamic process control applied to cryptocurrency in financial markets. The first problem is unstable cryptocurrency rate stabilization based on modal CM impulse process control. The second problem is constrained external and internal disturbances suppression during CM impulse processes control based on the invariant ellipsoid method. The third problem is minimizing the generalized variance of CM node coordinates and controls for stabilizing coordinates at given levels.

The chapters of the monograph have been structured to provide an easy-to-follow introduction to the topics that are addressed, including the most relevant references, so that anyone interested in this field can get started in the area.

This book may be useful for researchers and students who are interested in recent developments of modern control systems, robust adaptive systems, optimal control, fuzzy control, motion control, identification, modeling, differential games, evolutionary optimization, reliability control, security control, intelligent robotics, and cyber–physical systems.

We would like to express our deep appreciation to all authors for their contributions as well as to reviewers for their timely and interesting comments and suggestions. We certainly look forward to working with all contributors again in nearby future.

July 31, 2021

Editors:

Yuriy P. Kondratenko, Prof., Dr.Sc., Corresponding Academician of the Royal European Academy of Doctors – Barcelona 1914 (Spain), Petro Mohyla Black Sea National University, Ukraine

Vsevolod M. Kuntsevich, Prof., Dr.Sc., Academician of the National Academy of Sciences of Ukraine, Space Research Institute of the National Academy of Sciences and State Space Agency, Ukraine

Arkadii A. Chikrii, Prof., Dr.Sc., Academician of the National Academy of Sciences of Ukraine, Glushkov Institute of Cybernetics of the National Academy of Sciences, Ukraine

Vyacheslav F. Gubarev, Prof., Dr.Sc., Corresponding Member of the National Academy of Sciences of Ukraine, Space Research Institute of the National Academy of Sciences and State Space Agency, Ukraine

List of Figures

List of Tables

List of Contributors

Ashhepkova, Natalja, *Department of Mechanotronics, Oles Honchar Dnipro National University, Ave. Gagarin, 72, Dnipro, Ukraine, 49010; E-mail: ashchepkova.ftf.dnu@gmail.com, ORCID: http://orcid.org/0000 -0002-1870-1062*

Atamanyuk, I., *Warsaw University of Life Sciences, Nowoursynowska 166, 02-787 Warsaw, Poland and Mykolayiv National Agrarian University, Georgy Gongadze street, Mykolaiv, 9, 54020, Ukraine; E-mail:atamanyuk@mnau.edu.ua*

Baranovska, Lesia, *Institute of Applied System Analysis of the National Technical University of Ukraine "Igor Sikorsky Kyiv Polytechnic Institute", 37 Peremohy av., Kyiv, 03056, Ukraine; E-mail: lesia@baranovsky.org*

Belousov, A.A., *V.M. Glushkov Institute of Cybernetics of National Academy of Sciences of Ukraine, Kyiv, Ukraine; E-mail: belousov@nas.gov.ua*

Chikrii, A., *Glushkov Institute of Cybernetics of the National Academy of Sciences of Ukraine, 40 Glushkov Avenue, 03187, Kyiv, Ukraine; E-mail: g.chikrii@gmail.com*

Chikrii, G.Ts., *Glushkov Institute of Cybernetics of the National Academy of Sciences of Ukraine, 40 Glushkov Avenue, 03187, Kyiv-187, Ukraine; E-mail: g.chikrii@gmail.com*

Denisov, S., *Faculty of Computer Science and Cybernetics, Taras Shevchenko National University of Kyiv, Kyiv, Ukraine; E-mail: denisov.univ@gmail.com*

Gomolka, Zbigniew, *University of Rzeszow, Department of Computer Engineering, Rzeszow, Poland; E-mail: zgomolka@ur.edu.pl*

Gubarev, V., *Space Research Institute, National Academy of Sciences of Ukraine and State Space Agency of Ukraine, 40, b. 4/1 Glushkov av., Kyiv, 03187, Ukraine*

Holovatenko, Illya, *Department of Information Systems and Technologies, National Technical University of Ukraine "Igor Sikorsky Kyiv Polytechnic Institute", Kyiv, Ukraine; E-mail: illyaholovatenko@gmail.com*

Kantsedal, Heorhii, *Institute of Applied System Analysis of National Technical University of Ukraine "Igor Sikorsky Kyiv Polytechnic Institute", 37 Peremohy av., Kyiv, 03056, Ukraine; E-mail: romanenko.viktorroman@gmail.com*

Klyushin, D.A., *Taras Shevchenko National University of Kyiv, Prospekt Glushkova 4D, 03680, Kyiv, Ukraine; E-mail: dokmed5@gmail.com*

Kondratenko, Yuriy P., *Petro Mohyla Black Sea National University, Intelligent Information Systems Dept., Mykolayiv, 54003, Ukraine; E-mail: y_kondrat2002@yahoo.com*

Koshevoy, Nicolay, *Department of Intelligent Visual Systems and Engineering of Quality, National Aerospace University "Kharkov Aviation Institute", st. Chkalov, 17, Kharkov, Ukraine, 61070, E-mail: kafedraapi@ukr.net, ORCID: http:// orcid.org/0000-0001-9465-4467*

Kozlov, Oleksiy, *Petro Mohyla Black Sea National University, 10 68th Desantnykiv st., Mykolayiv, 54003, Ukraine; E-mail: kozlov_ov@ukr.net*

Kuleshyn, V.V., *V.M. Glushkov Institute of Cybernetics of National Academy of Sciences of Ukraine, Kyiv, Ukraine; E-mail: v.v.kuleshin@gmail.com*

Lebedev, D.O., *Department of Computerized Control Systems, National Shipbuilding University, Geroev Ukraine ave. 9, 54025, Mykolayiv, Ukraine; E-mail: dns19944@gmail.com*

Lyashko, S.I., *Taras Shevchenko National University of Kyiv, Prospekt Glushkova 4D 03680, Kyiv, Ukraine; E-mail: lyashko.serg@gmail.com*

Melnychuk, S., *Space Research Institute, National Academy of Sciences of Ukraine and State Space Agency of Ukraine, 40, b. 4/1 Glushkov av., Kyiv, 03187, Ukraine; E-mail: melnychuk89s@gmail.com*

Miliavskyi, Yurii, *Institute of Applied System Analysis of National Technical University of Ukraine "Igor Sikorsky Kyiv Polytechnic Institute", 37 Peremohy av., Kyiv, 03056, Ukraine; E-mail: yuriy.milyavsky@gmail.com*

Mordukhovich, Boris S., *Department of Mathematics, Wayne State University, Detroit, Michigan 48202, USA; E-mail: aa1086@wayne.edu*

Pysarenko, Andrii, *Department of Information Systems and Technologies, National Technical University of Ukraine "Igor Sikorsky Kyiv Polytechnic Institute", Kyiv, Ukraine; E-mail: andrew.pisarenko@gmail.com*

Romanenko, Viktor, *Institute of Applied System Analysis of National Technical University of Ukraine "Igor Sikorsky Kyiv Polytechnic Institute", 37 Peremohy av., Kyiv, 03056, Ukraine; E-mail: romanenko.viktorroman@gmail.com*

Roshanineshat, Arash, *The University of Arizona, Electrical & Computer Engineering Dept., Tucson, AZ, 85721, USA; E-mail: aroshanineshat@email.arizona.edu*

Salnikov, N., *Space Research Institute, National Academy of Sciences of Ukraine and State Space Agency of Ukraine, 40, b. 4/1 Glushkov av., Kyiv, 03187, Ukraine; E-mail: sergvik@ukr.net*

Semenov, V., *Faculty of Computer Science and Cybernetics, Taras Shevchenko National University of Kyiv, Kyiv, Ukraine; E-mail: semenov.volodya@gmail.com*

Simon, Dan, *Cleveland State University, Electrical Engineering and Computer Science Dept., Cleveland, OH, 44115, USA; E-mail: d.j.simon@csuohio.edu*

Skakodub, Oleksandr, *Petro Mohyla Black Sea National University, 10 68th Desantnykiv st., Mykolayiv, 54003, Ukraine; E-mail: aleksandrskakodub1996@gmail.com*

Solesvik, M., *Western Norway University of Applied Sciences, Inndalsveien 28, 5063 Bergen, Norway; E-mail: marina.solesvik@hvl.no*

Solovchuk, K., *Poltava Scientific Research Forensic Center of the MIA of Ukraine, Rybalskiy lane, 8, Poltava, 36011, Ukraine; E-mail: solovchuk_ok@ukr.net*

Timchenko, V.L., *Department of Computerized Control Systems, National Shipbuilding University, Geroev Ukraine ave. 9, 54025, Mykolayiv, Ukraine; E-mail: vl.timchenko58@gmail.com*

Tymoshenko, A.A., *Taras Shevchenko National University of Kyiv, Prospekt Glushkova 4D, 03680, Kyiv, Ukraine; E-mail: inna-andry@ukr.net*

Vedel, Y., *Faculty of Computer Science and Cybernetics, Taras Shevchenko National University of Kyiv, Kyiv, Ukraine; E-mail: yana.vedel@gmail.com*

Vyshenskyy, V.I., *V.M. Glushkov Institute of Cybernetics of National Academy of Sciences of Ukraine, Kyiv, Ukraine; E-mail: vyshenskyy@ukr.net*

Yugay, L.P., *Dr.Sc.(Math), Professor, Department of Natural Sciences, Uzbek State University of Physical Culture and Sport, Chirchik City, Republic of Uzbekistan, Tel: +998-90-187-1478(m); E-mail: yugailp@mail.ru*

Zbrutsky, Alexander, *Department of Aircraft Control Systems, National Technical University of Ukraine "Igor Sikorsky Kyiv Polytechnic Institute", Str. Botkin, 1, Kyiv, Ukraine, 03056, E-mail: zbrutsky@cisavd.kpi.ua, ORCID: http://orcid.org/0000-0002-2206-7148*

Zhiteckii, L., *International Research and Training Center for Information Technologies and Systems of the National Academy of Science of Ukraine and Ministry of Education and Sciences of Ukraine, Acad. Glushkova av., 40, Kyiv, 03187, Ukraine; E-mail: leonid_zhiteckii@i.ua*

List of Notations and Abbreviations

$\bar{\nu}_0$	linear speed of point O_0 relative to the origin of the inertial coordinate system
$\bar{\Omega} = (\Omega_x\ \Omega_y\ \Omega_z)^T$	angular velocity of the $CX_cY_cZ_c$ coordinate system relative to the $AXYZ$ inertial coordinate system
$\bar{A}(t), \bar{B}(t), \bar{C}(t)$	linearized matrices $\mathbf{A}_X$, $\mathbf{B}_X$ and $\mathbf{C}_X$
$\bar{F}_v$	the main vector of external forces acting on the v-th point of the system in the inertial coordinate system
$\bar{M}^M$	the main vector of the moment of external forces and forces in the hinge acting on the manipulator, relative to the origin of the inertial coordinate system
$\bar{M}_0^M$	the main vector of the moment of forces in the hinge acting on the manipulator, relative to the point O_1
$\bar{p}_{o1i}$	radius vector of the ith elementary particle of the AMR platform relative to the point O_1
$\bar{p}_{o1j}$	radius vector of the jth elementary particle of the manipulator relative to the point O_1
$\bar{p}_{oc}^m$	radius vector of the center of mass of the manipulator relative to the point O_0
$\bar{p}_{ol.M}$	radius vector of the center of mass of the manipulator of point M relative to O_1
$\bar{p}_{oo1}$	radius vector of point O_1 relative to the center of mass of the platform of point C
$\bar{r}_C$	radius vector of point C relative to the origin of the inertial coordinate system
$\bar{v}_j$	linear speed of the jth point of the manipulator relative to the origin of the inertial coordinate system
v_{EF}	engine failure velocity
V_1	decision-making velocity
$\bar{w}_1 = (w_{1x}\ w_{1y}\ w_{1z})^T$	angular velocity of the $O_1X_1Y_1Z_1$ coordinate system relative to the $CX_cY_cZ_c$ coordinate system

$\dfrac{\tilde{d}_0}{dt}$	operator of local derivative in the $CX_cY_cZ_c$ coordinate system
$\dfrac{\tilde{d}_1}{dt}$	local derivative operator in the $O_1X_1Y_1Z_1$ coordinate system
$\dfrac{d}{dt}$	operator of the derivative in the inertial $AXYZ$ coordinate system
α	rudder angle
η	parametric noise
ω	angular speed
ψ	roll angle
θ	pitch angle
ε	error value
φ	yaw (course) angle
m	mass of the system
m_i	mass of the ith elementary particle of the AMR platform
m_j	mass of the jth elementary particle of the manipulator
m_M	mass of the manipulator
$t_{ij}^s{-}i$	th switching moments for jth state coordinates
$\mathbf{A}$	damping coefficient matrix
$\mathbf{A_X}$	matrix parameters that depends on the coordinates (6×6)
$\mathbf{A}_m, \mathbf{B}_m, \mathbf{C}_m$	matrices of the reference mathematical model
$\mathbf{B_X}$	matrix of control parameters that depends on the coordinates (6×6)
$\mathbf{C_X}$	matrix of coefficients of perturbations that depends on the coordinates (6×6)
$\mathbf{E}$	error vector
$\mathbf{F}$	vector of external disturbances (6×1)
$\mathbf{F}_{cont}$	measured and controlled disturbances
$\mathbf{g}$	gravity acceleration vector (6×1)
$\mathbf{I}$	unit matrix
$\mathbf{Q}, \mathbf{G}_1, \mathbf{G}_2$	matrices of weighting coefficients
$\mathbf{R}$	space of variables
$\mathbf{S}$	matrices of sensitivity functions
$\mathbf{U}$	control vector (6×1)
$\mathbf{U}_{max}$	maximum control constraints
$\mathbf{U}_{cor}$	corrective robust control
$\mathbf{U}_{opt}$	optimal control

$\mathbf{W}_{cor}(p)$	matrix transfer function of robust correction
$\mathbf{W}_o(p)$	matrix transfer function of the physical object
$\mathbf{W}_{opt}(p)$	matrix transfer function of the optimal mathematical model
$\mathbf{W}_{sens}(p)$	matrix transfer function of the sensors
$\mathbf{X}$	state coordinate vector of moving object (6×1)
$\mathbf{X}_0, \mathbf{X}_k$	vectors of initial and final conditions
$f(t)$	external disturbance
$i = 0,\ldots, m$	order of the derivatives of state coordinate $\mathbf{X}$
J	optimality functional
$j = 1,\ldots, n$	order of mathematical model's moving object
L	Lagrange function
m	maximum order of the positive or negative derivatives determined by control constraints
N	quadrotor UAV rotor thrust
Q	quadratic functional
T_j	control end time for jth state coordinates
U	quadrotor UAV controls
V_w	wind speed
$W_f(p)$	transfer function of irregular waves impact
ACO	Ant colony optimization
AFM	Aircraft flight manual
AI	Artificial intelligence
AMR	Autonomous mobile robot
ASDA	Accelerate-stop distance available
ASD	Accelerate-stop distance
EIV	Errors-in-variables
FC	Fuzzy controller
FS	Fuzzy system
I.I.D.	Independent and identically-distributed
ICCRSU	Initial conditions and control requirements setting unit
LT	Linguistic terms
M	Manipulator
MF	Membership functions
NP	Target point
Obs	Obstacle
QUAV	Quadrotor unmanned aerial vehicle
RB	Rule base
SCU	Switch control unit

SNR	Signal-to-noise ratio
SP	Start position
SVD	Singular value decomposition
SW	Switching key

Part I
Advances in Theoretical Research of Control Systems

1

Control of Moving Object Groups in a Conflict Situation

A. Chikrii

Glushkov Institute of Cybernetics of the National Academy of Sciences
of Ukraine, 40 Glushkov Avenue, 03187, Kyiv, Ukraine
E-mail: g.chikrii@gmail.com

Abstract

This chapter is devoted to an overview of the results concerning the differential games of pursuit–evasion with many participants on the opposing sides. According to L. S. Pontryagin, such problems are the pinnacle of the theory of conflict-controlled processes. The basic method of research is the method of resolving functions. Here the problems with phase constraints, problems of group and successive pursuit as well as the problems of interaction of groups of participants are considered. Sufficient and sometimes the necessary conditions for the completion of the game in a finite time are given. In this case, quasi and stroboscopic strategies are used. The results are used to solve several model examples from the book of R. Isaacs illustrating game situations.[1]

Keywords: Differential game, guaranteed result, Pontryagin's first direct method, resolving function, extremal aiming rule, set-valued mapping and its selections, stroboscopic strategy, group pursuit, state constraints, successive pursuit.

[1]This work was partially supported by National Research Foundation of Ukraine. Grant No 2020.02/ 0121

1.1 Introduction

Several effective mathematical methods have been developed in the theory of dynamic games (differential games, conflict-controlled processes). In some cases, they provide optimal results. From a practical point of view, the concept of a guaranteed result is more appropriate. The game problems of pursuit–evasion are the most interesting.

A survey of results in the theory of evasion is contained in [1]. The works [2–6] provide an idea of the methods of pursuit. The last one describes the method of resolving functions. This method theoretically substantiates the rule of parallel pursuit and a more general method of pursuit along a beam, well known to engineers engaged in rocket and space technology [7].

The mathematical apparatus for substantiating various schemes of the method of resolving functions are the methods of applied nonlinear analysis, in particular, the technique of set-valued mappings and their selections [8, 9]. The creation and development of the method [6] were stimulated by a special 'minmaxmin' function, introduced by the author in the theory of evasion, and the chapter [10], which laid the foundation for group pursuit problems [6, 10–13]. Subsequently, this method was used to study the traveling salesman problems, namely, the problems of successive pursuit [6–14], and the problems with state constraints [2, 6, 15].

The method of resolving functions is based on the natural accumulative principle. Reaching in the sum a certain value of the total winnings means the game termination, that is, the fact that the trajectory of the conflict-controlled process has hit a given terminal set.

The versatility of the method makes it possible to cover in a unified scheme a wide range of dynamic processes which is proceeding in conditions of conflict and uncertainty. These include, first of all, functional-differential systems, stochastic processes, as well as processes with partial and fractional derivatives [16–23].

1.2 Function $\omega(n,\ \nu)$, Encirclement by Pshenihnyi, Scheme of the Method of Resolving Functions

In [24] the following function was introduced:

$$\omega(n,\ \nu) = \min_{\|p_i\|=1}\ \max_{\|v\|=1}\ \min_{i=1,\ \ldots,\ \nu}\ |(p_i,\ v)|,\ p_i \in R^n,\ v \in R^n,$$

$$n \geq 2,\ \ \nu \geq 2,$$

Because of the importance of this function, we analyze in detail the case of 'simple motions'.

The dynamics of a group of ν pursuers and single evader is described by the equations, respectively,

$$\dot{x}_i = u_i, \quad \|u_i\| \leq 1, \quad i = 1, \ldots, \nu, \quad x_i(0) = x_i^0, \quad x_i \in R^n, \tag{1.1}$$

$$\dot{y} = v, \quad \|v\| \leq a, \quad y(0) = y^0, \quad y \in R^n.$$

That the evader is captured by the ith pursuer means that $x_i = y$. It was shown that under the condition $\omega(n, \nu) \cdot a > 1$ the straight motion of the evader in a suitable direction provides avoiding meeting with the pursuers for any initial states.

The next important step after the introduction of the function $\omega(n, \nu)$ was made by B.N. Pshenichnyi [10]. Let us put $a = 1$ and analyze the problem from the point of view of a group of pursuers. In the expression for the function $\omega(n, \nu)$ we set

$$p_i = \frac{x_i^0 - y^0}{\|x_i^0 - y^0\|}, \quad i = 1, \ldots, \nu,$$

and remove the outer minimum and the modulus.

One can easily see that the condition $\max\limits_{\|v\|=1} \min\limits_{i=1,\ldots,\nu} (p_i, v) > 0$ is equivalent to the inclusion

$$y^0 \in intco\{x_1^0, \ldots, x_\nu^0\}, \tag{1.2}$$

This inclusion shows that at the initial moment the position of the evader belongs to the interior of the convex hull spanned over the initial positions of the evaders. The condition (1.2) is usually referred to as the condition for encirclement by Pshenichnyi [10].

Pshenichnyi's theorem. Let in the group pursuit–evasion problem (1.1) $a = 1$. Then, for the group of pursuers to catch the evader in a finite time, it is necessary and sufficient that the initial positions of the players satisfy inclusion (1.2).

The process of group pursuit of a surrounded evader is realized using the parallel approach strategy [7]. If the initial position of the evader lies on the boundary or outside the convex hull spanned over the initial positions of the pursuers, then there always exists a hyper-plane, generally speaking, loosely separating the evader from the pursuers of the group, and the motion along the normal to this hyper-plane in the corresponding direction insures the evader avoiding meeting with each pursuer from the group.

The method of resolving functions [25] proved its efficiency in solving problems of group and successive pursuit. Now we outline its basic scheme in the one-on-one case. To encompass wider classes of functional-differential systems in a unified scheme, we consider a general representation of the solution.

Let a conflict-controlled process be described by a generalized quasi-linear system [25],

$$z(t) = g(t) + \int_0^t \Omega(t, \tau) \phi(u(\tau), v(\tau)) d\tau, \ t \geq 0, \qquad (1.3)$$

where $z(t) \in R^n$, the function $g(t)$, $g : R_+ \to R^n$, $R_+ = \{t : t \geq 0\}$, is measurable and bounded at $t>0$, matrix function $\Omega(t, \tau)$, $t \geq \tau \geq 0$, is t measurable and summable in τ for each $t \in R_+$. The control block is given by function $\varphi(u, v)$, $\varphi : U \times V \to R^n$, which is jointly continuous on the direct product of nonempty sets U and V from R^n. Controls of the players $u(\tau)$, $u : R_+ \to U, v(\tau)$, $v : R_+ \to V$, are measurable functions of time.

We assume that the terminal set M^* has a cylindrical form

$$M^* = M^0 + M, \qquad (1.4)$$

where M^0 is a linear subspace from R^n, $M \in K(L)$, L is the orthogonal complement to M^0 in R^n, $K(L)$ is a set of nonempty compact sets from L.

We take the side of the first player (the pursuer) and define his strategies under the assumption that the evader applies an arbitrary measurable control.

Let the game (1.3), (1.4) is evolving on the interval [0, T]. The pursuer's control at the moment t is selected based on the information about $g(T)$ and $v_t(\cdot)$, i.e. either $u(t) = u(g(T), v_t(\cdot)) \in U$, where $v_t(\cdot) = \{v(s) : s \in [0, t]\}$ is the evader's control prehistory, or $u(t) = u(g(T), v(t)) \in U$.

Control $u(t) = u(g(T), v_t(\cdot))$ realizes quasi-strategy [5], and the counter-control–counter-strategy $u(t) = u(g(T), v(t))$, which is prescribed by a stroboscopic strategy of O. Hayek [26].

Let π be the orthogonal projector acting from R^n to L. We set $\varphi(U, v) = \{\varphi(u, v) : u \in U\}$, and consider the set-valued mappings

$$W(t, \tau, v) = \pi \Omega(t, \tau) \varphi(U, v), W(t, \tau) = \bigcap_{v \in V} W(t, \tau, v)$$

on the sets $\Delta \times V$ and Δ, $\Delta = \{(t, \tau) : 0 \leq \tau \leq t < \infty\}$.

Pontryagin's condition. The set-valued mapping $W(t, \tau)$ is closed-valued and has nonempty direct images on the set Δ.

Given the properties of parameters of the process (1.3), (1.4), the mapping $\varphi(U, v)$, $v \in V$, is continuous in the Hausdorff metric [9], the mapping $W(t, \tau, v)$ is measurable in τ, $\tau \in [0, t]$, and closed in v, $v \in V$. Therefore, the set-valued mapping $W(t, \tau)$ is measurable in $\tau \in [0, t]$. From Pontryagin's condition and the measurable choice theorem [9], it follows that for any $t > 0$ there exists at least one τ measurable selection $\gamma(t, \tau)$, $\gamma(t, \tau) \in W(t, \tau), (t, \tau) \in \Delta$.

Let us denote

$$\xi(t, g(t), \gamma(t, \cdot)) = \pi g(t) + \int_0^t \gamma(t, \tau) \, d\tau.$$

We introduce into consideration the set-valued mapping

$$Q(t, \tau, v) == \sup\{\alpha \geq 0 : [W(t, \tau, v) - \gamma(t, \tau)] \quad (1.5)$$
$$\bigcap \alpha [M - \xi(t, g(t), \gamma(t, \cdot))] \neq \phi\},$$

$$Q : \Delta \times V \to 2^{R_+}.$$

Its support function in the direction +1 is

$$\alpha(t, \tau, v) = \sup\{\alpha \geq 0 : \alpha \in Q(t, \tau, v)\}, (t, \tau) \in \Delta, v \in V. \quad (1.6)$$

We call this function the resolving function [6]:

$$\alpha_X(p) = \sup\{\alpha \geq 0 : \alpha p \in X\}, p \in R^n, 0 \in X, X = \bar{X}.$$

Since Pontryagin's condition is satisfied, then the set-valued mapping $Q(t, \tau, v)$ has nonempty closed direct images on the set $\Delta \times V$. In the case $\xi(t, g(t), \gamma(t, \cdot)) \in M$ we have $Q(t, \tau, v) = [0, \infty)$, and, therefore, $\alpha(t, \tau, v) = +\infty$ for $\tau \in [0, t]$, $v \in V$. This corresponds to Pontryagin's first direct method [6]. Using the theorems on the characterization and the inverse image [9], it can be shown that the set-valued mapping $Q(t, \tau, v)$ is $L \times B$-measurable for the set (τ, v), $\tau \in [0, t]$, $v \in V$, and the resolving function $\alpha(t, \tau, v)$ is joint $L \times B$-measurable in (τ, v), by the support function theorem [9].

Let V_* be the set of measurable functions with values in V. We consider the set

$$T(g(\cdot), \gamma(\cdot, \cdot)) = \left\{ t > 0 : \inf_{v(\cdot) \in V_*} \int_0^t \alpha(t, \tau, v(\tau)) \, d\tau \geq 1 \right\}. \quad (1.7)$$

The function $\alpha(t, \tau, v)$ is $L \times B$-measurable, consequently, it is super-positionally measurable [9, 27] and the integral in (2.7) has sense. If the inequality in (1.7) does not hold for all $t > 0$, then we set $T(g(\cdot), \gamma(\cdot, \cdot)) = \varnothing$.

Theorem 1.1 Suppose that Pontryagin's condition holds for the conflict-controlled process (1.3), (1.4), the set M is convex and for function $g(\cdot)$ and some selection $\gamma(\cdot, \cdot)$

$$T \in T(g(\cdot), \gamma(\cdot, \cdot)) \neq \varnothing.$$

Then the trajectory of the process (1.3) can be brought by the pursuer to the terminal set (1.4) at a time T using a suitable quasi-strategy.

Moreover, if the mapping $Q(t, \tau, v)$ is convex-valued, then the process of approach can be realized in the class of stroboscopic strategies [25].

There are several modifications of the method of resolving functions related to various aspects of the game problem. These are a scheme for bringing the trajectory to a given point of the set M [6], a variant of the pursuit method when the pursuer employs only stroboscopic strategies [25], a case when the function $\alpha(t, \tau, v)$ is t- independent that makes it feasible to terminate the game before the estimated time at the account of the evader's mistakes [6], as well as the functional form of Pontryagin's first direct method [25, 27]. Various forms of the pursuer advantage–Pontryagin's condition, bring essential changes to the method scheme. For oscillatory processes, the Pontryagin integral condition is sometimes used [6]. For the problems of soft meeting (the instantaneous coincidence of the players' geometric coordinates and velocities), an effective result is achieved by using the principle of time dilation [28], which is based on the modified Pontryagin's condition.

The method of resolving functions makes it feasible to study in a unified scheme the group pursuit problems, the problems with phase constraints as the problems with static "recumbent" pursuers [6, 15], the game problems of successive approach—traveling salesman problems [6, 14], the pursuit–evasion problems under the interaction of groups [6], the problem of multiple captures [11, 12], and the problem of pursuit of the formation [12]. The form of solution representation [1] allows extending the method to the nonlinear case.

Not so long ago, the schemes for two types of upper and lower resolving functions were created [29, 30], focused on the case when $0 \notin Q(t, \tau, v)$ and Pontryagin's condition does not hold. Also, the matrix resolving function was introduced. This makes it possible not only to attract the set $M -$

$\xi(t, \ g(t), \ \gamma(t, \cdot))$ to the intersection from (2.5) in the cone spanned on it, but to rotate it as well [31].

1.3 Group Pursuit of a Moving Object

When a group of ν pursuers and a single evader are evolved in the group pursuit, the following effects can be observed. Each of the pursuers from the group cannot independently catch the evader in a finite time, but by acting together they can achieve the desired goal [6, 11–13, 15]. Another case is when each pursuer can individually intercept the evader, but acting together in a group they can reach the desired goal faster.

Formally, the problem setting should contain the dynamics (1.3) and a terminal set–the set of ν cylindrical sets M_i^*, $i = 1, ..., \nu$, of the form (1.4). To emphasize the presence of several pursuers, we consider the conflict-controlled process,

$$z_i(t) = g_i(t) + \int_0^t \Omega_i(t, \ \tau) \, \varphi_i(u_i(\tau), \ v(\tau)) \, d\tau, i = 1, \ ..., \ \nu, \ t \geq 0, \ (1.8)$$

where $z_i(t) \in R^{n_i}$, $g_i : R_+ \to R^{n_i}$, functions $g_i(t)$ are measurable and bounded in $t > 0$, matrix functions $\Omega_i(t, \ \tau)$, $t \geq \tau \geq 0$, are measurable in t and summable in τ for each $t \in R_+$. Functions $\varphi_i(u_i, \ v)$, $\varphi_i : U_i \times V \to R^{n_i}$, are jointly continuous in their variables, and U_i, V are nonempty compacts. The admissible controls of the players are measurable functions. The pursuers apply quasi or stroboscopic strategies [5, 6, 26]. The terminal set has the form

$$M_i^* = M_i^0 + M_i, i = 1, \ ..., \ \nu, \qquad (1.9)$$

where M_i^0 are linear subspaces from R^{n_i}, and $M_i \in K(L_i)$, L_i are orthogonal complements to M_i^0 in R^{n_i}.

Let us denote by z_i a pair $(x_i, \ y)$, where x_i and y are the states of the ith pursuer and the evader, respectively. Since the goal of the group of pursuers consists in the capture of the evader by at least one of the pursuers, then vector z_i hitting the set M_i^* means the capture of the evader by the ith pursuer. . .

Let π_i be the operators of orthogonal projection from R^{n_i} to L_i, and $\varphi_i(U_i, \ v) = \{\varphi_i(u_i, \ v) : u_i \in U_i\}$, $v \in V$. Let us take into consideration the forms of the set-valued mappings, respectively:

$$W_i(t, \ \tau, \ v) = \pi_i \Omega_i(t, \ \tau) \, \varphi_i(U_i, \ v), W_i(t, \ \tau) = \bigcap_{v \in V} W_i$$

$$(t,\ \tau,\ v),\ t \geq \tau \geq 0, i = 1,\ ...,\ v.$$

Pontryagin's condition for a group of pursuers. The set-valued mappings $W_i(t,\ \tau),\ (t,\ \tau) \in \Delta,\ i = 1,\ ...,\ \nu$, are closed-valued and have nonempty direct images.

Since the set-valued mappings $W_i(t,\ \tau),\ i = 1,\ ...,\ \nu$, are measurable in τ and closed-valued, then there exist τ measurable selections $\gamma_i(t,\ \tau),\ \gamma_i(t,\ \tau) \in W_i(t,\ \tau),\ (t,\ \tau) \in \Delta$, analogously to the case $\nu = 1$.

We set

$$\xi_i(t,\ g_i(t),\ \gamma_i(t,\cdot)) = \pi_i g_i(t) + \int_0^t \gamma_i(t,\ \tau)\, d\tau, i = 1,\ ...,\ \nu,$$

and introduce numerical set-valued mappings

$$Q_i(t,\ \tau,\ v) =$$
$$= \{\alpha_i \geq 0 : [W_i(t,\ \tau,\ v) - \gamma_i(t,\ \tau)] \bigcap \alpha_i [M_i - \xi_i(t,\ g_i(t),\ \gamma_i(t,\cdot))] \neq \varnothing\},$$
$$Q_i : \Delta \times V \to 2^{R+}.$$

Below are given their support functions in the direction +1:

$$\alpha_i(t,\ \tau,\ v) = \sup\{\alpha_i \geq 0 : \alpha_i \in Q_i(t,\ \tau,\ v)\},\ \ t \geq \tau \geq 0,$$

$$v \in V,\ i = 1,\ ...,\ \nu.$$

Let us set

$$g^\nu(\cdot) = (g_1(\cdot)\ ...\ g_\nu(\cdot)), \gamma^\nu(\cdot,\cdot) = (\gamma_1(\cdot,\cdot)...\gamma_\nu(\cdot,\cdot)),$$

and take a closer look at the set

$$T^\nu(g^\nu(\cdot),\ \gamma^\nu(\cdot,\cdot)) = \left\{ t > 0 : \inf_{v(\cdot) \in V_*} \max_{i = \overline{1,\nu}} \int_0^t \alpha_i(t,\ \tau,\ v(\tau)) d\tau \geq 1 \right\}.$$
$$(1.10)$$

If for some index $i \in \{1,\ ...,\ \nu\}$ and time instant $t > 0$ inclusion $\xi_i(t,\ g_i(t),\ \gamma_i(t,\cdot)) \in M_i$, has a place, then at the time instant t the ith pursuer can catch the evader, i.e. $Q_i(t,\ \tau,\ v) = [0,\ \infty),\ \alpha_i(t,\ \tau,\ v) = +\infty$, for all $\tau \in [0,\ t],\ v \in V$, and the capture is realized in the class of stroboscopic strategies in compliance with the scheme of the first direct method [3].

In the case the inequality in (1.10) is not fulfilled, we put $T^\nu(g^\nu(\cdot),\ \gamma^\nu(\cdot,\cdot)) = \phi$.

Theorem 1.2. Let for the group pursuit problem (1.8), (1.9) an analog of Pontryagin's condition be satisfied, the sets M_i, $i = 1, ..., \nu$, are convex and suppose that for function $g^\nu(\cdot)$ there exist measurable selections, $\gamma^\nu(\cdot, \cdot)$, $\gamma_i(t, \tau) \in W_i(t, \tau)$, $(t, \tau) \in \Delta$, $i = 1, ..., \nu$, and an integer $T^\nu \in T^\nu(g^\nu(\cdot), \gamma^\nu(\cdot, \cdot)) \neq \phi$.

Then there is a quasi-strategy of the pursuers such that at least one of the trajectories (1.8) can be brought to the corresponding set at the time instant T^ν for any admissible control of the evader.

By the way of illustration, below is given a simple example of the group pursuit with the same control domains of the players

$$\dot{z}_i = u_i - v, z_i \in R^{n_i} = R^n, n \geq 2, \|u_i\| \leq 1, \|v\| \leq 1, z_i(0) = z_i^0,$$
(1.11)

and the captures described by the sets:

$$M_i^* = M_i^0 = M_i = \{z_i : z_i = 0\}, \ i = 1, ..., \nu.$$
(1.12)

Here $L_i = R^n$, π_i are identity transformation operators $\pi_i : R^n \to R^n$, described by the n-dimensional unit matrix E.

Here in the presentation of solution (1.8) $g_i(t) = z_i^0$, $\Omega(t, \tau) = E$, $\varphi_i(u_i, v) = u_i - v$, $i = 1, ..., \nu$, see (1.8), and the Pontryagin's condition for a group of pursuers holds: $W_i(t, \tau) = S * S = 0$, $S = \{p : \|p\| < 1\}$, with $\gamma_i(t, \tau) = 0$.

Here $\alpha_i(t, \tau, v) = \alpha_i(z_i^0, v) = \max\{\alpha \geq 0 : -\alpha z_i^0 \in S - v\}$, the resolving functions show up as the greatest positive roots of the quadratic equations $\| v - \alpha_i z_i^0 \| = 1$, $i = 1, ..., \nu$, and have the form

$$\alpha_i(z_i^0, v) = \frac{(z_i^0, v) + \sqrt{(z_i^0, v)^2 + \| z_i^0 \|^2 (1 - \|v\|^2)}}{\| z_i^0 \|^2}, \ i = 1, ..., \nu.$$

The encirclement by Pshenichnyi [10] for the initial states $z^0 = (z_1^0, ..., z_\nu^0)$ can be expressed in similar forms [6], namely:

1) $\delta(z^0) = \min\limits_{\|v\| \leq 1} \max\limits_{i=1,...,\nu} \alpha_i(z_i^0, v) > 0$;

2) $\sigma(z^0) = \min\limits_{\|v\| \leq 1} \max\limits_{i=1,...,\nu} \left(\dfrac{z_i^0}{\| z_i^0 \|}, v \right) > 0$;

3) $0 \in intco \left\{ \dfrac{z_i^0}{\| z_i^0 \|} \right\}$.

Below is given the statement that provides necessary and sufficient conditions for the successful termination of the group pursuit–evasion and the pursuit at the moment

$$T_0^\nu = \min T^\nu(z^0, \, 0).$$

Corollary 1.1. If $0 \in intco \left\{ \frac{z_i^0}{\| z_i^0 \|} \right\}$, where z^0 is the initial state of the process (1.11), then the group pursuit can be terminated at the time T_0^ν and the following estimate is valid: $T_0^\nu \leq \frac{\nu}{\delta(z^0)}$. Herewith, if $t_* = t_*(v(\cdot))$ is the time of control switching, i.e. a zero of the control function

$$1 - \max_{i=1,\ldots,\nu} \int_0^t \alpha_i(z_i^0, \, v(\tau)) \, d\tau,$$

then the controls of the pursuers, realizing time T_0^ν on the active section $[0, \, t_*]$, are

$$u_i(\tau) = v(\tau) - \alpha_i(z_i^0, \, v(\tau)) z_i^0, \quad i = 1, \, \ldots, \, \nu, \qquad (1.13)$$

and on the passive one $(t_*, \, T_0^\nu]$ $u_i(\tau) = v(\tau)$ for indices i, satisfying the equality

$$\int_0^{t_*} \alpha_i(z_i^0, \, v(\tau)) \, d\tau = \max_{i \in \overline{1,\nu}} \int_0^{t_*} \alpha_i(z_i^0, \, v(\tau)) \, d\tau,$$

and for the rest of the indices, the controls of the pursuers at the interval $(t_*, \, T_0^\nu]$ can be arbitrary admissible. If

$$0 \bar{\in} intco \left\{ \frac{z_i^0}{\| z_i^0 \|} \right\},$$

then the escape is possible.

Corollary 1.2. Let in the group pursuit problem (1.11)

$$M_i^0 = \{0\}, M_i^* = M_i = \varepsilon_i S, \quad \varepsilon_i > 0, \quad i = 1, \, \ldots, \, \nu.$$

Then, if

$$0 \in int \; co \, \{z_i^0 + \varepsilon_i S\},$$

then at least one of the pursuers can catch the evader in a finite time. If

$$0 \; \bar{\in} \; int \; co \, \{z_i^0 + \varepsilon_i S\},$$

then the escape is possible at all semi-infinite intervals of time.

Simple matrix. The dynamics are given by the equations

$$\dot{z}_i = a_i z_i + u_i - v, a_i < 0, z_i \in R^n, \|u_i\| \leq 1, \|v\| \leq 1, i = 1, \, ..., \, \nu.$$

The goal of the pursuer's group is the exact capture of the evader by at least one of the pursuers (1.12). The result has the form of necessary and sufficient conditions [10], which is similar to Corollary 1 and realizes a parallel approach (1.13) on the active section, with the only difference that the following estimate is fulfilled for the guaranteed time

$$T^\nu(z^0, \, 0) \leq \frac{1}{a_*} \ln \left(1 + \frac{a_* \cdot \nu}{\delta(z^0)}\right), \quad a_* = - \max_{i=\overline{1,\nu}} a_i.$$

Second-order objects.
$$\dot{z}_i^1 = z_i^2,$$

$$\dot{z}_i^2 = - a_1 z_i^2 - a_2 z_i^1 + u_i - v, \|u_i\| \leq 1, \quad \|v\| \leq 1, i = 1, \, ..., \, \nu. \quad (1.14)$$

The terminal set is described by the sets

$$M_i^* = M_i^0 = M_i = \{z_i = (z_i^1, \, z_i^2) : z_i^1 = 0\}, \quad i = 1, \, ..., \, \nu,$$

where z_i^1 is the difference between the geometric coordinates of the pursuer x_i and the evader y, $z_i^1 = x_i - y$, and z_i^2 is the difference of their speeds, $z_i^2 = \dot{x}_i - \dot{y}$.

Because of the equivalence of the Equation (1.14) to the second-order equations:

$$\ddot{x}_i + a_1 \dot{x}_i + a_2 x_i = u_i,$$

$$\ddot{y} + a_1 \dot{y} + a_2 y = v,$$

the characteristic polynomial takes the form

$$\lambda^2 + a_1 \lambda + a_2 = 0. \quad (1.15)$$

We will restrict ourselves to the case of real roots and present the results in the form of consequences of the general method scheme for classical examples [13, 15].

The pursuers and the evader are material points [5]
Let $a_1 = a_2 = 0$ in (1.14). Then the roots of the polynomial (3.8) are $\lambda_1 = \lambda_2 = 0$.

If $0 \in intco\{z_i^*\}$, where $z_i^* = \begin{cases} z_i^{01}, & z_i^{02} = 0, \\ z_i^{02}, & z_i^{02} \neq 0, \end{cases}$ then $T^\nu(z^0, 0) < +\infty$.

Pontryagin's test case [3]
Let $a_2 = 0, a_1 > 0$ in (1.14). Then $\lambda_1 = -a_1, \lambda_2 = 0$.

If $0 \in intco\{z_i^*\}$, where $z_i^* = a_1 z_i^{01} + z_i^{02}$, then $T^\nu(z^0, 0) < +\infty$.

If the objects are of a different type, then a_1 is changed for a_1^i, and in the encirclement, the condition z_i^* should be put equal to $z_i^{01} + \frac{1}{a_1^i} z_i^{02}$. Then the result holds [13, 15].

1.4 Non-fixed Time of Game Termination

The scheme of the method of resolving functions allows, under certain assumptions, to obtain the conditions for the game termination not only at the moment but also no later than in a certain time, using the evader's mistakes [6]. Let the motion of a group of controlled objects $z = (z_1, ..., z_\nu)$ in space R^n be described by quasi-linear equations

$$\dot{z}_i = A_i z_i + \varphi_i(u_i, v), i = 1, ..., \nu, \tag{1.16}$$

$z_i \in R^{n_i}, n = n_1 + ... + n_\nu, A_i$ are square matrices of the order $n_i, u_i \in U_i$, $v \in V, U_i, V$ are nonempty compact sets, functions $\varphi_i(u_i, v)$ are jointly continuous in their variables.

The terminal set consists of cylindrical sets (1.9), where $M_i, i = 1, ..., \nu$, are convex compact sets, As usual, π_i are the ortho–projector acting from R^{n_i} to L_i, con $X = \{x \in X : \lambda x \in X \; \forall \lambda > 0\}$, $\overline{con} X$ is the cone closure.

Condition 1.1. For the fixed point $z \in R^n$,
$\overline{con}(M_i - \pi_i e^{tA_i} z_i) \bigcap \pi_i e^{(t-\tau)A_i} \varphi_i(U_i, v) \neq \phi, i = 1, ..., \nu, (t, \tau) \in \Delta, v \in V$.

Such points are given, let us introduce the resolving functions
$\alpha_i(t, \tau, z_i, v) = \sup\{\alpha_i \geq 0 : \pi e^{(t-\tau)A_i} \varphi_i(U_i, v) \bigcap \alpha[M_i - \pi_i e^{tA_i} z_i] \neq \phi\}$,

$$i = 1, ..., \nu, (t, \tau) \in \Delta, v \in V,$$

and set

$$T^{\nu}(z) = \min \left\{ t \geq 0 : \inf_{v(\cdot) \in V_*} \max_{i=\overline{1,\nu}} \int_0^t \alpha_i(t, \tau, z_i, v(\tau)) \, d\tau \geq 1 \right\}.$$

Theorem 1.3. Suppose that for the process (1.16), (1.9) $M_i = \{0\}$, $\pi_i A_i = A_i \pi_i$, $i = 1, ..., \nu$, the initial position z^0 satisfies Condition 1.1, and $T^{\nu}(z^0) < +\infty$.

Then the group pursuit can be terminated starting from the initial state z^0 no later than in time $T^{\nu}(z^0)$ in the class of stroboscopic strategies.

Let us put $A_i = \{0\}$, $i = 1, ..., \nu$. Then

$$\dot{z}_i = \varphi_i(u_i, v), i = 1, ..., \nu. \tag{1.17}$$

Now we consider the set-valued mappings

$$W_i(z_i, v) = \pi_i \varphi_i(U_i, v) \bigcap \overline{\mathrm{con}}(M_i - \pi_i z_i),$$

$$\tilde{W}_i(z_i, v) = \mathrm{co} \{\pi_i \varphi_i(U_i, v)\} \bigcap \overline{\mathrm{con}}(M_i - \pi_i z_i),$$

and for $z_i \in R^{n_i} \backslash M_i^*$ the sets

$$W_i = \{z_i : W_i(z_i, v) \neq \phi \forall v \in V\},$$
$$\tilde{W}_i = \{z_i : \tilde{W}_i(z_i, v) \neq \phi \forall v \in V\}.$$

The resolving functions have the forms

$$\bar{\alpha}_i(z_i, v) = \max \{\alpha \geq 0 : \pi_i \varphi_i(U_i, v) \bigcap$$

$$\alpha[M_i - \pi_i z_i] \neq\neq \phi\}, \ z_i \in W_i, \ v \in V,$$

$$\tilde{\alpha}_i(z_i, v) = \max \{\alpha \geq 0 : \mathrm{co} [\pi_i \varphi_i(U_i, v)] \bigcap \alpha[M_i - \pi_i z_i] \neq\neq \phi\},$$

$$z_i \in \tilde{W}_i, \ v \in V, \ i = 1, ...\nu.$$

We set

$$\tilde{T}^{\nu}(z) = \min \{t \geq 0 : \inf_{v(\cdot) \in V_*} \max_{i=\overline{1,\nu}} \int_0^t \bar{\alpha}_i(z_i, v(\tau)) d\tau \geq 1\}.$$

Let us introduce the summary functions

$$\bar{\alpha}(z) = \inf_{v \in V} \max_{i=\overline{1,\nu}} \bar{\alpha}_i(z_i, v), \tag{1.18}$$

$$\tilde{\alpha}(z) = \inf_{v \in V} \max_{i=\overline{1,\nu}} \tilde{\alpha}_i(z_i, v).$$

Theorem 1.4. Let $z_i^0 \in W_i$, $i = 1, ..., \nu$, for the initial state of the process (1.17) z^0 and $\bar{\alpha}(z^0) > 0$. Then the group pursuit problem can be solved in the class of counter-controls no later than in time $\tilde{T}^\nu(z^0)$ and $\tilde{T}^\nu(z^0) \leq \frac{\nu}{\bar{\alpha}(z^0)}$.

If, otherwise, for z^0, $z_i^0 \in \tilde{W}_i$, $i = 1, ..., \nu$, and $\tilde{\alpha}(z^0) = 0$, then the problem of evasion from a group of pursuers can be solved under the constant control of the evader, furnishing minimum in (1.18).

Let be

$$\dot{z}_i = u_i - v, z_i \in R^n, v \in V, u_i \in \partial coV, i = 1, ..., \nu. \tag{1.19}$$

Here V is a compact set from R^n, ∂coV is the boundary of the set coV. The terminal set consists of convex compact sets $M_1, ..., M_\nu$ from R^n.

Now we deduce necessary and sufficient conditions, ensuring solvability of the group pursuit problem (1.19) for given initial states.

Let us set

$$\widehat{W}_i(z_i, \nu) = [\partial coV - \nu] \bigcap \overline{con}[M_i - z_i],$$

$$\widehat{W}_i = \{z_i : \widehat{W}_i(z_i, \nu) \neq \phi \forall \nu \in V\}, i = 1, ..., \nu,$$

$$\widehat{\alpha}_i(z_i, \nu) = \max\{\alpha \geq 0 : [\partial coV - \nu] \bigcap [M_i - z_i] \neq \phi\},$$

$$z_i \in \widehat{W}_i, \ z_i \bar{\in} M_i, \ v \in V,$$

$$\widehat{\alpha}(z) = \inf_{v \in V} \max_{i=\overline{1,\nu}} \widehat{\alpha}_i(z_i, v),$$

$$\hat{T}^\nu(z) = \min \left\{ t \geq 0 : \inf_{v(\cdot)} \max_{i=\overline{1,\nu}} \int_0^t \widehat{\alpha}_i(z_i, v(\tau)) d\tau \geq 1 \right\}.$$

We now take into consideration the support function and the support set for an arbitrary compact set X from the space R^n, respectively.

$$C_X(p) = \max_{x \in X}(x, p), \ p \in R^n,$$

$$H(X; p) = \{x \in X : (p, x) = C_X(p)\}, \ p \neq 0.$$

In the case the images of the mapping $H(X;\ p)$ consist of single points for any $p \in R^n$, they say that X is a strictly convex compact.

The set X is called a compact with smooth boundary if $H(X;\ p_1) \cap H(X;p_2) = \phi\ \forall p_1,\ p_2,\ p_1 \neq p_2,\ \|p_1\| = \|p_2\| = 1$.

Lemma 1.1. Let for the process (1.19) the set V be a strictly convex compact with smooth boundary. Then $\widehat{\alpha}\ (z^0) > 0$ if and only if

$$0 \in intco \left[\bigcup_{i=\overline{1,\ldots,\nu}} (M_i - z_i^0) \right]. \tag{1.20}$$

It should be emphasized that in the case that V is a strongly convex compact set, the inequality $\widehat{\alpha}\ (z^0) > 0$ leads to inclusion (1.20), but the opposite, generally speaking, is not true. If V is a compact set with a smooth boundary, then inclusion (1.20) implies inequality $\widehat{\alpha}\ (z^0) > 0$,the opposite is generally not true.

Theorem 1.5. Let for the conflict-controlled process (1.19) V be a strictly convex compact set with a smooth boundary. Then, if $0 \in intco \left[\bigcup_{i=\overline{1,\ldots,\nu}} (M_i - z_i^0) \right]$, then the group pursuit can be terminated starting from the initial state z^0 no later than in the time $\widehat{T}^\nu(z^0)$. If, otherwise, $0 \bar{\in} intco \left[\bigcup_{i=\overline{1,\ldots,\nu}} (M_i - z_i^0) \right]$, then starting from the initial state z^0 it is possible to avoid meeting the group of pursuers.

1.5 The Group Pursuit. Linear State Constraints

In what follows, we show that the group pursuit problem under linear state constraints on the evader's state can be reduced to the equivalent group pursuit problem without state constraints with a larger number of pursuers [6, 13, 15]. Consider the conflict-controlled process (1.16)

$$\dot{z}_i = A_i z_i + \varphi_i(u_i,\ v), z_i \in R^{n_i}, u_i \in U_i,\ v \in V, z_i(0) = z_i^0, i = 1,\ \ldots,\ \nu.$$

The terminal set consists of cylindrical sets

$$M_i^* = M_i^0 + M_i, i \in N_\nu = \{1,\ \ldots,\ \nu\}. \tag{1.21}$$

Here M_i^0 are linear subspaces from R^{n_i}, M_i are nonempty convex compact sets from L_i, the orthogonal complements to M_i^0 in R^{n_i}, and $M_i = \{a_i\}$, $i \in N_\nu \backslash N_k$, $k \leq \nu$, and a_i are vectors from L_i, $\dim L_i = 1$.

Consequently, for $i \in \{k+1, ..., \nu\}$, $\dim M_i^0 = n_i - 1$ and sets M_i^* appear as affine manifolds. We assume that the pursuers use quasi-strategies. We denote by π_i, the operator of orthogonal projecting from R^{n_i} to L_i, $i \in N_\nu$. Let us introduce the set-valued mappings

$$W_i(t, \tau) = \pi_i e^{A_i t} \varphi_i(U_i, v), i \in N_\nu.$$

For $i \in N_k$, we set $W_i(t) = \bigcap_{v \in V} W_i(t, v)$, $t \geq 0$.

Condition 1.2. Direct images of mappings $W_i(t)$ are nonempty for all $i \in$ $\in N_k$, $t \geq 0$.

We fix some measurable selections $\gamma_i(t)$, $\gamma_i(t) \in W_i(t)$, $i \in N_k$, and set

$$\xi_i(t, z_i^0, \gamma_i(\cdot)) = \pi_i e^{t A_i} z_i^0 + \int_0^t \gamma_i(\tau) \, d\tau.$$

Let us define the resolving functions for $i \in N_k$

$$\alpha_i(t, \tau, z_i^0 v, \gamma_i(\cdot)) = \sup \{\alpha \geq 0 : [W_i(t - \tau, v) - \gamma_i(t - \tau)] \bigcap$$
$$\bigcap \alpha[M_i - \xi_i(t, z_i^0, \gamma_i(\cdot))] \neq \phi\}.$$

They enjoy all the properties provided earlier. Consider the case $i \in N_\nu \backslash N_k$.

Condition 1.3. Set-valued mappings $W_i(t, v)$, $i \in N_\nu \backslash N_k$, are continuous single-valued functions $\omega_i(t, v)$.

Let us set

$$\xi_i(t, z_i^0) = \pi_i e^{t A_i} z_i^0, i \in N_\nu \backslash N_k,$$

and write the resolving functions for $i = k+1, ..., \nu$:

$$\alpha_i^*(t, \tau, z_i^0 v) = \begin{cases} \frac{(p_i, \omega_i(t-\tau, v))}{(p_i, a_i - \xi_i(t, z_i^0))}, & p_i \in L_i, \ \|p_i\| = 1, \ a_i \neq \xi_i(t, z_i^0), \\ \|\omega_i(t - \tau, v)\| + \frac{1}{t}, & a_i = \xi_i(t, z_i^0). \end{cases}$$

It is supposed that they are continuous in t, $t > 0$, $\gamma(\cdot) = (\gamma_1(\cdot), ..., \gamma_\nu(\cdot))$.

Let us take into consideration the function

$$T_*^\nu(z^0, \gamma(\cdot)) = \min \left\{ t \geq 0 : \inf_{v(\cdot)} \max \left[\max_{i \in N_k} \int_0^t \alpha_i(t, \tau, z_i^0, v(\tau), \right. \right.$$

$$\gamma_i(\cdot))\, d\tau,\ \max_{i\in N_\nu\backslash N_k} \int_0^t \alpha_i^*(t,\ \tau,\ z_i^0,\ v(\tau))\, d\tau\ \Bigg]\geq 1\ \Bigg\}.$$

If the inequality in braces is not satisfied, then we set $T_*^\nu(z^0,\ \gamma(\cdot)) = +\infty$.

Theorem 1.6. Let the parameters of the conflict-controlled process (4.1), (5.1) meets Conditions 1.2 and 1.3 and let for the initial state z^0 and some measurable selections

$$\gamma(t),\ \gamma_i(t)\in W_i(t), i=1,\ ...,\ k, T_*^\nu(z^0,\ \gamma(\cdot)) < +\infty.$$

Then, the trajectory of the process (1.16) can be brought from the initial state z^0 to the corresponding set M_i^* at the moment $T_*^\nu(z^0,\ \gamma(\cdot))$ for at least one i, $i\in N_\nu$, using quasi-strategies.

By way of illustration, let us study the linear problem of control of a group of pursuers when the evader cannot leave some open polyhedral set.

The dynamics of the pursuers and the evader have the forms, respectively:

$$\dot{x}_i = C_i x_i + u_i, x_i \in R^{r_i},\ u_i \in U_i, i = 1,\ ...,\ k, \tag{1.22}$$

$$\dot{y} = By + v, y \in R^s, v \in V,$$

C_i are quadratic matrices of order r_i, and B is a matrix of the order s, U_i, V are nonempty compact sets. The motion of the evader is bounded by the constraints

$$G = \{y \in R^s : (p_i,\ y) < l_i,\ p_i \in R^s,\ \|p_i\| = 1,\ i = k+1,\ ...,\ \nu\} \neq \varnothing. \tag{1.23}$$

We see that the terminal set consists of cylindrical sets M_i^*, $i = 1,\ ...,\ k$, from the spaces $R^{n_i} = R^{r_i} \times R^s$.

We say that at some moment the pursuit is completed if at this moment at least one of the pursuers catches the evader, i.e. for some $i \in N_k$ $z_i = (x_i,\ y) \in M_i^*$, or the evader is forced to violate the state constraints, that is, $(p_i,\ y) = l_i$ for some $i \in N_\nu\backslash N_k$.

For $i \in N_\nu\backslash N_k$ we set

$$z_i = y,\ A_i = B,\ \varphi_i(u,\ v) = v,$$

$$M_i^0 = \{z_i : (p_i,\ z_i) = 0\}, M_i = \{a_i : a_i = l_i p_i\}, M_i^* = \{z_i : (p_i,\ z_i) = l_i\}.$$

Thus, the group pursuit problem (1.22) under the state constraints (1.23) is reduced to a group pursuit problem (1.16), (1.21) without constraints.

The above-outlined scheme can be used to study the game problem with arbitrary state constraints being convex sets [6].

Let us analyze in detail group pursuit for 'simple motions'

$$\dot{x}_i = u_i, \|u_i\| \leq 1, x_i \in R^s, s \geq 2, i = 1, ..., k, \dot{y} = v \|v\| \leq 1, y \in R^s, \tag{1.24}$$

under the state constraints (1.23).

The group pursuit terminates if $\|x_i - y\| \leq \varepsilon_i, \varepsilon_i \geq 0$, for some $i \in N_k$, or the evader hits the boundary of set G, i.e. for some $i \in N_\nu \backslash N_k$ the equality, $(p_i, y) = l_i$ takes place. The game problem (1.24), (1.23) is easily reduced to a differential group pursuit game. Let us put $z_i = x_i - y. i \in N_k$. Then

$$\dot{z}_i = u_i - v, \|u_i\| \leq 1, \|v\| \leq 1, M_i^* = \{z_i : \|z_i\| \leq \varepsilon_i\}, i = 1, ..., k. \tag{1.25}$$

We set $z_i = y., i \in N_\nu \backslash N_k$. Then we have

$$\dot{z}_i = v, \|v\| \leq 1, M_i^* = \{z_i : (p_i, z_i) = l_i\}, i = k+1, ..., \nu. \tag{1.26}$$

It is evident that the analog of Pontryagin's condition for a group is fulfilled, the selections $\gamma_i(t) \equiv 0, i \in N_k, t \geq 0$. Hence, the functions $\xi_i(t, z_i^0, 0) = z_i^0, \xi_i(t, z_i^0) - z_i^0, i \in N_\nu \backslash N_k, \omega_i(t, v) = v$, and the resolving functions take the forms

$$\alpha_i(t, \tau, z_i^0, v, 0) = \rho_i(z_i^0, v) =$$

$$= \frac{(v, z_i^0) + \varepsilon_i + \sqrt{((v, z_i^0) + \varepsilon_i)^2 + (\|z_i^0\|^2 - \varepsilon_i^2)(1 - \|v\|^2)}}{\|z_i^0\|^2 - \varepsilon_i^2}, i \in N_k,$$

$$\alpha_i^*(t, \tau, z_i^0, v) = \rho_i(z_i^0, v) = \frac{(p_i, v)}{l_i - (p_i, z_i^0)}, z_i^0 \neq a_i = l_i p_i, i \in N_\nu \backslash N_k.$$

The termination time of the group pursuit game (1.25), (1.26) is defined as the smallest positive root of the equation

$$\inf_{v(\cdot)} \max_{i \in N_\nu} \int_0^t \rho_i(z_i^0, v(\tau)) \, d\tau = 1. \tag{1.27}$$

In the sequel, we provide results concerning problems of group pursuit–evasion in various forms, in their number, in the form of encirclement about the initial states [6, 12].

Let vectors b_1, ..., b_ν, $\|b_i\| = 1$, belong to space R^s. They form a positive basis if for each vector $z \in R^s$ there exist positive numbers σ_i, $i = 1$, ..., ν, such that $z = \sum_{i=1}^{\nu} \sigma_i b_i$. Let us put

$$\beta_i(b_i,\ v) = (b_i,\ v) + \sqrt{(b_i,\ v)^2 + 1 - \|v\|^2}, i = 1,\ ...,\ k,$$

$$\beta_i(b_i,\ v) = (b_i,\ v), i = k + 1,\ ...,\ \nu,$$

$$\beta(b_1,\ ...,\ b_\nu) = \min_{\|v\| \leq 1} \max_{i=\overline{1,\nu}} \beta_i(b_i,\ v).$$

Lemma 1.2. The following statements are equivalent:

1) vectors b_1, ..., b_ν form a positive basis;
2) vectors b_1, ..., b_ν form a Carathéodory set, i.e. $0 \in int$ $co\{b_1,\ ...,\ b_\nu\}$;
3) $\overline{con}\{b_1,\ ...,\ b_\nu\} = R^s$;
4) $\min_{\|v\|=1} \max_{i=\overline{1,\nu}} (b_i,\ v) > 0$;
5) $\beta(b_1,\ ...,\ b_\nu) > 0$.

Theorem 1.7. Let $z^0 = (z_1^0,\ ...,\ z_\nu^0)$ be the initial state of the process (1.25), (1.26). If there is a number $i \in N_k$, such that $\varepsilon_i > 0$, or $k \geq s$, and

$$0 \in int \{co\ [z_1^0 + \varepsilon_1 S,\ ...,\ z_k^0 + \varepsilon_k S] + con\ (p_{k+1},\ ...,\ p_\nu)\}, \qquad (1.28)$$

then there exists a finite root of equation (1.27).

Corollary 1.3. Let z^0 be the initial state of the process (1.25), (1.26). Then, if $0 \bar{\in} int$ $co\{z_1^0 + \varepsilon_1 S,\ ...,\ z_k^0 + \varepsilon_k S,\ p_{k+1},\ ...,\ p_\nu\}$, then the escape is possible on set $\bar{G}$.

Corollary 1.4. Let for the process (1.25), (1.26) the set G be a polyhedron (a bounded set), and one of the following conditions hold:

1) for some $i \in N_k \varepsilon_i > 0$;
2) $k \geq s$.

Then the group pursuit problem (1.25), (1.26) with state constraint $\bar{G}$ can be solved in a finite time for any initial states.

Remark. If for the process (1.25), (1.26) $\varepsilon_i = 0$ for $i \in N_k$, $k < s$, then even in the case the inclusion $0 \in int$ $co\{z_1^0,\ ...,\ z_k^0,\ p_{k+1},\ ...,\ p_\nu\}$ is fulfilled, the escape on a set $\bar{G}$ is possible. This follows from the results from [6].

Below we analyze some classical problems with state constraints from the book [2], which can be solved based on Theorem 1.7 and its corollaries [6].

Lion versus man [2]. Let $k = 1$, $\varepsilon_1 > 0$, the pursuer and the evader move according to equations (1.24), $\nu - 1 > s$, set G be bounded and given by inequalities (1.23).

Then $con\{p_2, ..., p_\nu\} = R^s$ and inclusion (1.28) is satisfied for all initial positions. Hence, capture is always possible. What is more, the result holds in the case G be an arbitrary convex compact.

Rat cornered [2]. Let motion of a cat and a rat be described by the equations (1.24), respectively, with $k = 1$, $\varepsilon_1 > 0$, a set G be a convex cone K in space R^s. Then the barrier cone of a set G, $K_G = -K^*$, where K^* is a cone conjugate to K, given the form of the support function

$$C_K(p) = \begin{cases} 0, & p \in -K^*, \\ +\infty, & p\bar{\in} - K^*. \end{cases}$$

Consequently, if there are vectors p_i, $i = 2, ..., \nu$, $p_i \in -K^*, \|p_i\| = 1$, such that $y^0 \in int\{x^0 + \varepsilon_1 S + +con[p_2, ..., p_\nu]\}$, then the capture can be performed in a finite time.

Patrolling the corridor [2]. Let $k = 1$, $s = 2$, $\varepsilon_1 > 0$. State constraints are two parallel straight lines on the plane having the normals p_2, p_3. Without loss of generality, we can assume that these straight lines are parallel to the abscissa axis. We see that p_2 and p_3 are orthogonal to the latter. Therefore, the condition, which is sufficient for the capture (1.28), takes the form $| x_1^0 - y_1^0 | < \varepsilon_1$, where x_1^0, y_1^0 are the first coordinates of the initial states of the pursuer and the evader, respectively.

Death line game [2]. Let $k = 1$, $\varepsilon_1 > 0$, $s = 2$. The game takes place on a half-plane–below the abscissa axis, which is the line of death for the evader, and the normal p_2 coincides in direction with the axis of ordinates. The inequalities $| y_1^0 - x_1^0 | < \varepsilon_1, y_2^0 > x_2^0$ separate the initial positions of the pursuer and the evader, starting from which the capture is possible in a finite time.

'Simple' group pursuit. Let $k = \nu \geq s$, $\varepsilon_i \geq 0$. Then sufficient and necessary condition for capture in the group pursuit problem (1.25) is

$$0 \in int \ co \{z_1^0 + \varepsilon_1 S, ..., z_k^0 + \varepsilon_k S\}..$$

1.6 Principle of Shortest Broken Line in Successive Pursuit

Now we proceed to the analysis of the problem of successive capturing of a group of evaders by a single pursuer. Various schemes have been developed with a fixed and non-fixed moment of capture of each of the evaders. The proposed schemes [6, 14] are especially effective in the case of 'simple motions' with the beforehand programmed choice of the order of capture. Then the pursuit strategy is the parallel approach and, consequently, the total capture time depends only on the evader's controls. In so doing, extremum of the functional is attained on constant controls of evaders, and the infinite-dimensional problem of maximizing the total pursuit time is reduced to a finite-dimensional conditional optimization problem.

Let a conflict-controlled process be described by the quasi-linear equation

$$\dot{z}_j = A_j z_j + \varphi_j(u, v_j), z_j \in R^{n_j}, u \in U, v_j \in V_j, j = 1, \ldots, \mu, \quad (1.29)$$

A_j are square matrices of the order n_j, u, v_j are the control parameters of the pursuer and evaders belonging to the nonempty compacts U and V_j, respectively, functions $\varphi_j(u, v_j)$ are jointly continuous in their variables.

The terminal set consists of cylindrical sets

$$M_j^* = M_j^0 + M_j, M_j^* \subset R^{n_j}, j = \overline{1, \mu}, \quad (1.30)$$

where M_j^0 are linear subspaces from R^{n_j}, and M_j are convex compact sets from the orthogonal complements L_j to M_j^0 in R^{n_j}.

The goal of the pursuer is to bring in turn all the trajectories $z_j(t)$, $j = \overline{1, \mu}$, to the corresponding sets M_j^* in the least total time. The pursuer is faced with a double problem, dealing with control and enumerating, which can in no way be divided and studied separately. Let us denote by z_j a pair (x, y_j), where x is the pursuer state and y_j-the evaders. Then inclusion $z_j \in M_j^*$ signifies the capture of the jth evader by the pursuer. We suppose that the pursuer applies quasi-strategies.

We state that the successive pursuit can be completed, starting from the initial state, $z^0 = (z_1^0, \ldots, z_\mu^0)$ no later than at the time T, if there is a quasi-strategy of the pursuer such that for any controls of evaders $v_T(\cdot)$ there exist moments t_j, $0 < t_j \leq T$, $j = \overline{1, \mu}$, such $z_j(t_j) \in M_j^*$ for all $j = \overline{1, \mu}$. It should be noted that the values of the function $v_j(t)$ are used only on time intervals $[0, t_j]$. In the case the order of bringing trajectories $z_j(t)$ to the corresponding sets M_j^* is chosen beforehand, i.e. is fixed at the initial moment and then does not change, the number of possible options is $\mu\_!$

We denote by Λ the set of all possible orders for capturing the evaders. Let us assume that at the initial moment the pursuer selects some order l, $l \in \Lambda$, and then sticks it. Without loss of generality, we determine by $l = \{1, 2, \ldots, \mu\}$ the order of bringing the trajectories to corresponding sets.

Let us embark on minimization of the total time. We denote by π_j be the operator of orthogonal projecting from R^{n_j} to L_j. We set

$$W^j(t, \, v_j) = \pi_j e^{A_j t} \varphi_j(U, \, v_j), W^j(t) = \bigcap_{v_j \in V_j} W^j(t, \, v_j),$$

$$j = 1, \, ..., \, \mu, t \geq 0.$$

Condition 1.4. The direct images of set-valued mappings $W^j(t)$ are not empty and closed for all $j = 1, ..., \, \mu, t \geq 0$.

Since the mappings $W^j(t)$ are measurable and closed-valued, then there exist measurable selections $\gamma_j(t)$, $\gamma_j(t) \in W_j(t)$, $t \geq 0$, $j = 1, \, ..., \, \mu$ [9]. We set

$$\xi_j(t, \, z_j, \, \gamma_j(\cdot)) = \pi_j e^{A_j t} z_j + \int_0^t \gamma_j(\tau) \, d\tau.$$

Then the resolving functions take the forms

$\alpha_j(t, \, \tau, \, z_j, \, v_j, \, \gamma_j(\cdot)) = \sup \{\alpha \geq 0 : [W^j(t - \tau, \, v_j) - \gamma_j(t - \tau)] \bigcap \alpha[M_j - \xi_j(t, \, z_j, \, \gamma_j(\cdot))] \neq \phi\}, j = 1, \, ..., \, \mu, t \geq \tau \geq 0.$

We see that they enjoy all the above-listed properties.

Now we determine recursively the moments of trajectories (1.29) hitting the corresponding sets (1.30) in order l. The following relations are true:

$$t_j = t_j(z_j(t_{j-1}), \, v_j(\cdot)) = t_j(z_1^0, \, ..., \, z_j^0, \, v_1(\cdot), \, ..., \, v_j(\cdot)),$$

Here

$$t_j = \min \left\{ t \geq t_{j-1} : \int_{t_{j-1}}^{t_j} \alpha_j(t, \, \tau, \, z_j(t_{j-1}), \, v_j(\tau), \, \gamma_j(\cdot)) \, d\tau \geq 1 \right\},$$

$$j = 1, \, ..., \, \mu.$$

It is evident that the total time of capturing all the evaders coincides with the moment of capturing the last evader and, consequently, is equal to t_μ.

Let us denote $z^0 = (z_1^0, \, ..., \, z_\mu^0)$, $v(\cdot) = (v_1(\cdot), \ldots, v_\mu(\cdot))$, and by

$$T_\mu^l(z^0, \, v(\cdot)) = t_\mu(z_1^0, \, ..., \, z_\mu^0, \, v_1(\cdot), \, ..., \, v_\mu(\cdot))$$

–the total capture time.

This time depends on the order of capturing the evaders and its optimization in the case of the programmed order of capture consists in maximizing in $v(\cdot)$ and minimizing concerning the order, i.e.

$$T_\mu(z^0, \gamma(\cdot)) = \min_{l \in \Lambda} \sup_{v(\cdot)} T_\mu^l(z^0, \, v(\cdot)),$$

$\gamma(\cdot) = (\gamma_1(\cdot), \ ..., \ \gamma_\mu(\cdot))$ are the fixed forehand selections.

Theorem 1.8 Let Condition 1.4 be satisfied, and there exist measurable selections of set-valued mappings $\gamma_j(t) \in W^j(t)$, $t \geq 0$, $j = 1, ..., \mu$, such that $T_\mu(z^0, \ \gamma(\cdot)) < +\infty$.

Then, starting from the initial state z^0, the successive pursuit can be terminated at the time instant $T_\mu(z^0, \ \gamma(\cdot))$, using quasi-strategies.

To apply the modified method scheme with non-fixed termination time to solving the successive pursuit problem using stroboscopic strategies requires significantly more burdensome constraints on the process parameters. It is suited to the case of 'simple motions'. For the sake of simplicity and geometric descriptiveness, we restrict ourselves to the case of plane motions. Let be

$$\dot{z}_j = u - v_j, z_j \in R^2, z_j(0) = z_j^0,$$

$$\|u\| \leq 1, \| \ v_j \ \| \leq \beta < 1, j = 1, \ ..., \ \mu, \qquad (1.31)$$

$$z_j = x - y_j, \ \dot{x} = u, \ \dot{y}_j = v_j.$$

It is easy to observe that each of the pursuers has an advantage over each of the evaders, otherwise, the problem has no sense. The goal of the pursuer is to bring in turn all trajectories $z_j(t)$ to the origin in a finite time, i.e. $M_j^* : z_j = 0$.

As before, it is assumed that the system with the lower index has priority in the order of bringing. We denote

$$t_j = \min \{t > 0 : z_j(t) = 0\} \qquad (1.32)$$

the moment of the first hitting the set M_j^*. by the trajectory $z_j(t)$.

We have

$$M_j^0 = M_j = \{0\}, L_j = R^2, \pi_j = I, \ A_j = \{0\},$$

$$\varphi_j(u, \ v_j) = u - v_j, j = 1, \ ..., \ \mu.$$

Then

$$W^j(t, \ v_j) = S - v_j, W^j(t) = (1 - \beta) \, S, j = 1, \ ..., \ \mu, t \geq 0.$$

Since $\beta < 1$, then Condition 1.4 is fulfilled. We select $\gamma_j(t) \equiv 0$, $j = 1, \ ..., \ \mu$. Then $\xi_j(t, \ z_j, \ 0) = z_j$, and the resolving functions are

$$\alpha_j(t, \tau, z_j, v_j, 0) = \alpha_j(z_j, \ v_j)$$

$$= \frac{(v_j, \ z_j) + \sqrt{(v_j, \ z_j)^2 + \| \ z_j \ \|^2 \, (1 - \| \ v_j \ \|^2)}}{\| \ z_j \ \|^2}, j = 1, \ ..., \ \mu.$$

Therefore, the pursuer's control, which brings the *j*th system (1.31) to zero, takes the form

$$u(\tau) = v_j(\tau) - \alpha_j(z_j(t_{j-1}), \ v_j(\tau)) \, z(t_{j-1}), \ \tau \in [t_{j-1}, \ t_j), \qquad (1.33)$$

and the moments t_j are defined by expression (1.32).

This control law (1.33) applies the strategy of parallel pursuit and is closely related with the Apollonian circle [6], centered at the point, called the Apollonius point. If $\mu = 1$, then the Apollonian circle is the locus of the capture points of the evader, under the condition the evader is moving with maximum speed along a straight line, and the pursuer implements the strategy of parallel pursuit (1.33).

As was shown in [6] in the case of 'simple motions' of the players and $\mu \neq 1$, in order to maximize the total time of capturing all the evaders, when the pursuer sticks the strategy of parallel pursuit under fixed capture order, each of the evaders should move with maximum speed along a straight line. Let us call by the attainable circle of the *j*th evader the set

$$y_j^0 + \beta l_{j-1} \partial \, S.$$

Each point of this circle and point $x(t_{j-1})$ possesses an own Apollonius circle with the corresponding center. Herewith, the centers of the Apollonius circles, corresponding to the points of the attainable circle of the *j*th evader, also form a circle (the circle of centers). Imagine that at each point of the circle of centers the corresponding Apollonius circle is constructed. Then the outer envelope of these circles is exactly the locus of the points of the capture of player y_j. The equation of this envelope in both polar and Cartesian coordinates is presented in [6].

Hence, it is shown that if the order of captures is fixed, the pursuers apply the law of parallel pursuit and the evaders are moving rectilinearly at maximum speed, then the total capture time is functional depending only on the constant controls of the evaders, and its maximization presents the nonlinear finite-dimensional optimization problem [6].

To select the order of captures ensuring the least total time, the principle of the shortest broken line was developed [6]. By the shortest broken line is called a broken line of minimal length, starting from the initial position of the pursuer and one at a time passing through all the initial positions of the evaders. In the case of 'simple motions', under certain assumptions, the principle of the shortest broken line provides the smallest total capture time [6].

For practical applications see [32].

1.7 Conclusion

An overview of contemporary research methods of the problem of decision making and control in conflict conditions with the participation of moving object groups is presented. The general problem of the game interaction of groups with the help of decomposition is divided into separate problems of group and successive pursuit with subsequent iterative repetition.

Methods for solving the mentioned problems, based on the method of resolving functions, are presented. Their effectiveness is demonstrated, in particular, on test examples from the monograph by R. Isaacs.

In so doing, the presence of phase coordinates for the state of objects is taken into account, and the case of a non-fixed time of the game termination is also highlighted. The situation of the environment by Pshenichnyi and the principle of the shortest broken line make the results of group interaction geometrically descriptive.

References

[1] A. A. Chikrii, 'Escape Problem for Control Dynamic Objects', Journal of Automation and Information Sciences, 1997, vol. 29, No. 6, pp. 71–82.

[2] R. F. Isaacs, 'Differential Games: Their Scope, Nature and Future', J. Opt. Theory Appl., 1969, No 3, p. 283–295.

[3] L. S. Pontryagin, 'Linear Differential Games of Pursuit', Matem. Sbornik, 1980, 112, No 3, p. 307–330. (in Russian).

[4] B. N. Pschenichnyi, 'Structure of Differential Games', Dokl. Akad. Nauk. SSSR, 1969, 184, No 2. (in Russian).

[5] N. N. Krasovskii, A. I. Subbotin, 'Positional Differential Games', Nauka, Moscow, 1974 (in Russian).

[6] A. A. Chikrii, 'Conflict controlled processes', Naukova Dumka, Kiev, 1992 (in Russian).

[7] A. S. Locke, 'Guidance', D. Van Nostrand Company, Inc., Princeton, 1955.

[8] B. S. Mordukhovich, 'Approximation Methods in Problems of Optimization and Control', Nauka, Moscow, 1988 (in Russian).

[9] J. Warga, 'Optimal Control of Differential and Functional Equations', Nauka, Moscow, 1977 (in Russian).

[10] B. N. Pschenitchnyi, 'Simple pursuit by several objects', Kibernetika, 1976, No. 3, pp. 145–146. (in Russian).

[11] N. L.Grigorenko, 'Differential Games of Pursuit by Several Objects' Izdat. Gos. Univ., Moscow, 1983. (in Russian).

[12] N. N. Petrov, N.N. Petrov, 'On Differential Game "Kazaki - Razboyniki"', Diff. Uravn., 1983, 19, No 8, p. 1366–1374 (in Russian).

[13] A. A. Chikrii, 'Methods of Group Pursuit', Stochastic Optimization, Int. Conf., Edited by V.Arkin, A. Shiraev, R. Wets, Springer–Verlag, 1986, pp. 632–640.

[14] A. A. Chikrii, S. F. Kalashnikova, L. A. Sobolenko, 'A numerical method for the solution of the successive pursuit problem', Cybernetics, 1988, vol. 24, No. 1, pp. 53–59.

[15] B. N. Pshenichnyi, A. A. Chikrii, I. S. Rappoport, 'An efficient method of solving differential games with many pursuers', Soviet Math. Dokl., 1981, vol. 23, No. 1, pp. 104–109.

[16] G. Ts. Dzyubenko, B. N. Pschenichnyi, 'Discrete differential Games with information lag', Cybernetics, 1972, vol. 8, No. 6, pp. 947–952.

[17] M. S. Nikolskiy, 'Linear Differential Games of Pursuit with Information Time Lag', Soviet Math. Dokl., 1971, 197, No. 5, pp. 1018–1021. (in Russian).

[18] L. A.Vlasenko, A. D. Myshkis, A. G. Rutkas, 'On a class of differential equations of parabolic type with impulsive action', Differential Equations, 2008, **44**, No. 2, pp. 231–244.

[19] A. A. Chikrii, 'Game Dynamic Problems for Systems with Fractional Derivatives', ≪ Pareto-Optimality, Game Theory and Equilibria ≫, 2008, vol.17, New York, Springer, pp. 349–387.

[20] L. A. Vlasenko, A. G. Rutkas, 'Stochastic impulse control of parabolic systems of Sobolev type', Differential Equations, 2011, **47**, No. 10, pp. 1498–1507.

[21] V. A. Pepelyaev, Al. A. Chikrii, K. A. Chikrii, 'On nonstationary problems of motion control in conflict situation', Journal of Automation and Information Sciences, 2019, 51, No. 7, pp. 55–66.

[22] A. N. Khimich, K. A. Chikrii, 'Game approach problems for dynamic processes with impulse controls', Cybernetics and Systems Analysis, 2009, 45, No 1, pp. 123–140.

[23] I. Yu. Krivonos, Al. A. Chikrii, K. A. Chikrii, 'Some modifications to the pontryagin's condition in nonstationary quasilinear games', Journal of Automation and Information Sciences, 2013, **45**, No. 11, pp. 15–21.

[24] A. A. Chikrii, 'Linear problem of avoiding several pursuers', Engineering Cybernetics, 1976, 14, No 4, pp. 38–42.

[25] A. A Chikrii, I. S. Rappoport, 'Systems analysis method of resolving functions in the theory of conflict-controlled processes', Cybernetics and Systems Analysis, 2012, 48, No 4, pp. 512–531.

[26] O. Hajek, 'Pursuit games', New York: Academic Press, 1975.

[27] M. S Nikolskiy, 'The first direct L. S. Pontryagin method in differential games', Izdatelstvo MGU, Moscow, 1984.

[28] G. T. Chikrii, 'On the problem of approach for damped oscillations', Journal of Automation and Information Sciences, 2009, **41**, No. 10, pp.1–9.

[29] A. G. Nakonechnyi, S. O. Mashchenko, V. K. Chikrii, 'Motion control under conflict condition', Journal of Automation and Information Sciences, 2018, **50**, No. 1, pp. 54–75.

[30] A. A. Chikrii, V. K. Chikrii, 'Image Structure of Multivalued Mappings in Game Problems of Motion Control', Problemi upravlenia i informatiki, 2016, No 2, pp. 65–78.

[31] A. O. Chikrii, G. T. Chikrii, 'Matrix Resolving Functions in Dynamic Games of Approach', Cybernetics and System Analysis, 2014, 50, No. 2, pp. 201–217.

[32] Y. P. Kondratenko, V. M. Kuntsevich, A. A. Chikrii, V. F. Gubarev, 'Advanced Control Systems: Theory and Applications', River Publishers, Series in Automation, Control and Robotics, 2021.

2

Applications of Variational Analysis to Controlled Sweeping Processes

Boris S. Mordukhovich

Department of Mathematics, Wayne State University, Detroit,
Michigan 48202, USA.
E-mail: aa1086@wayne.edu.

Abstract

We present recent developments and applications of variational analysis and generalized differentiation to rather new classes of optimal control problems governed by discontinuous differential inclusions that arise in the so-called sweeping processes and the like. Such problems are particularly important for numerous applications to automatics control systems, engineering design, hysteresis, etc. To study highly challenging control problems of this type, we develop appropriate constructions of the method of discrete approximations, which allow us to investigate various theoretical and numerical aspects of optimal control for discontinuous differential systems. In particular, we establish new necessary optimality conditions for such problems including appropriate extensions of the Euler–Lagrange conditions and the Pontryagin maximum principle along with the optimality conditions of the novel type specific for the problems under consideration. New applications to practical engineering and mechanical models, robotics, traffic equilibria, etc. are also discussed.

Keywords: Automatic control systems, optimal control, sweeping processes, hysteresis, variational analysis, generalized differentiation, discrete approximations, optimality conditions, Euler–Lagrange and Hamiltonian inclusions, Pontryagin's maximum principle.

2.1 Introduction and Discussions

Automatic control systems and related problems of control engineering design are among the most important practical models that frequently appear in applications. It has been well recognized that appropriate methods of optimal control and differential games play a major role in theoretical investigations and solving of such problems; see, e.g., the books [13, 23, 28, 29, 31, 32, 42, 44] and the references therein. Among advanced techniques in the study of challenging issues appearing in automatic control systems, we highlight methods of modern variational analysis and generalized differentiation.

Variational analysis has been well recognized as an active and fruitful area of mathematics sciences that, on one hand, concerns the study of optimization-related problems and, on the other hand, applies optimization, perturbation, and approximation ideas to the analysis of a broad range of problems that may not be of a variational nature. This area can be considered as an outgrowth of the classical calculus of variations, optimal control theory, and mathematical programming, where the focus is on optimization of functions relative to various constraints and on the solution sensitivity with respect to perturbations.

A characteristic feature of modern variational analysis is a broad involvement of objects with nonsmooth structures (nondifferentiable functions, sets with nonsmooth boundaries, and set-valued mappings), which appear naturally and frequently in the framework of optimization, equilibrium, and control problems while being often generated by the usage of variational principles and techniques. Nonlinear systems and variational principles in applied sciences also give rise to nonsmooth structures and motivate the developments of new forms of analysis that rely on generalized differentiation. Starting with nonsmooth convex functions, subgradient mappings are inevitably set-valued and thus require tools of set-valued analysis for their study and applications.

The last two decades have witnessed rapid progress in developing and applications of variational analysis and generalized differentiation. Starting with the fundamental monograph by Rockafellar and Wets [45] devoted to finite-dimensional aspects of variational analysis and their applications to constrained optimization, an enormous amount of publications on the subject appeared and great many applications were developed to various branches of mathematical sciences as well as to engineering, economics, biology, environmental and behavioral sciences, operations research, computer sciences, robotics, etc. We refer the reader to [45] and the author's monographs

[36, 37], with the vast bibliographies and commentaries therein, for the genesis of ideas, basic constructions and results, and applications of variational analysis and generalized differentiation in finite and infinite dimensions. In particular, the second volume of [36] contains a rather comprehensive account of the methods and results of variational analysis applied to nonsmooth controlled systems governed by ordinary differential, functional differential, and partial differential equations and Lipschitzian differential inclusions with a variety of applications.

New trends in variational analysis largely relate to the investigation and solving of meaningful problems, which often come from applications. A broad class of such problems are governed by discontinuous differential inclusions introduced originally by Moreau [40, 41] in the form of the *sweeping process*

$$\dot{x}(t) \in -N(x(t); C(t)) \quad a.e. \quad t \in [0, T] \quad with \quad x(0) := x_0 \in C(0), \quad (2.1)$$

where 'a.e.' means as usual 'almost everywhere' in the sense of Lebesgue measure on $[0, T]$, and where $N(x; C)$ stands for the normal cone of convex analysis defined by

$$N(x; C) := \{v \in \mathbb{R}^n | \langle v, y - x \rangle \leq 0, \ y \in C\} \ if \ x \in C \ and$$
$$N(x; C) := \varnothing \ if \ x \notin C \qquad (2.2)$$

for the continuously moving convex set $C = C(t)$ at the point $x = x(t)$. The primal motivation of [41] mainly came from problems of elastoplasticity, but over the years the sweeping process and its modifications have been developed in dynamical system theory with many applications to various areas of mechanics, engineering, robotics, economics, traffic equilibria, etc.; see, e.g., the excellent recent survey [8] with the references therein. Independently, basic theory and important engineering applications of sweeping processes were largely developed (not under this name) in the Soviet literature by Krasnosel'skii, Pokrovskii, and their followers for *systems of hysteresis*; see the English book [30] with many references to the previous publications in Russian.

Optimization and control problems for sweeping processes were formulated much latter. A primal reason for this situation is that the Cauchy problem for the sweeping process (1) has a *unique* solution, and so there is nothing to optimize. To the best of our knowledge, optimal control problems for sweeping processes were first defined with controls functions acting in additive perturbations as in Edmond and Thibault [22]. However, the focus of [22], as well as of subsequent publication in this direction (see, e.g., Tolstonogov [48] and related chapters), was on the existence of optimal controls and relaxation procedures, while not on deriving necessary optimality conditions that have always been at the core of the calculus of variation and optimal control.

Necessary optimality conditions for controlled sweeping processes were first established less than ten years ago in Colombo et al. [16] (see also [17]) for a novel class of problems with control functions acting in the moving set formalized as $C(t) = C(u(t))$ on $[0, T]$. Another type of controlled sweeping processes was introduced in Brokate and Krejčcí [9], with deriving necessary optimality conditions, in the setting where control functions entered a linear ODE system adjacent to the sweeping dynamics. First necessary optimality conditions for sweeping control problems with control actions in dynamic perturbations were derived in Cao and Mordukhovich [11]. After these initial works, and especially in the very recent years, optimal control theory for various types of controlled sweeping processes has been strongly developed in many publications with establishing new necessary optimality conditions such that some of them do not have any analogs in the classical theory of optimal control; see, e.g., [2, 4, 10, 12, 15–20, 26, 52] among other chapters.

It is important to emphasize that all the aforementioned types of optimal control problems dealing with the sweeping differential inclusions (2.1) are highly different from optimal control systems governed by smooth ordinary differential equations as in the classical monograph by Pontryagin et al. [42] and by their nonsmooth counterparts, as well as by set-valued problems governed by *Lipschitzian* differential inclusions of the type

$$\dot{x}(t) \in F(x(t)) \quad a.e. \quad t \in [0, T], \tag{2.3}$$

where $F(x)$ is a Lipschitz continuous set-valued mapping (for simplicity we consider here only autonomous systems); see, e.g., the books [14, 35, 36, 43, 46, 50] for more details and references. It is clear that the latter formalism covers the ODE control systems

$$\dot{x}(t) = f(x(t), u(t)), \quad u(t) \in U \quad a.e. \quad t \in [0, T], \tag{2.4}$$

where the control set U in (4) may depend also of the state $U = U(x)$ and thus reflects *feedback laws*. The sweeping inclusions (2.1) are dramatically different from (2.3), and hence from (2.4), due to highly non-Lipschitzian (actually discontinuous) structure of the sweeping dynamics (2.1), which does not allow us to employ the conventional techniques developed earlier in optimal control theory.

Another, even more essential difference between (2.1) and the dynamics in (2.3) and (2.4) is that in the case of controlled moving sets $C(t) = C(u(t))$ in (2.1), the right-hand site in (2.1) is *not fixed* as in (2.3) and in (2.4) with $F(x) := f(U(x))$. Thus optimizing (2.1) is actually a problem of *shape optimization*.

The main goal of this chapter is to discuss some recent results in dynamic optimization of controlled sweeping processes and some of their applications. We focus on the following two types of the sweeping dynamics: with controls acting in moving sets and with controls entering additive perturbations. These two classes of sweeping optimal controls and the necessary optimality conditions obtained for them are essentially different from each other.

Nevertheless, we derive necessary optimality conditions for both of these classes of sweeping control problems by using the *method of discrete approximations* and advanced tools of *generalized differentiation* in variational analysis. Recall that the discrete approximation approach to obtain optimality conditions for continuous-time variational problems goes back to Euler in the classical calculus of variations. This method was used by Pshenichnyi [43] to optimize *Lipschitzian* differential inclusions governed by set-valued mappings with convex graphs and also with convex values under rather restrictive assumptions. Those restrictions have been overcome for convex-valued Lipschitzian differential inclusions, with the usage of generalized differentiation developed by the author (see Section 2), in the books by Mordukhovich [35] and Smirnov [46]; see also the references therein. Then the author [36] established, by employing again the method of discrete approximations, refined optimality conditions for Lipschitzian differential inclusions without any convexity assumptions. Below we present major results on necessary optimality conditions obtained by appropriate developments of the method of discrete approximations for the aforementioned classes of optimal control problems for *sweeping differential inclusions* of highly *discontinuous* dynamics.

The rest of the chapter is organized as follows. In Section 2 we review basic tools of *first-order and second-order generalized differentiation* in variational analysis that is broadly used in the formulations and proofs of the main necessary optimality conditions given below.

Section 3 addresses optimal control problems for sweeping processes with control functions entering *moving sets* that generate the sweeping dynamics. Here we establish necessary optimality conditions of generalized *Euler–Lagrange and Hamiltonian types*, which yield a new version of the *Weierstrass–Pontryagin maximization condition* in terms of the *modified*

Hamiltonian while an expected form of the Pontryagin maximum principle *fails*.

Section 4 deals with controlled sweeping processes, where control functions appear in additive *perturbations of the dynamics*. A variety of necessary optimality conditions obtained for such systems include, among other relationships of the new type, an appropriate version of the *Pontryagin maximum principle* in the class of measurable controls. The concluding Section 5 presents selected applications of the achieved results and discuses some perspectives of the ongoing and future research.

Throughout the chapter we use the standard notation of variational analysis and optimal control; see, e.g., the books [36, 45, 50].

2.2 Generalized Differentiation of Variational Analysis

This section overviews the basic generalized differential constructions, which play a central role of many aspects of variational analysis and its applications. These constructions were mainly introduced by the author in different times and then were largely developed by him and his collaborators along with other researchers in this field. The reader can find more details, comments, and references in the books [35–36, 37, 45].

Employing the geometric approach to generalized differentiation [35–37], we start with our basic concept of the normal cone to a locally closed set that induces the corresponding generalized differentiability notions for nonsmooth functions and (single-valued and set-valued) mappings. Note that our discrete approximation technique requires considering normals to nonconvex sets even in the case of sweeping processes generated by convex moving sets as in (2.1). The (basic, limiting, Mordukhovich) *normal cone* to a nonempty locally closed set $\Omega \subset \mathbb{R}^n$ at $\bar{x} \in \mathbb{R}^n$ is defined by

$$
N(\bar{x}; \Omega) := \begin{cases} \{v \in \mathbb{R}^n \mid \exists x_k \to \bar{x}, \ \alpha_k \geq 0, \ w_k \in \Pi(x_k; \Omega), \ \alpha_k(x_k - w_k) \\ \qquad\qquad\qquad\qquad\qquad\qquad\qquad\qquad\qquad \to v \ if \ \bar{x} \in \Omega, \\ \varnothing \qquad\qquad\qquad\qquad\qquad\qquad\qquad\qquad\qquad otherwise, \end{cases}
$$
$$(2.5)$$

where $\Pi(x; \Omega)$ stands for the nonempty Euclidean projector of x onto Ω. If Ω is convex, the normal cone (2.5) reduces to the normal cone of convex analysis (2.2), but in general the multifunction $x \rightrightarrows N(x; \Omega)$ is nonconvex-valued while satisfying a *full calculus* together with the associated subdifferential of extended-real-valued functions and coderivative of set-valued mappings considered below. Such a calculus is due to *variational/extremal principles* of variational analysis; see [36, 37, 45] for more details.

Given a set–valued mapping $F : \mathbb{R}^n \rightrightarrows \mathbb{R}^q$ and a point $(\bar{x}, \bar{y}) \in gph\, F$ from its graph

$$gph\, F := \{(x, y) \in \mathbb{R}^n \times \mathbb{R}^q | \ y \in F(x))\},$$

the *coderivative* $D^* F(\bar{x}, \bar{y}) : \mathbb{R}^q \rightrightarrows \mathbb{R}^n$ of F at $(\bar{x}, \bar{y})$ is defined by

$$D^* F(\bar{x}, \bar{y})(u) := \{v \in \mathbb{R}^n | \ (v, -u) \in N((\bar{x}, \bar{y}); \ gph\, F)\}, \quad u \in \mathbb{R}^q, \tag{2.6}$$

where $\bar{y}$ is omitted in the notation when $F : \mathbb{R}^n \to \mathbb{R}^q$ is single-valued. If F is C^1-smooth around $\bar{x}$, we have $D^* F(\bar{x})(v) = \{\nabla F(\bar{x})^* v\}$ via the adjoint Jacobian matrix. In general, the coderivative (2.6) is a positively homogeneous set-valued mapping satisfying extended calculus rules and providing fullcharacterizations of the fundamental *well-posedness* properties in variational analysis concerning Lipschitzian stability, metric regularity, and linear openness (covering); see [36, 45].

Given an extended-real-valued function $\phi : \mathbb{R}^n \to \overline{\mathbb{R}} := (-\infty, \infty]$ finite at $\bar{x}$, i.e., with $\bar{x} \in dom\, \phi$, the (first-order) *subdifferential* of ϕ at $\bar{x}$ is introduced geometrically by

$$\partial\phi(\bar{x}) := \{v \in \mathbb{R}^n | \ (v, -1) \in N((\bar{x}, \phi(\bar{x})); \ epi\, \phi)\} \tag{2.7}$$

via the normal cone (2.5) to the epigraphical set $epi\, \phi := \{(x, \alpha) \in \mathbb{R}^{n+1} | \ \alpha \geq \phi(x)\}$. If $\phi(x) := \delta_\Omega(x)$, the indicator function of a set Ω equal to 0 for $x \in \Omega$ and to ∞ otherwise, we get $\partial\phi(\bar{x}) = N(\bar{x}; \Omega)$. Fixing $\bar{v} \in \partial\phi(\bar{x})$, the *second-order subdifferential* (or generalized Hessian) $\partial^2\phi(\bar{x}, \bar{v}) : \mathbb{R}^n \rightrightarrows \mathbb{R}^n$ of ϕ at $\bar{x}$ relative to $\bar{v}$ is defined as the coderivative of the first-order subdifferential by

$$\partial^2\phi(\bar{x}, \bar{v})(u) := (D^*\partial\phi)(\bar{x}, \bar{v})(u), \quad u \in \mathbb{R}^n, \tag{2.8}$$

When ϕ is twice continuoisly differentiable around $\bar{x}$, then (2.8) reduces to the classical (symmetric) Hessian matrix

$$\partial^2\phi(\bar{x})(u) = \{\nabla^2\phi(\bar{x})u\} \ \ for\ all \ \ u \in \mathbb{R}^n.$$

In what follows, we also use partial versions of the above subdifferential constructions for functions of two variables $\phi : \mathbb{R}^n \times \mathbb{R}^m \to \overline{\mathbb{R}}$. Define the *partial first-order subdifferential* mapping for $\phi(x, w)$ with respect to x by

$$\partial_x\phi(x, w) := \{ \ set\ of\ subgradients \ \ v \in \mathbb{R}^n \ of \ \ \phi_w := \phi(\cdot, w) \ \ at \ \ x\}$$
$$= \partial\phi_w(x)$$

and then, for $(\bar{x}, \bar{w}) \in \ dom \ \phi$ and $\bar{v} \in \partial_x \phi(\bar{x}, \bar{w})$, introduce the *partial second-order subdifferential* of ϕ with respect to x at $(\bar{x}, \bar{w})$ relative to $\bar{v}$ by

$$\partial_x^2 \phi(\bar{x}, \bar{w}, \bar{v})(u) := (D^* \partial_x \phi)(\bar{x}, \bar{w}, \bar{v})(u) \quad for \ all \ \ u \in \mathbb{R}^n. \tag{2.9}$$

If ϕ is twice continuously differentiable around $(\bar{x}, \bar{w})$, we have

$$\partial^2 \phi(\bar{x}, \bar{w})(u) = \{(\nabla_{xx}^2 \phi(\bar{x}, \bar{w})^* u, \nabla_{xw}^2 \phi(\bar{x}, \bar{w})^* u)\} \quad for \ all \ \ u \in \mathbb{R}^n.$$

Consider now the following class of *parametric constraint systems*

$$S(w) := \{x \in \mathbb{R}^n \mid \psi(x, w) \in \Theta\}, \quad w \in \mathbb{R}^m, \tag{2.10}$$

generated by a vector function $\psi : \mathbb{R}^n \times \mathbb{R}^m \to \mathbb{R}^s$ and a set $\Theta \subset \mathbb{R}^s$, and then associate with (2.10) the *normal cone mapping* $N : \mathbb{R}^n \times \mathbb{R}^m \rightrightarrows \mathbb{R}^n$ defined by

$$N(x, w) := N(x; S(w)) \quad for \ \ x \in S(w). \tag{2.11}$$

It is easy to see that the mapping N in (2.11) admits the representation

$$N(x, w) = \partial_x \phi(x, w) \quad with \quad \phi(x, w) := (\delta_\Theta \circ \psi)(x, w) \tag{2.12}$$

via the composition of ψ and the indicator function δ_Θ of the set Θ. We get from (2.12) and the second-order subdifferential construction (2.9) that

$$\partial_x^2 \phi(\bar{x}, \bar{w}, \bar{v})(u) = D^* N(\bar{x}, \bar{w}, \bar{v})(u) \quad for \ any \ \ \bar{v} \in N(\bar{x}, \bar{w}) \ \ and \ \ u \in \mathbb{R}^n.$$

The following second-order calculus result can be derived from [39, Theorem 3.1] applied to the composition in (2.12). This computation foprmula plays an important role in the our applications to establishing optimality conditions for controlled sweeping processes.

Theorem 2.13 (second-order subdifferential chain rule). *Let ψ be C^2-smooth around $(\bar{x}, \bar{w})$ with the partial Jacobian matrix $\nabla_x \psi(\bar{x}, \bar{w})$ of full rank. Then for each $\bar{v} \in N(\bar{x}, \bar{w})$ there exists a unique vector $\bar{p} \in N_\Theta(\psi(\bar{x}, \bar{w})) := N(\psi(\bar{x}, \bar{w}); \Theta)$ satisfying*

$$\nabla_x \psi(\bar{x}, \bar{w})^* \bar{p} = \bar{v}$$

and such that the coderivative of the normal cone mapping is calculated for all $u \in \mathbb{R}^n$ by

$$D^* N(\bar{x}, \bar{w}, \bar{v})(u) = \begin{bmatrix} \nabla_{xx}^2 \langle \bar{p}, \psi \rangle (\bar{x}, \bar{w}) \\ \nabla_{xw}^2 \langle \bar{p}, \psi \rangle (\bar{x}, \bar{w}) \end{bmatrix} u + \nabla \psi(\bar{x}, \bar{w})^* D^* N_\Theta(\psi(\bar{x}, \bar{w}), \bar{p})$$

$(\nabla_x \psi(\bar{x}, \bar{w}) u).$

2.3 Dynamic Optimization via Controlled Moving Sets

The results of this section are mostly based on the recent chapter [26], which addresses a general class of *sweeping processes with controlled moving* sets defined as inverse images of closed subsets of finite-dimensional spaces under nonlinear differentiable mappings that depend on both state and control variables. Such problems naturally arise in the study of and applications to *hysteresis* and *rate-independent systems*.

The problem under consideration here is formulated as follows:

$$minimize\ J[x,u] := \varphi(x(T)) + \int_0^T \ell(t, x(t), u(t), \dot{x}(t), \dot{u}(t))dt \quad (2.14)$$

over absolutely continuous control actions $u(\cdot)$ and the corresponding absolutely continuous trajectories $x(\cdot)$ of the sweeping differential inclusion

$$\dot{x}(t) \in f(t, x(t)) - N(g(x(t)); C(t, u(t))) \quad a.e. \quad t \in [0, T],$$
$$x(0) = x_0 \in C(0, u(0)), \quad (2.15)$$

where the controlled moving set is given by

$$C(t, u) := \{x \in \mathbb{R}^n | \psi(t, x, u) \in \Theta\}, \quad (t, u) \in [0, T] \times \mathbb{R}^m, \quad (2.16)$$

with $f : [0, T] \times \mathbb{R}^n \to \mathbb{R}^n$, $g : \mathbb{R}^n \to \mathbb{R}^n$, $\psi : [0, T] \times \mathbb{R}^n \times \mathbb{R}^m \to \mathbb{R}^s$, and $\Theta \subset \mathbb{R}^s$. The sets $C(t, u)$ may not be convex, and the normal cone in (2.15) is understood in the sense of (2.5). It follows from (2.15) due to the normal cone definition that the formulated optimal control problem contains the *pointwise constraints* on both *state and control functions* given by

$$\psi(t, g(x(t)), u(t)) \in \Theta \quad for\ all \quad t \in [0, T].$$

Such constraints have been realized among the *most challenging* ones even in standard optimal control theory for smooth ordinary differential equations.

Let us now formulate our *standing assumptions* in this section:

(H1) There is $L_f > 0$ such that $\|f(t, x) - f(t, y)\| \le L_f\|x - y\|$ for all $x, y \in \mathbb{R}^n$, $t \in [0, T]$ and the mapping $t \mapsto f(t, x)$ is a.e. continuous on $[0, T]$ for each vector $x \in \mathbb{R}^n$.

(H2) There is $L_g > 0$ such that $\|g(x) - g(y)\| \le L_g\|x - y\|$ for all $x, y \in \mathbb{R}^n$.

(H3) For each $(t, u) \in [0, T] \times \mathbb{R}^m$, the mapping $\psi_{t,u}(x) := \psi(t, x, u)$ is twice continuously differentiable around the reference points with the surjective derivative $\nabla\psi_{t,u}(x)$ satisfying

$$\|\nabla\psi_{t,u}(x) - \nabla\psi_{t,v}(x)\| \le L_\psi\|u - v\|$$

with the uniform Lipschitz constant L_ψ. Also we assume that the mapping $t \mapsto \psi(t, x)$ is a.e. continuous on $[0, T]$ for each $x \in \mathbb{R}^n$ and $u \in \mathbb{R}^m$.

(H4) There are a number $\tau > 0$ and a mapping $\vartheta : \mathbb{R}^n \times \mathbb{R}^n \times \mathbb{R}^n \times \mathbb{R}^m \to \mathbb{R}^m$ locally Lipschitz continuous and uniformly bounded on bounded sets such that for all $t \in [0, T]$, $\bar{v} \in N(\psi_{(t,\bar{u})}(\bar{x}); \Theta)$, and $x \in \psi_{(t,u)}^{-1}(\Theta)$ with $u := \bar{u} + \vartheta(x - \bar{x}, x, \bar{x}, \bar{u})$ there exists $v \in N(\psi_{(t,u)}(x); \Theta)$ satisfying $\|v - \bar{v}\| \leq \tau \|x - \bar{x}\|$.

(H5) The objective functions $\varphi : \mathbb{R}^n \to \overline{\mathbb{R}} := [-\infty, \infty)$ and $\ell(t, \cdot) : \mathbb{R}^{2(n+m)} \to \overline{\mathbb{R}}$ in (2.14) are bounded from below and lower semicontinuous around a given feasible solution to (P) for a.e. $t \in [0, T]$, while the running costs ℓ is a.e. continuous in t and is uniformly majorized by a summable function on $[0, T]$.

(H6) The set Θ in (2.16) has nonempty interior and locally closed around the reference points.

Assumption (H4) is technical being the most restrictive. As shown in [26], it is automatically satisfied in the case of *polyhedral* moving sets

$$C(t) := \{x \in \mathbb{R}^n | \langle u_i(t), x \rangle \leq w_i(t), \ i = 1, \dots, m\}$$

controlled by both $u_i(t)$ and $w_i(t)$ as in [17], and also in various nonconvex settings.

Our approach to the study of the sweeping control problem formulated in (2.14)–(2.16) is based on the *method of discrete approximations* containing the following *major steps*:

Step I: Construct a *well-defined family* of discrete approximations involving a finite-difference replacement of the derivative $\dot{x}$ in (2.15). The key point of this step is to verify the possibility to approximate any *feasible* trajectory of (2.15) by feasible trajectories of discrete-time systems in a topology yielding the a.e. convergence of the extended discrete derivatives alonmg a subsequence. We address not only *qualitative* aspects of well-posedness but also *quantitative* ones with estimating error bounds, convergence rates, etc., which are of undoubted numerical values. Achieving these goals would lead us then to the strong $W^{1,2}$-norm approximation of a given *local minimizer* for the continuous-time problem under consideration by a sequence of optimal solutions to the discrete-time problems that are piecewise linearly extended to the whole interval $[0, T]$.

Step II: For each fixed step of discretization, the approximating discrete-time problems can be reduced to nondynamic problems of *mathematical programming* with various constraints, including increasingly many *geometric constraints*. The latter problems are *finite-dimensional* of increasing dimensions. Powerful tools of generalized differentiation in variational analysis overviewed in Section 2 can be employed for deriving *discrete-time necessary optimality conditions* in the approximation problems, even without any Lipschitzian and convexity assumptions, by using appropriate *calculus rules*. However, dealing with the *graphical structure* of the geometric constraints in the approximation problems requires robust generalized differential constructions enjoying full calculus that should be subtle and small enough to be efficiently applied to graphical sets. Note to this end, note that applying the *convexified* normal cone and related constructions by Clarke [14] to graphs of mappings often gives us the whole space or its subspace of maximal dimension; see [36, 45]. On the other hand, the nonconvex generalized differential constructions from Section 2 satisfy all the required properties and can be successfully used.

Step III: The concluding step in deriving necessary conditions for optimal solutions of (2.14)–(2.16) is the passage to the limit from those for discrete-time problems obtained in Step II with employing the strong convergence of discrete optimal solutions obtained in Step I. This final step is highly involved, since it requires justifying an appropriate convergence of *dual* arcs as trajectories of *adjoint differential inclusions*. In this way we arrive at the *necessary optimality conditions for sweeping control problems* presented below.

Let us now formulate the notion of local minimizers for which we derive necessary optimality conditions in what follows.

Definition 2.17 (local minimizers for controlled sweeping processes). *Let the pair $(\bar{x}(\cdot), \bar{u}(\cdot))$ be feasible to problem (2.14)–(2.16) under the imposed standing assumptions. We say that $(\bar{x}(\cdot), \bar{u}(\cdot))$ be a* local $W^{1,2} \times W^{1,2}$-*minimizer for this problem if $\bar{x}(\cdot) \in W^{1,2}([0,T]; \mathbb{R}^n)$, $\bar{u}(\cdot) \in W^{1,2}([0,T]; \mathbb{R}^m)$, and*

$$J[\bar{x}, \bar{u}] \leq J[x, u] \quad for \ all \quad x(\cdot) \in W^{1,2}([0,T]; \mathbb{R}^n) \quad and$$
$$u(\cdot) \in W^{1,2}([0,T]; \mathbb{R}^m)$$

sufficiently close to $(\bar{x}(\cdot), \bar{u}(\cdot))$ in the corresponding $W^{1,2}$-norm topology.

Note that this notion, while being rather specific for the class of control problems under consideration, occupies an *intermediate position* between the conventional notions of *weak* and *strong* local minimizers in variational problems; see [36, Chapter 6] for more discussions in the framework of Lipschitzian differential inclusions. Observe also the choice of the class of feasible solutions to problem (2.14)–(2.16) is due to the known existence theorems for the sweeping dynamics; compare, e.g., [12, 33, 48].

As has been well understood in the calculus of variations and optimal control, starting with the pioneering chapters by Bogolyubov [6] and Young [51], finishing limiting procedures in variational arguments require some *relaxation stability* under the velocity convexification, which in fact holds automatically in fairly general settings; see [14, 36, 47, 50] for Lipschitzian differential inclusions and [22, 48] for non–Lipschitzian frameworks of sweeping processes.

Following this idea, let us define the mapping

$$F = F(t, x, u) := f(t, x) - N(g(x); C(t, u)) \tag{2.18}$$

and formulate the *relaxed optimal control problem* for (2.14)–(2.16) as follows:

$$minimize \ \widehat{J}[x, u] := \varphi(x(T)) + \int_0^T \widehat{\ell}_F(t, x(t), u(t), \dot{x}(t), \dot{u}(t)) dt, \tag{2.19}$$

where $\widehat{\ell}(t, x, u, \cdot, \cdot)$ is defined as the largest l.s.c. convex function majorized by $\ell(t, x, u, \cdot, \cdot)$ on the convex closure of the set F in (2.18) with $\widehat{\ell} := \infty$ otherwise. Then we say that the pair $(\bar{x}(\cdot), \bar{u}(\cdot))$ is a *relaxed local* $W^{1,2} \times W^{1,2}$-*minimizer* for (2.14)–(2.16) if in additions to the conditions of Definition 2.17 we have $J[\bar{x}, \bar{u}] = \widehat{J}[\bar{x}, \bar{u}]$. Observe that, contrary to the original sweeping control system in (2.14)–(2.16), the convexified structure of the relaxed dynamic optimization problem (2.19) allows us to establish the *existence* of global optimal solutions in the prescribed classes of controls and trajectories; cf. [12, 33, 48].

As we see, there is no difference between the original and relaxed problems if the normal cone in (2.18) is convex and the integrand ℓ in (2.14) is convex with respect to velocity variables. On the other hand, the nonatonomicity of Lebesgue's measure on $[0, T]$ and the differential inclusion structure of the sweeping process (2.15) are instrumental to conclude, due to the classical Lyapunov convexity theorem [34] (see also [5, 27]) that in many important situations local minimizer of the types under consideration is also a relaxed one. Without delving into details here, we just mention that the possibility to derive such a *local relaxation stability* from [48, Theorem 4.2] for *strong* local minimizers of problem (2.14)–(2.16), provided that the controlled moving sets $C(t, u)$ in (2.16) are convex and continuous.

Now we present our first major result establishing generalizesd the *Euler–Lagrange formalism* for the class of sweeping control problems (2.14)–(2.16). The reader can see that, besides extended conditions of the Euler–Lagrange and transversality types, the obtained theorem contains the necessary optimality conditions of novel types, which are specific for the controlled sweeping dynamics. The proof of this theorem, which follows the lines in the proof of [26, Theorem 4.3], realizes the steps (I)–(III) of the discrete approximation method and machinery of variational analysis with the particular usage of the second-order calculations from Theorem 2.13. For simplicity the theorem is formulated in the case of (2.14)–(2.16) where $g(x) := x$, $f := 0$ while ψ and ℓ do not depend on t. Recall the the symbol 'co' stands for the convex hull of the set in question.

Theorem 3.2 (necessary conditions for controls in moving sets). *Let* $(\bar{x}(\cdot), \bar{u}(\cdot))$ *be a relaxed local $W^{1,2} \times W^{1,2}$-minimizer local minimizer for problem (2.14)–(2.16). In addition to the standing assumptions, suppose that* $\psi = \psi(x, u)$ *is C^2-smooth with respect to both variables while the cost functions φ and ℓ are locally Lipschitzian around the corresponding components of the optimal solution. Then there exist a multiplier $\lambda \geq 0$, an adjoint arc $p(\cdot) = (p^x, p^u) \in W^{1,2}([0, T]; \mathbb{R}^n \times \mathbb{R}^m)$, a signed vector measure $\gamma \in C^*([0, T]; \mathbb{R}^s)$, as well as pairs $(w^x(\cdot), w^u(\cdot)) \in L^2([0, T]; \mathbb{R}^n \times \mathbb{R}^m)$ and $(v^x(\cdot), v^u(\cdot)) \in L^\infty([0, T]; \mathbb{R}^n \times \mathbb{R}^m)$ with*

$$(w^x(t), w^u(t), v^x(t), v^u(t)) \in \mathrm{co}\,\partial\ell(\bar{x}(t), \bar{u}(t), \dot{\bar{x}}(t), \dot{\bar{u}}(t)) \quad a.e. \quad t \in [0, T]$$

satisfying the following necessary optimality conditions:

- PRIMAL–DUAL DYNAMIC RELATIONSHIPS:

$$\dot{p}(t) = \lambda w(t) + \begin{bmatrix} \nabla^2_{xx}\langle \eta(t), \psi\rangle(\bar{x}(t), \bar{u}(t)) \\ \nabla^2_{xw}\langle \eta(t), \psi\rangle(\bar{x}(t), \bar{u}(t)) \end{bmatrix}$$

$$(-\lambda v^x(t) + q^x(t)) \quad a.e. \quad t \in [0, T], \quad q^u(t) = \lambda v^u(t) \quad a.e. \quad t \in [0, T],$$

where $\eta(\cdot) \in L^2([0, T]; \mathbb{R}^s)$ is a uniquely determined function from

$$\dot{\bar{x}}(t) = -\nabla_x\psi(\bar{x}(t), \bar{u}(t))^*\eta(t) \quad a.e. \quad t \in [0, T]$$

with $\eta(t) \in N(\psi(\bar{x}(t), \bar{u}(t)); \Theta)$, and where $q : [0, T] \to \mathbb{R}^n \times \mathbb{R}^m$ is a function of bounded variation on $[0, T]$ with its left-continuous representative given, for all $t \in [0, T]$ except at most a countable subset, *by*

$$q(t) = p(t) - \int_{[t,T]} \nabla \psi(\bar{x}(\tau), \bar{u}(\tau))^* d\gamma(\tau).$$

- MEASURED   CODERIVATIVE   CONDITION: *Considering   the*] *t-dependent outer limit*

$$\underset{|B| \to 0}{Lim\,sup} \frac{\gamma(B)}{|B|}(t) := \Big\{ y \in \mathbb{R}^s \, | \, \exists \ sequence \ B_k \subset [0,1] \ \ with$$

$$t \in IB_k, \ |B_k| \to 0, \ \frac{\gamma(B_k)}{|B_k|} \to y \Big\}$$

over Borel subsets $B \subset [0,1]$ *with the Lebesgue measure* $|B|$, *for a.e.* $t \in [0,T]$ *we have*

$$D^* N_\Theta(\psi(\bar{x}(t), \bar{u}(t)), \eta(t))(\nabla_x \psi(\bar{x}(t), \bar{u}(t))(q^x(t) - \lambda v^x(t))) \cap$$

$$\underset{|B| \to 0}{Lim\,sup} \frac{\gamma(B)}{|B|}(t) \neq \varnothing.$$

- TRANSVERSALITY CONDITION AT THE RIGHT ENDPOINT:

$$-(p^x(T), p^u(T)) \in \lambda(\partial\varphi(\bar{x}(T)), 0) + \nabla\psi(\bar{x}(T), \bar{u}(T)) N_\Theta((\bar{x}(T), \bar{u}(T)).$$

- MEASURE   NONATOMICITY   CONDITION: Whenever $t \in [0,T)$ with $\psi(\bar{x}(t), \bar{u}(t)) \in int\,\Theta$ there is a neighborhood V_t of t in $[0,T]$ such that $\gamma(V) = 0$ for any Borel subset V of V_t.

Nontriviality condition:

$$\lambda + \underset{t \in [0,T]}{sup} \|p(t)\| + \|\gamma\| \neq 0 \ \ with \ \ \|\gamma\| := \underset{\|x\|_{C([0,T])} = 1}{sup} \int_{[0,T]} x(s) d\gamma.$$

Suppose in addition that $\eta(T)$ is well defined and that $\theta = 0$ is the only vector for which

$$\theta \in D^* N_\Theta(\psi(\bar{x}(T), \bar{u}(T)), \eta(T))(0),$$
$$\nabla\psi(\bar{x}(T), \bar{u}(T))^*\theta \in \nabla\psi(\bar{x}(T), \bar{u}(T)) N_\Theta(\bar{x}(T), \bar{u}(T)).$$

Then the above necessary optimality conditions hold with the stronger nontriviality

$$\lambda + mes\{t \in [0,T] | \, q(t) \neq 0\} + \|q(0)\| + \|q(T)\| > 0.$$

Looking at the necessary optimality conditions obtained in Theorem 3.2, we do not see there any maximization condition of the Weierstrass–Pontryagin type, which is expected for continuous-time variational problems. However, the new class of the sweeping control problems (2.14)–(2.16) with controlled moving sets is largely different from conventional classes of problems in dynamic optimization. Thus the reader should not be surprised by the *failure* of the expected extension of the Pontryagin maximum principle for problems of this type. We'll construct below a (counter)example illustrating this phenomenon, but first present necessary optimality conditions of the *novel Hamiltonian type* that contain a modified (not expected a priori) form of the maximum principle.

To proceed, consider the constraint set Θ in (2.16) given by

$$\Theta = h^{-1}(\mathbb{R}^l_-) := \{z \in \mathbb{R}^s | \, h(z) \in \mathbb{R}^l_-\}, \qquad (2.21)$$

where $h : \mathbb{R}^s \to \mathbb{R}^l$ is a C^2-smooth mapping. Define the collection of active indexes

$$I(x, u) := \{i \in \{1, \dots, s\} | \, \psi_i(x, u) = 0\}.$$

It follows from the standing assumption (H3) that for each $v \in -N(x; C(u))$ there exists a unique collection $\{\alpha_i\}_{i \in I(x,u)}$ with $\alpha_i \leq 0$ and $v = \sum_{i \in I(x,u)} \alpha_i [\nabla_x \psi(x, u)]_i$. Given $\nu \in \mathbb{R}^s$, define the vector $[\nu, v] \in \mathbb{R}^n$ by

$$[\nu, v] := \sum_{i \in I(x,u)} \nu_i \alpha_i [\nabla_x \psi(x, u)]_i$$

and introduce the *modified Hamiltonian* function

$$H_\nu(x, u, p) := \sup\{\langle [\nu, v], p \rangle | \, v \in -N(x; C(u))\},$$
$$(x, u, p) \in \mathbb{R}^n \times \mathbb{R}^m \times \mathbb{R}^n. \qquad (2.22)$$

Observe that in the linear case $h(z) := Az - b$ in (2.21), the full rank assumption in (H3) corresponds to the classical *linear independence constraint qualification* (LICQ).

The following major results can be derived from Theorem 2.20 by using the *precise computation* of the second-order construction $D^* N_{\mathbb{R}^s_-}$.

Theorem 3.3 (novel Hamiltonian formalism and maximum principle for optimal control in moving sets). *Consider the sweeping control problem* (2.14)–(2.16) *in the frameworks of Theorem 2.20 with the set Θ given by (2.21), where $h : \mathbb{R}^s \to \mathbb{R}^l$ is C^2-smooth around the relaxed local*

$W^{1,2} \times W^{1,2}$-*minimizer* $\bar{z}(t) := (\bar{x}(t), \bar{u}(t))$ *for all* $t \in [0, T]$. *Suppose that the matrix* $\nabla h(\bar{z}(t))$ *is of full rank for all* $t \in [0, T]$. *Then, in addition to the necessary optimality conditions of Theorem 2.20, the new maximum condition*

$$\langle [\nu(t), \dot{\bar{x}}(t)], q^x(t) - \lambda v^x(t) \rangle = H_{\nu(t)}(\bar{x}(t), \bar{u}(t), q^x(t) - \lambda v^x(t)) = 0 \ \ a.e.$$
$$t \in [0, T] \tag{20}$$

holds in terms of the modified Hamiltonian (2.22), where $\nu : [0, T] \to \mathbb{R}^s$ is taken from

$$\nu(t) \in D^* N_{\mathbb{R}^s_-}(h(\psi(\bar{x}(t), \bar{u}(t))), \mu(t))(\nabla_x \psi(\bar{x}(t), \bar{u}(t))(q^x(t) - \lambda v^x(t))) \cap$$
$$\limsup_{|B| \to 0} \frac{\gamma(B)}{|B|}(t)$$

for a.e. $t \in [0, T]$ with a measurable vector function $\mu : [0, T] \to \mathbb{R}^l$ satisfying

$$\mu(t) \in N_{\mathbb{R}^l_-}(h(\psi(\bar{x}(t), \bar{u}(t)))) \ \ and$$
$$\eta(t) = \nabla h(\psi(\bar{x}(t), \bar{u}(t))^* \mu(t) \ \ a.e. \ \ t \in [0, T].$$

Discussions on the Maximum Principle. Observe that our form of the new Hamiltonian (2.22) is different from the *conventional Hamiltonian* function

$$H(x, p) := \sup\{\langle p, v \rangle | \ v \in F(x)\} \tag{2.25}$$

used in optimal control of Lipschitzian differential inclusions $\dot{x} \in F(x)$ that extends the usual Hamiltonian functions in classical mechanics and optimal control. We show below in Example 2.26 that the maximum principle via the standard Hamiltonian (2.25) *fails* for the sweeping control problem in (2.14)–(2.16). The main reason for this is that (2.25) does not reflects *implicit* state constraints, which do not appear for Lipschitzian problems while being an essential part of the sweeping dynamics.

Observe also that the new maximum principle form (2.24) incorporates vector measures appearing through the measured coderivative condition of Theorem 2.20. The fact that measures naturally arise in descriptions of necessary optimality conditions in optimal control problems with state constraints has been first realized by Dubovitskii and Milyutin [21] and since that has been largely employed in optimal control. There are interesting connections between our form of the maximum principle for controlled sweeping processes and the Hamiltonian formalism in models of contact and nonsmooth mechanics; see, e.g., [7].

Here is the promised example, which shows the failure of the conventional maximum principle for sweeping optimal processes with controlled moving sets and also illustrates the fulfillment of the new one obtained in Theorem 2.23.

Example 3.4 (failure of the maximum principle for sweeping processes with controlled moving sets). Consider problem (2.14)–(2.16) with $u = (a, b) \in \mathbb{R}^6$ as $a = (a_1, a_2) \in \mathbb{R}^4$ and $b = (b_1, b_2) \in \mathbb{R}^2$, $x = (x_1, x_2) \in \mathbb{R}^2$, $x_0 = (1, 1)$, $T = 1$,

$$\varphi(x) = \frac{\|x\|^2}{2} \quad and \quad \ell(t, x, u, \dot{x}, \dot{u}) := \frac{1}{2}(\dot{b}_1^2 + \dot{b}_2^2).$$

The moving sets $C(t, u)$ are defined by

$$C(t, u) := \{x \in \mathbb{R}^2 \mid \langle a_i(t), x \rangle \leq b_i(t), \ i = 1, 2\}.$$

Fix the a-controls as $\bar{a}_1 \equiv (1, 0)$ and $\bar{a}_2 \equiv (0, 1)$, and then deduce from Theorem 2.20 the following relationships on $[0, 1]$ involving only the b-components of controls together with the corresponding state and adjoint functions:

1. $w(\cdot) = 0$, $v^x(\cdot) = 0$, $v^b(\cdot) = (\dot{b}_1(\cdot), \dot{b}_2(\cdot))$
2. $\dot{\bar{x}}_i(t) \neq 0 \Rightarrow q_i^x(t) = 0$, $i = 1, 2$
3. $p^b(\cdot)$ is constant with nonnegative components, and $-p_i^x(\cdot) = \lambda \bar{x}(1) + p_i^b(\cdot)\bar{a}_i$ are also constant for both $i = 1, 2$.
4. $q^x(t) = p^x - \gamma([t, 1])$, $q^b(t) = \lambda \bar{b}(t) = p^b + \gamma([t, 1])$ for a.e. $t \in [0, 1]$.
5. $\lambda + \|q(0)\| + \|p(1)\| \neq 0$ with $\lambda \geq 0$.

Observe first that the pair $\bar{x}(t) = (1, 1)$ and $\bar{b}(t) = 0$ on $[0, 1]$ satisfies the necessary conditions with $p_1^x = p_2^x = -1$, $p_1^b = p_2^b = \gamma_1 = \gamma_2 = 0$, and $\lambda = 1$. The conventional Hamiltonian function (2.25) reads now as

$$H(x, b, p) = \sup\{\langle p, v \rangle \mid v \in -N(x; C((1, 0), (0, 1)), b)\},$$

and we get by the direct calculation that

$$\begin{aligned}
H(\bar{x}(t), \bar{b}(t), q^x(t) &- \lambda v^x(t)) \\
&= H((1, 1), (1, 1), (-1, -1)) \\
&= \sup\{\langle(-1, -1), v\rangle \mid v \in -N((1, 1); C(((1, 0), (0, 1)), (1, 1)))\} \\
&= \sup\{\langle(-1, -1), v\rangle \mid v_1 \leq 0, \ v_2 \leq 0\} = \infty,
\end{aligned}$$

while $\langle \dot{\bar{x}}(t), q^x(t) - \lambda v^x(t) \rangle = 0$. Thus the conventional maximum principle fails in this example. At the same time, the new maximum condition (2.24) from Theorem 2.23 holds trivially with the choice of $\nu(t) = 0$ for all $t \in [0,1]$ therein.

2.4 Sweeping Processes with Controlled Dynamics

In this section we consider a different type of controlled sweeping processes, where measurable control functions enter dynamical perturbations. Such problems can be viewed as far-going extensions of the standard class of ODE optimal control and in fact reduce to the latter in the trivial (for sweeping processes) case where the moving sets are absent. Employing again the method of discrete approximations together with advanced tools of variational analysis, we derive necessary optimality conditions for sweeping control problems considered here that contain appropriate extensions of the Pontryagin maximum principle along novel conditions of the types discussed in Section 3 for problems with controlled moving sets. The material of the section is mainly based on the recent chapter [19], where the reader can find more details.

Here we consider the problem:

$$\text{minimize} \quad J[x,u] := \varphi(x(T)) \tag{2.27}$$

over pairs $(x(\cdot), u(\cdot))$ of *measurable* controls $u(t)$ and absolutely continuous trajectories $x(t)$ on the fixed time interval $[0,T]$ satisfying the controlled sweeping differential inclusion

$$\dot{x}(t) \in -N(x(t);C) + g(x(t), u(t)) \quad a.e. \quad t \in [0,T], \quad x(0) := x_0 \in C \subset \mathbb{R}^n, \tag{2.28}$$

subject to the pointwise/hard *control constraints* given by

$$u(t) \in U \subset \mathbb{R}^d \quad a.e. \quad t \in [0,T]. \tag{2.29}$$

The set C in (2.28) is a *convex polyhedron* given by

$$C := \bigcap_{i=1}^{m} C_i \quad \text{with} \quad C_i := \{x \in \mathbb{R}^n | \langle x_i^*, x \rangle \leq c_i\}. \tag{2.30}$$

Observe that the sweeping differential inclusion (2.28) and the polyhedral description (2.30) automatically yield the *pointwise state constraints*

$$\langle x_i^*, x(t) \rangle \leq c_i \quad for \ all \ \ t \in [0,T].$$

It is said that a feasible pair $(\bar{x}(\cdot), \bar{u}(\cdot))$ for (2.27)–(2.30) is a $W^{1,2} \times L^2$-*local minimizer* for this problem if there exists a positive number ε such that $J[\bar{x}, \bar{u}] \leq J[x, u]$ for all feasible pairs $(x(\cdot), u(\cdot))$ satisfying the localization condition

$$\int_0^T \left(\|\dot{x}(t) - \dot{\bar{x}}(t)\|^2 + \|u(t) - \bar{u}(t)\|^2 \right) dt < \varepsilon.$$

Our *standing assumptions* in this section are as follows:

(A1) The set U is compact and convex in $\mathbb{R}^d$, while $g(x, U)$ is convex in $\mathbb{R}^n$.

(A2) The cost function $\varphi : \mathbb{R}^n \to \mathbb{R}$ in (2.27) is C^1-smooth around $\bar{x}(T)$.

(A3) The perturbation mapping $g : \mathbb{R}^n \times \mathbb{R}^d \to \mathbb{R}^n$ in (2.28) is C^1-smooth around $(\bar{x}(\cdot), \bar{u}(\cdot))$ and satisfies the sublinear growth condition

$$\|g(x, u)\| \leq \beta(1 + \|x\|) \quad for \ all \ \ u \in U \ \ with \ some \ \ \beta > 0.$$

(A4) The vertices x_i^* of (2.30) satisfy the linear independence constraint qualification

$$\left[\sum_{i \in I(\bar{x})} \alpha_i x_i^* = 0, \ \alpha_i \in \mathbb{R} \right] \Rightarrow [\alpha_i = 0 \ \ for \ all \ \ i \in I(\bar{x})\}$$

along the trajectory $\bar{x} = \bar{x}(t)$ as $t \in [0, T]$, where $I(\bar{x}) := \{i \in \{1, \ldots, m\} \mid \langle x_i^*, \bar{x} \rangle = c_i\}$.

Note that the usage of the relaxation procedure similar the one described in Section 3 allows us to significantly relax the convexity assumptions in (A1).

Furnishing all the steps (I)–(III) of the method of discrete approximations as discussed in Section 3 and using the generalized differential constructions reviewed in Section 2, we arrive the following set of necessary optimality conditions for the sweeping control problem (2.27)–(2.30). Note that an important role in this device is played by explicit coderivative calculation of the mapping on the right-hand side of (2.28) obtained in [25].

Theorem 2.31 (necessary conditions for sweeping processes with controlled dynamics). *Let $(\bar{x}(\cdot), \bar{u}(\cdot))$ be a $W^{1,2} \times L^2$-local minimizer for problem (2.27)–(2.30) under the assumptions in (A1)–(A4), where $\bar{u}(\cdot)$ is of bounded variation (BV) with a right continuous representative on $[0, T]$. Then there exist a multiplier $\lambda \geq 0$, a measure $\gamma = (\gamma_1, \ldots, \gamma_n) \in C^*([0, T]; \mathbb{R}^n)$ as well as adjoint arcs $p(\cdot) \in W^{1,2}([0, T]; \mathbb{R}^n)$ and $q(\cdot) \in BV([0, T]; \mathbb{R}^n)$ such that $\lambda + \|q(t)\|_{L^\infty} + \|p(T)\| > 0$ and the following conditions are satisfied:*

- PRIMAL VELOCITY REPRESENTATION:

$$-\dot{\bar{x}}(t) = \sum_{i=1}^{m} \eta_i(t)x_i^* - g(\bar{x}(t), \bar{u}(t)) \quad for \ a.e. \ \ t \in [0,T], \qquad (2.32)$$

where $\eta^i(\cdot) \in L^2([0,T]; \mathbb{R}_+)$ being uniquely determined by (2.32) and well defined at $t = T$.

- ADJOINT SYSTEM:

$$\dot{p}(t) = -\nabla_x g(\bar{x}(t), \bar{u}(t))^* q(t) \quad for \ a.e. \ \ t \in [0,T],$$

where the dual arcs $q(\cdot)$ and $p(\cdot)$ are connected by the equation

$$q(t) = p(t) - \int_{(t,T]} d\gamma(\tau)$$

that holds for all $t \in [0,T]$ except at most a countable subset.

- MAXIMIZATION CONDITION:

$$\langle \psi(t), \bar{u}(t) \rangle = \max\{\langle \psi(t), u \rangle \mid u \in U\} \quad with \ \ \psi(t) := \nabla_u g(\bar{x}(t), \bar{u}(t))^*$$

$$q(t) \quad for \ a.e. \ \ t \in [0,T].$$

- COMPLEMENTARITY CONDITIONS:

$$\langle x_i^*, \bar{x}(t) \rangle < c_i \Rightarrow \eta_i(t) = 0 \ \ and \ \ \eta_i(t) > 0 \Rightarrow \langle x_i^*, q(t) \rangle = c_i$$

for a.e. $t \in [0,T]$ including $t = T$ and for all $i = 1, \ldots, m$.

- RIGHT ENDPOINT TRANSVERSALITY CONDITIONS:

$$-p(T) = \lambda \nabla \varphi(\bar{x}(T)) + \sum_{i \in I(\bar{x}(T))} \eta_i(T)x_i^* \quad with$$

$$\sum_{i \in I(\bar{x}(T))} \eta_i(T)x_i^* \in N(\bar{x}(T); C).$$

- MEASURE NONATOMICITY CONDITION: If $t \in [0,T)$ and $\langle x_i^*, \bar{x}(t) \rangle < c_i$ for all $i = 1, \ldots, m$, then there is a neighborhood V_t of t in $[0,T]$ such that $\gamma(V) = 0$ for all the Borel subsets V of V_t.

Observe that the *maximization condition* of Theorem 2.31 provided an appropriate linearized version of the maximum principle for the class of optimal control problems under consideration satisfying the imposed assumptions.

2.5 Some Applications

In this section we briefly review selected applications of the obtained results and also discuss some potential avenues for future developments.

Elastoplasticity and Systems of Hysteresis. Consider the model of this type discussed in [3], but without any control actions. Let Q be a closed convex subset of the $\frac{1}{2}n(n+1)$-dimensional space of symmetric $n \times n$ tensors with nonempty interior. Using the notation of [3], define the strain tensor $= \{\}_{i,j}$ by $:= {}^e + {}^p$, where ^{e} is the elastic strain and ^{p} is the plastic strain. The elastic strain ^{e} depends on the stress tensor $\sigma = \{\sigma\}_{i,j}$ linearly, i.e., ${}^e = A^2\sigma$, where A is a constant symmetric positive-definite matrix. The *principle of maximal dissipation* says that

$$\langle {}^p(t), z \rangle \leq \langle {}^p(t), \sigma(t) \rangle \quad for \ all \ \ z \in Q. \tag{2.33}$$

As follows from [3], the variational inequality (2.33) is equivalent to the *sweeping process*

$$\dot{\zeta}(t) \in -N(\zeta(t); C(t)), \ \zeta(0) = A\sigma(0) - A^{-1}(0) \in C(0),$$

where $\zeta(t) := A\sigma(t) - A^{-1}(t)$ and $C(t) := -A^{-1}(t) + AQ$. This can be rewritten in the frame of the sweeping optimal control problem with *controlled moving sets* considered in Section 3 where $x := \zeta$, $u := $, $\psi(x, u) := x + A^{-1}u$, and $\Theta := AQ$. Thus we can apply Theorems 2.20 and 2.31 to this class of hysteresis operators for the general elasticity domain Q. In this way we cover various elastoplasticity models including those with the Drucker-Prager, Mohr-Coulomb, Tresca, and von Mises yield criteria; see [1]. More details with precise computations, examples, and discussions can be found in [26].

Optimal Control of Sweeping Processes in Somew Models of Robotics. Among important dynamical models arising in robotics, the *mobile robot model with obstacles* has been well recognized and investigated. It is shown in [24] that the dynamics of this model can be adequately described as a sweeping process. An *optimal control* version of the mobile robot model with obstacles has been recently developed in [18] as a sweeping control problem with *controlled dynamics* studied in Section 4. This model concerns n mobile robots ($n \geq 2$) identified with safety disks in the plane of the same radius R. The goal of each robot is to reach the target by the shortest path during a fixed time interval $[0, T]$ while avoiding the other $n - 1$ robots that are treated by this robot as obstacles.

To formalize the model, consider the configuration vector $x = (x^1, \ldots, x^n) \in \mathbb{R}^{2n}$, where $x^i \in \mathbb{R}^2$ is the center of the safety disk i with coordinates $(\|x^i\| \cos \theta_i, \|x^i\| \sin \theta_i)$. This means that the trajectory $x^i(t)$ of the i-robot/obstacle admits the representation

$$\bar{x}^i(t) = (\|\bar{x}^i(t)\| \cos \theta_i(t), \|\bar{x}^i(t)\| \sin \theta_i(t)) \quad for \quad i = 1, \ldots, n,$$

where the angle θ_i signifies the corresponding direction. According to the model dynamics, at the moment of contacting the obstacle (one or more) the robot in question keeps its velocity and pushes the other robots in contact to go to the target with the same velocity and then to maintain their constant velocities until reaching either other obstacles or the end of the process at the final time $t = T$. In this framework, the constant direction θ_i of x^i is the smallest positive angle in standard position formed by the positive x-axis and Ox^i with taking the origin as the target point.

The optimal control problem corresponding to the described model is formulated as:

$$minimize \quad J[x, u] := \frac{1}{2} \|x(T)\|^2,$$

over the controlled sweeping dynamics

$$\begin{cases} -\dot{x}(t) \in N(x(t); C) - g(x(t), u(t)) \quad for \ a.c. \ \ t \in [0, T], \\ x(0) = x_0 \in C, \ u(t) \in U \ \ a.e. \ on \ [0, T], \end{cases}$$

where the controlled additive perturbations of the dynamics are modeled by

$$g(x(t), u(t)) = (s_1 u^1(t) \cos \theta_1(t), s_1 u^1(t) \sin \theta_1(t), \ldots, s_n u^n(t) \cos \theta_n(t),$$

$$s_n u^n(t) \sin \theta_n(t)),$$

where the generating polyhedral set $C \subset \mathbb{R}^n$ is given by

$$C := \{x \in \mathbb{R}^n \mid \langle x_*^j, x \rangle \le c_j, \ j = 1, \ldots, n-1\} \quad with$$

$$x_*^j := e_j - e_{j+1}, \ c_j = -2R$$

for $j = 1, \ldots, n-1$ with $(e_1, \ldots, e_n)$ being the orths in $\mathbb{R}^n$, and where the control set U is specified in [18] in particular settings.

Applying the necessary optimality conditions of Theorem 2.31 to this problem leads to constructive algorithms of solving of the mobile mobile model that we completely implemented in [18] to find optimal solutions in some interesting cases, while more serious practical applications require more elaborations.

Traffic Equilibria and Crowd Motion Models. There are several practically important models of traffic equilibria and related areas, where the dynamics can be formulated as a sweeping process. One of the most interesting classes of models of this type is known under the name of *crowd motion models*. In what follows we consider some *control* versions of such problems, which can be efficiently investigated and partly solved by using the necessary optimality conditions discussed above.

The original motivation for crowd motion models came the modeling of local interactions between participants in order to describe the dynamics of pedestrian traffic. However, by now such models have been successfully used to investigate more general classes of problems in engineering, operations research, economics, etc.

The *microscopic form* of the crowd motion model is based on the following *two postulates*. On one hand, each individual has a *spontaneous* velocity that he/she intends to implement in the absence of other participants. On the other hand, the *actual* velocity must be taken into account. The latter one is incorporated via a projection of the spontaneous velocity into the set of admissible/feasible velocities, i.e., those which do not violate certain nonoverlapping constraints. A mathematical description of the uncontrolled microscopic crowd motion model is given [49] in the *sweeping process* form with the subsequent usage therein for numerical simulations and various applications.

In the *corridor version* of the crowd motion model, the crucial nonoverlapping condition is written in the form

$$Q_0 = \{x = (x_1, \ldots, x_n) \in \mathbb{R}^n, \ x_{i+1} - x_i \geq 2R\},$$

where $n \geq 2$ indicates the number of participants identified with rigid disks of the same radius R in a corridor. The actual velocity field is described via the projection operator by

$$\dot{x}(t) = \Pi(U(x); C_x) \ \text{for a.e.} \ t \in [0, T], \quad x(0) = x_0 \in Q_0,$$

with spontaneous velocities $U(x) = (U_0(x_1), \ldots, U_0(x_n))$, $x \in Q_0$, satisfying

$$U(x) \in N_x + \dot{x}(t) \ \text{for a.e.} \ t \in [0, T], \quad x(0) = x_0,$$

where N_x stands for the normal cone to Q_0 at x. Since all the participants exhibit the same behavior and want to reach the exit by the shortest path, their spontaneous velocities can be represented as follows:

$$U(x) = (U_0(x_1), \ldots, U_0(x_n)) \quad with \quad U_0(x) = -s\nabla D(x),$$

where $D(x)$ denotes the distance between the position $x = (x_1, \ldots, x_n) \in Q_0$ and the exit, and where $s \geq 0$ stands for the speed. By taking into account the aforementioned postulate that in the absence of other participants each participant tends to remain his/her spontaneous velocity until reaching the exit, the (uncontrolled) dynamic perturbations in this model are described by

$$g(x) = -(s_1, \ldots, s_n) \in \mathbb{R}^n \quad for \ all \quad x = (x_1, \ldots, x_n) \in Q_0,$$

where s_i denotes the speed of the participant $i \in \{1, \ldots, n\}$. To control the actual speed of all the participants in the presence of the nonoverlapping condition, we suggest in [11] to involve control functions $u(\cdot) = (u_1(\cdot), \ldots, u_n(\cdot))$ into perturbations as follows:

$$g(x(t), u(t)) = (s_1 u_1(t), \ldots, s_n u_n(t)), \quad t \in [0, T].$$

To optimize the sweeping dynamics by using controls $u(\cdot)$, consider the cost functional

$$minimize \ \ J[x, u] := \frac{1}{2}(\|x(T)\|^2 + \int_0^T \|u(t)\|^2 dt) \qquad (2.34)$$

the meaning of which is to *minimize the distance* of all the participants to the exit at the origin together with minimizing the *energy* of feasible controls $u(\cdot)$.

The described sweeping optimal control problem falls into the category of dynamic optimization problems with *controlled dynamics*, which were considered in Section 4. The necessary optimality conditions for such problems obtained in Theorem 2.31 led us in [11] to developing an efficient algorithmic procedure to determine optimal controls and trajectories in the general case of finitely many participants and solve the problem analytically in the cases where $n = 2, 3$. More recent results in this direction allow us to do calculations for a fairly large number of participants in practical settings that appear in real applications.

In the *planar version* of the microscopic crowd motion model [49], the overlapping condition is not polyhedral anymore while being represented by

$$Q := \{x \in \mathbb{R}^{2n} | \ D_{ij}(x) \geq 0 \ \ for \ all \ \ i \neq j\},$$

where $D_{ij}(x) := \|x_i - x_j\| - 2R$ is the signed distance between the disks i and j of the same radius R identified with $n \geq 2$ participants on the plane. The corresponding optimal control problem formulated and investigated in [12] is described via the sweeping dynamics as follows: minimize the cost functional (2.34) over the controlled sweeping process

$$\begin{cases} -\dot{x}(t) \in N(x(t); C(t)) + g(x(t), u(t)) & for\ a.e.\ \ t \in [0, T], \\ C(t) := C + \bar{v}(t),\ \|\bar{v}(t)\| = r \in [r_1, r_2] \ \ on\ \ [0, T],\ x(0) = x_0 \in C(0), \end{cases}$$

where the initial data and constraints are given by

$$g(x(t), u(t)) := (s_1 u_1(t) \cos \theta_1(t), s_1 u_1(t) \sin \theta_1(t), \dots, s_n u_n(t) \cos \theta_n(t),$$

$$s_n u_n(t) \sin \theta_n(t)), \bar{v}_{i+1}(t) = \bar{v}_i(t) := \left(\frac{r}{\sqrt{2n}}, \frac{r}{\sqrt{2n}}\right), \quad i = 1, \dots, n-1,$$

$$C := \{x = (x_1, \dots, x_n) \in \mathbb{R}^{2n} | \ h_{ij}(x) \geq 0 \ \ for\ all\ \ i \neq j \ \ as$$

$$i, j = 1, \dots, n\}$$

with the functions $h_{ij}(x) := D_{ij}(x) = \|x_i - x_j\| - 2R$ and with

$$x(t) - \bar{v}(t) \in C \ \ for\ all\ \ t \in [0, T].$$

Observe that the sweeping dynamic in the described planar version of the crowd motion model is generated by the set C, which is not a convex polyhedron, while belonging the class of *prox-regular* sets in the sense of [45]. The required extensions of necessary optimality conditions for this type of controlled sweeping processes were obtained in [12] by using the corresponding modification of the method of discrete approximations. The obtained results were applied therein to establishing efficient relationships which allowed us to determine optimal parameters, to develop numerical algorithms, and realize them in particular settings.

2.6 Conclusion

This chapter discusses recent applications of advanced constructions of variational analysis and generalized differentiation to challenging classes of optimal control problems governed by discontinuous differential inclusions that are known as sweeping processes. We present optimality conditions for such control systems obtained by employing the method of discrete approximations and then review the usage of the obtained conditions to solving some practical models. There is the intensive ongoing research on further aspects of the theory of controlled sweeping processes motivated by applications to new models arising in nanotechnology, marine surface vehicle modeling, electronics, etc.

Acknowledgments

This research was partially supported by the US National Science Foundation under grants DMS-1007132 and DMS-1512846, by the US Air Force Office of Scientific Research grant #15RT0462, and by the Australian Research Council under Discovery Project DP-190100555.

References

[1] B. Acary, O. Bonnefon, B. Brogliato, *Nonsmooth Modeling and Simulation for Switched Circuits*, Springer, Berlin, 2011.

[2] L. Adam, J.V. Outrata, 'On optimal control of a sweeping process coupled with an ordinary differential equation', *Discrete Contin. Dyn. Syst. Ser. B* **19** (2014), 2709–2738.

[3] S. Adly, T. Haddad, L. Thibault, 'Convex sweeping process in the framework of measure differential inclusions and evolution variational inequalities', *Math. Program.* **148** (2014), 5–47.

[4] C.E. Arroud and G. Colombo, 'A maximum principle of the controlled sweeping process', *Set-Valued Var. Anal.* **26** (2018), 607–629.

[5] R.J. Aumann, 'Integrals of set-valued functions', *J. Math. Anal. Appl.* **12** (1965), 1–12.

[6] N.N. Bogolyubov, 'Sur quelques méthodes nouvelles dans le Calculus des Variations', *Ann. Math. Pura Appl.* **7** (1929), 249–271.

[7] B. Brogliato, *Nonsmooth Mechanics: Models, Dynamics and Control*, 3rd edition, Springer, Cham, Switzerland, 2016.

[8] B. Brogliato, A. Tanwani, 'Dynamical systems coupled with monotone set-valued operators: formalisms, applications, well-posedness, and stability', *SIAM Rev.* **62** (2020), 3–129.

[9] M. Brokate, P. Krejčí, 'Optimal control of ODE systems involving a rate independent variational inequality', *Discrete Contin. Dyn. Syst. Ser. B* **18** (2013), 331–348.

[10] T.H. Cao, G. Colombo, B.S. Mordukhovich, D. Nguyen, 'Optimization of fully controlled sweeping processes', *J. Diff. Eqs.* **295** (2021), 138–186.

[11] T.H. Cao and B S. Mordukhovich, 'Optimal control of a perturbed sweeping process via discrete approximations', *Discrete Contin. Dyn. Sys. Ser. B*, **21** (2016), 3331–3358.

[12] T.H. Cao, B.S. Mordukhovich, 'Optimal control of a nonconvex perturbed sweeping process', *J. Diff. Eqs.* **266** (2019), 1003–1050.

[13] A.A. Chikrii, *Conflict Controlled Processes*, Kluwer Acad. Publ., Boston, 1997.

[14] F.H. Clarke, *Optimization and Nonsmooth Analysis*, Wiley, New York, 1983.

[15] G. Colombo and P. Gidoni, 'On the optimal control of rate-independent soft crawlers', *J. Math. Pure Appl.* **146** (2021), 405–417.

[16] G. Colombo, R. Henrion, N.D. Hoang, B.S. Mordukhovich, 'Optimal control of the sweeping process', *Dyn. Contin. Discrete Impuls. Syst. Ser. B* **19** (2012), 117–159.

[17] G. Colombo, R. Henrion, N.D. Hoang, B.S. Mordukhovich, 'Optimal control of the sweeping process over polyhedral controlled sets', *J. Diff. Eqs.* **260** (2016), 3397–3447.

[18] G. Colombo, B.S. Mordukhovich, D. Nguyen, 'Optimal control of sweeping processes in robotics and traffic flow models', *J. Optim. Theory Appl.* **182** (2019), 439–472.

[19] G. Colombo, B.S. Mordukhovich, D. Nguyen, 'Optimization of a perturbed sweeping process by discontinuous controls', *SIAM J. Control Optim.* **58** (2020), 2678–2709.

[20] M.d.R. de Pinho, M.M.A. Ferreira, G.V. Smirnov, 'Optimal control involving sweeping processes', *Set-Valued Var. Anal.* **27** (2019), 523–548.

[21] A.Y. Dubovitskii, A.A. Milyutin, 'Extremum problems in the presence of restrictions,' *USSR Comput. Math. Math. Phys.* **5** (1965), 1–80.

[22] J.F. Edmond, L. Thibault, 'Relaxation of an optimal control problem involving a perturbed sweeping process', *Math. Program.* **104** (2005), 347–373.

[23] R. Gabasov, F.M. Kirillova, *Qualitative Theory of Optimal Processes*, Marcel Dekker, New York, 1976.

[24] R. Hedjar, M. Bounkhel, 'Real-time obstacle avoidance for a swarm of autonomous mobile robots', *Int. J. Adv. Robot. Syst.* **11** (2014), 1–12.

[25] R. Henrion, J.V. Outrata, T. Surowiec, 'On the coderivative of normal cone mappings to inequality systems', *Nonlinear Anal.* **71** (2009), 1213–1226.

[26] N.D. Hoang, B.S. Mordukhovich, 'Extended Euler–Lagrange and Hamiltonian formalism in optimal control of sweeping processes with controlled sweeping sets', *J. Optim. Theory Appl.* **180** (2019), 256–289.

[27] A.D. Ioffe, V.M. Tikhomirov, *Theory of Extremal Problems*, North-Holland, Amsterdam, The Netherlands, 1979.

[28] Y.P. Kondratenko, A.A. Chikrii, V.F. Gubarev, J. Kacprzyk (eds.), *Advanced Control Techniques in Complex Engineering Systems: Theory and Applications*, Studies in Systems, Decision and Control, Vol. 203, Springer, Cham, Switzerland, 2019.

[29] Y.P. Kondratenko, V.M. Kunsevich, A.A. Chikrii, V.F. Gubarev (eds.), *Advanced Control Systems: Theory and Applications*, Series in Automation, Control and Robotics, River Publishers, Gistrub, Denmark, 2021.

[30] M.A. Krasnosel'skii, A.V. Pokrovskii, *Systems with Hysteresis*, Springer, Berlin, 1989.

[31] N.N. Krasovskii, A.I. Subbotin, *Game-Theoretical Control Problems*, Springer, New York, 1988.

[32] V.M. Kunsevich, V.F. Gubarev, Y.P. Kondratenko, D. Lebedev, V. Lysenko (eds.), *Control Systems: Theory and Applications*, Series in Automation, Control and Robotics, River Publishers, Gistrub, Denmark, 2018.

[33] M. Kunze, M. D. P. Monteiro Marques, "An introduction to Moreau's sweeping process," in: *Impacts in Mechanical Systems* (B. Brogliato, ed.), Lecture Notes in Phys. **551**, pp. 1–60, Springer, Berlin, 2000.

[34] A.A. Lyapunov, 'Sur les founctions-vecteurs complétement additives', *Izvest. Akad. Nauk SSSR, Ser. Mat.* **3** (1940), 465–478.

[35] B.S. Mordukhovich, *Methods of Approximations in Problems of Optimization and Control*, Nauka, Moscow, 1988.

[36] B.S. Mordukhovich, *Variational Analysis and Generalized Differentiation, I: Basic Theory, II: Applications*, Springer, Berlin, 2006.

[37] B.S. Mordukhovich, *Variational Analysis and Applications*, Springer, Cham, Switzerland, 2018.

[38] B.S. Mordukhovich, 'Optimal control of Lipschitzian and discontinuous differential inclusions with various applications', *Proc. Inst. Math. Mech. Azer. Acad. Sci.* **45** (2019), 52–74.

[39] B.S. Mordukhovich, R.T. Rockafellar, 'Second-order subdifferential calculus with applications to tilt stability in optimization', *SIAM J. Optim.* **22** (2012), 953–986.

[40] J.J. Moreau, 'Ra e par un convexe variable, I', *Trav. Semin. d'Anal. Convexe, Montpellier* **1** (1971), Exposé 15, 36 pp.

[41] J.J. Moreau, 'On unilateral constraints, friction and plasticity', in *New Variational Techniques in Mathematical Physics* (G. Capriz and G. Stampacchia, eds.), Proc. C.I.M.E. Summer Schools, pp. 173–322, Cremonese, Rome, 1974.

[42] L.S. Pontryagin, V.G. Boltyanskii, R.V. Gamkrelidze, E.F. Mishchenko, *The Mathematical Theory of Optimal Processes*, Wiley, New York, 1962.

[43] B.N. Pschenichnyi, *Convex Analysis and Extremal Problems*, Nauka, Moscow, 1980.

[44] B.N. Pschenichnyi, V.V. Ostapenko, *Differential Games*, Naukova Dumka, Kyiv, 1992.

[45] R.T. Rockafellar, R.J-B. Wets, *Variational Analysis*, Springer, Berlin, 1998.

[46] G. V. Smirnov, *Introduction to the Theory of Differential Inclusions*, American Mathematical Society, Providence, RI, 2002.

[47] A.A. Tolstonogov, *Differential Inclusions in a Banach Spaces*, Kluwer, Dordrecht, The Netherlands, 2000.

[48] A.A. Tolstonogov, 'Control sweeping processes', *J. Convex Anal.* **23** (2016), 1 099–1123.

[49] J. Venel, 'A numerical scheme for a class of sweeping process', *Numerische Mathematik*, **118** (2011), 451–484.

[50] R.B. Vinter, *Optimal Control*, Birkhaüser, Boston 2000.

[51] L.C. Young, 'Generalized curves and the existence of an attained absolute minimum in the calculus of variations', *C. R. Soc. Sci. Lett. Varsovie*, Cl. III, **30** (1937), 212–234.

[52] V. Zeidan, C. Nour, H. Saoud, 'A nonsmooth maximum principle for a controlled nonconvex sweeping process', *J. Diff. Eqs.* **269** (2020), 9531–9582.

3

Robust and Robustly-Adaptive Control of Some Noninvertible Memoryless Systems

L. Zhiteckii[1] and K. Solovchuk[2]

[1]International Research and Training Center for Information Technologies and Systems of the National Academy of Science of Ukraine and Ministry of Education and Sciences of Ukraine,
Acad. Glushkova av., 40, Kyiv, 03187, Ukraine
[2]Poltava Scientific Research Forensic Center of the MIA of Ukraine,
Rybalskiy lane, 8, Poltava, 36011, Ukraine
E-Mail: leonid_zhiteckii@i.ua; solovchuk_ok@ukr.net

Abstract

This chapter deals with the robust control of some uncertain multivariable memoryless (static) discrete-time systems in the presence of arbitrary unmeasurable bounded disturbances, whose bounds may be unknown, in general. The main feature of the systems to be controlled is that their gain matrices are noninvertible. Again, they are of not full rank. It is assumed that the elements of these matrices are exactly unknown. However, the bounded intervals to which they belong are *a priori* known. The nonadaptive robust control approach applicable to the case when the lengths of these intervals are relatively short is proposed. Such an approach is based on the so-called pseudoinverse model-based concept. The robustly-adaptive control utilizing a modification of inverse-model concept is designed to cope with large parametric uncertainty. The asymptotic properties of both nonadaptive and adaptive feedback control systems are derived. To support the theoretic study, numerical examples and simulation results are given.

Keywords: Boundedness, discrete time, estimation algorithm, feedback control system, robust adaptive controller, pseudoinverse model-based control, uncertainty.

3.1 Introduction

The problem of a perfect performance of control systems is an important problem from both theoretical and practical points of view. Within the framework of this actual problem, new approaches have been advanced by many researches. The latest results in this scientific area have been reported in numerous papers and generalized in several books including [1, 2]. Among them, the recent works [3–7] dealt with advanced multivariable control systems.

A long-standing problem of the optimal controller design for multivariable system in the presence of unmeasurable disturbances [8] has been solved using, in particular, the internal model control principle proposed in [9]. A perspective modification of this principle is the so-called model-inverse approach. Since the pioneering work [10], the problem of inversion of linear time-invariant multivariable systems has attracted an attention of several authors [11–14]. Recently, a significant progress in this research are has been achieved in [15, 16].

An inverse model-based approach to ensuring a perfect performance of linear multivariable control systems containing the memoryless (static) plants was first advanced in [17]. Regardless of this work, a similar approach was proposed in [18] to deal with some memoryless plants. However, the approach above mentioned is quite unacceptable if their gain matrices are either square but singular or nonsquare because they are noninvertible matrices.

It turned out that so-called pseudoinverse (generalized inverse) model-based concept advanced in the paper [19] can be exploited to cope with the possible noninvertibility of gain matrices. This fruitful concept was extended in [20, 21] to robust control of a noninvertible and uncertain plant with unmeasurable bounded disturbances.

The robust control theory gives a powerful tool to reject the unmeasured disturbances in the closed-loop control system with parametric and nonparametric uncertainty. The books [22–27] et al. provide insight into the results achieved in this theory to the beginning of the 2000s. A new estimation method applicable to a nonlinear discrete-time system with bounded disturbances has been advanced in the work [28]. Novel results with respect to a practical application of the robustness theory is reported, in particular, in recent paper [29].

Unfortunately, the robust control theory may not be employed if the initial parametric uncertainty is "wide" enough. Meanwhile, the adaptive approach gives some universal tool to deal with such a type of uncertainty. Foundations of this approach have been extended and generalized in the books [26, 30–35].

Difficulties that take place when adaptive control uses the point estimation algorithms are how to guarantee the stability (the boundedness) of the closed-loop control system [30, 31]. To overcome these difficulties, the so-called Frequency theorem given in [30, theorem 4.II.3] or the Key Technical Lemma given in [31, subsect. 6.2] are usually utilized to establish the boundedness properties of adaptive control systems. However, these tools become not applicable if the gain matrix is noninvertible; see [30, p. 242] and [31, p. 202]. Nevertheless, the problem of the adaptive control of nonsquare multivariable memoryless plants whose gain matrices have the full rank has been solved in the work [36].

This chapter extends the results of [21] concerning the robust nonadaptive approach to controlling uncertain noninvertible memoryless systems whose gain matrices have not full rank and generalizes the robustly-adaptive techniques proposed in [37 – 40] to deal with these systems.

3.2 Problem Statement

Let

$$y_n = Bu_{n-1} + v_n \qquad (3.1)$$

be the vector-valued difference equation of a static (memoryless) plant that is some linear multivariable discrete-time system to be stabilized. In this equation, $y_n \in R^m$, $u_n \in R^r$ and $v_n \in R^m$ represent the measured output, control input and unmeasured external disturbance vectors, respectively, at the nth time instant $(n = 1, 2, \dots)$ defined by $y_n = [y_n^{(1)}, \dots, y_n^{(m)}]^T$, $u_n = [u_n^{(1)}, \dots, u_n^{(r)}]^T$ and $v_n = [v_n^{(1)}, \dots, v_n^{(m)}]^T$,

$$B = \begin{pmatrix} b^{(11)} & \dots & b^{(1r)} \\ \dots & \dots & \dots \\ b^{(m1)} & \dots & b^{(mr)} \end{pmatrix} \qquad (3.2)$$

is an arbitrary time-invariant $m \times r$ gain matrix.

The case when the number of the control inputs $u_n^{(1)}, \dots, u_n^{(r)}$ is not less than two but does not exceed the number of the outputs $y_n^{(1)}, \dots, y_n^{(m)}$

meaning that

$$2 \le r \le m \tag{3.3}$$

is here considered. However, in contrast with problem stated in the work [36], the rank of B satisfies the strict inequality

$$\text{rank } B < r \tag{3.4}$$

implying that B is a not full rank matrix.

The following basic assumptions with respect to the gain matrix B and the sequences $\{v_n^{(i)}\} = v_0^{(i)}, v_1^{(i)}, \ldots$ ($i = 1, \ldots, m$) are introduced.

A1) The elements of B are all unknown. However, there are some interval estimates defined as

$$\underline{b}^{(ij)} \le b^{(ij)} \le \overline{b}^{(ij)}, \qquad i = 1, \ldots, m, \qquad j = 1, \ldots, r \tag{3.5}$$

where the upper and lower bounds $\underline{b}^{(ij)}$ and $\overline{b}^{(ij)}$, respectively, are assumed to be known.

A2) The rank of B satisfies Equation (3.4) giving that B is a noninvertible matrix.

A3) $v_n^{(i)}$s ($i = 1, \ldots, m$) are all the arbitrary scalar variables bounded in modulus according to

$$\left| v_n^{(i)} \right| \le \varepsilon^{(i)} < \infty$$

where $\varepsilon^{(i)}$s are constant. It is assumed that these upper bounds may be unknown, in general.

Let $y^0 = [y^{0(1)}, \ldots, y^{0(m)}]^T$ be a desired output vector ($y^{0(i)} \equiv$ const $\forall i = 1, \ldots, m$). Suppose that $|y^{0(1)}| + \cdots + |y^{0(m)}| \neq 0$ implying that, at least, one $y^{0(i)}$ of $y^{0(1)}, \ldots, y^{0(m)}$ is nonzero.

Define the output error vector

$$e_n = y^0 - y_n \tag{3.6}$$

with the components $e_n^{(i)} = y^{0(i)} - y_n^{(i)}$, i.e., $e_n = [e_n^{(1)}, \ldots, e_n^{(m)}]^T$.

The problem is to design the feedback controller guaranteeing the ultimate boundedness of the sequence $\{e_n\} = e_1, e_2, \ldots$, meaning

$$\lim_{n \to \infty} \sup \|e_n\| < \infty \tag{3.7}$$

provided

$$\lim_{n \to \infty} \sup \|u_n\| < \infty \tag{3.8}$$

Remark 3.1. Assumptions A1) – A3) differ from similar assumptions made in [36] only in that B is here the matrix of not full rank, and it is its essential feature.

Remark 3.2. Equation (3.8) is here introduced additionally since it may not be satisfied even if (3.7) takes place.

3.3 Preliminaries

Assume, for the time being, that B is a known noninvertible matrix of full rank (rank $B = r$), i.e., B represents some nonsquare matrix with $r < m$. In this case, the so-called pseudoinverse control

$$u_n = u_{n-1} + B^+ e_n \tag{3.9}$$

proposed by the authors in their previous works (see, e.g., [21]) ensures the minimum of the upper bound on the Euclidean norm $\|e_n\|_2$ of the output error vector of the closed-loop control system (3.1), (3.6), (3.9) with any bounded sequence $\{v_n\} = v_1, v_2, \dots$ Here the notation P^+ of any pseudoinverse (generalized inverse) matrix introduced in [41] and defined as

$$P^+ = \lim_{\delta \to 0} (P^T P + \delta I_r)^{-1} P^T \tag{3.10}$$

where I_r denotes the identity $r \times r$ matrix is used. Recall that if rank $P = r$ for some $P \in R^{m \times r}$, then the expression of P^+ is simplified [42, item 7.46] instead of (3.10) we have

$$P^+ = (P^T P)^{-1} P^T$$

This control system is designed as shown in Figure 3.1.

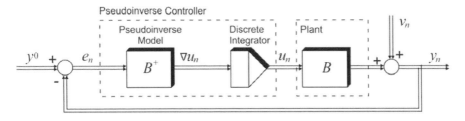

Figure 3.1　Configuration of the pseudoinverse model-based feedback control system without uncertainty

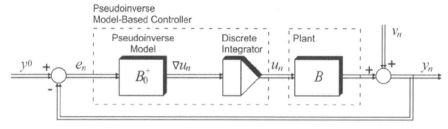

Figure 3.2   Configuration of the pseudoinverse model-based feedback control system with parametric uncertainty

In this control system, the variable $\nabla u_n := u_n - u_{n-1}$ produced by the pseudoinverse model represents the increment of the control action during one step determined as

$$\nabla u_n := B^+ e_n$$

whereas the signal u_n is formed as the sum $u_n = \sum_{k=1}^{n} \nabla u_k$.

Since B is indeed unknown, it needs to be replaced by an appropriate estimate $B_0 = (b_0^{(ij)})$, $i = 1, \ldots, m$, $j = 1, \ldots, r$, that is a fixed matrix (in the nonadaptive case) with rank $B_0 = r$ taken from the interval matrix set

$$\Xi = \{(b^{(ij)}): \quad b^{(ij)} \in \left[\underline{b}^{(ij)}, \overline{b}^{(ij)}\right], \quad i = 1, \ldots, m, \quad j = 1, \ldots, r\}$$
(3.11)

of admissible matrices B satisfying the constraints (3.5). Then, under the parameter uncertainty given as $B \in \Xi$, instead of (3.9), the control law becomes

$$u_n = u_{n-1} + B_0^+ e_n \tag{3.12}$$

Figure 3.2 represents the structure of the feedback control system containing the pseudoinverse model-based controller described by Equations (3.6), (3.12).

Introduce the matrix

$$\Delta := B_0 - B$$

which, by virtue of the fact that $\underline{b}^{(ij)} \le b^{(ij)} \le \overline{b}^{(ij)}$ and $\underline{b}^{(ij)} \le b_0^{(ij)} \le \overline{b}^{(ij)}$, is specified as

$$\Delta = \left(\delta^{(ij)}\right): \quad \delta^{(ij)} \in \left[\underline{\delta}^{(ij)}, \overline{\delta}^{(ij)}\right]$$

where $\underline{\delta}^{(ij)} = b_0^{(ij)} - \overline{b}^{(ij)}$ and $\overline{\delta}^{(ij)} = b_0^{(ij)} - \underline{b}^{(ij)}$.

Recall that if there are no disturbances ($v_n^{(i)} \equiv 0$ for all $i = 1, \ldots, m$), then an equilibrium state of the closed-loop control system (3.1), (3.6), (3.12) defined by the pair (u^e, y^e) in which $y^e = Bu^e$, where u^e represents the solution of the equation

$$B_0^+ Bu = B_0^+ y^0 \tag{3.13}$$

with respect to u, exist provided that rank $B_0 = r$ and the condition

$$q < 1 \tag{3.14}$$

where

$$q := \max_{\Delta \,:\, \delta^{(ij)} \in \left[\underline{\delta}^{(ij)},\, \bar{\delta}^{(ij)}\right]} \left\| B_0^+ \Delta \right\| \tag{3.15}$$

is satisfied [36, Lemma 7.1]. In this lemma it has been established that the condition (3.14) together with (3.15) is sufficient to achieve the ultimate boundedness of the sequences $\{u_n\}$ and $\{e_n\}$ if B is a matrix of full rank.

To establish the sufficient conditions under which the equilibrium state exist, the requirement that the $r \times r$ matrix $B_0^+ B$ of Equation (3.13) is nonsingular is utilized.

It turns out that when $y^0 \notin \aleph(B_0^+)$ and also $y^0 \notin \mathfrak{R}(B)$, where $\aleph(P)$ and $\mathfrak{R}(P)$ denote the null space and the range of a matrix P, respectively, then the equilibrium state may exist even if $\det B_0^+ B = 0$. To show this, consider an example.

Example 3.1. Let

$$B = \begin{pmatrix} 1 & -2 \\ 2 & -4 \\ -1 & 2 \end{pmatrix}, \qquad B_0 = \begin{pmatrix} 3 & -4.5 \\ 1 & -1.5 \\ -2 & 3 \end{pmatrix}$$

be some nonsquare matrices of the not full rank (rank B = rank B_0 = 1). By (3.10) we have

$$B_0^+ = \frac{1}{91} \begin{pmatrix} 6 & 2 & -4 \\ -9 & -3 & 6 \end{pmatrix}$$

Put $y^0 = [1, 3, 5]^T$. Such a choice of y^0 gives $y^0 \notin \aleph(B_0^+)$ and $y^0 \notin \mathfrak{R}(B)$. It can be verified that in this case, Equations (3.13) will have the infinite number of solutions $u^e = [u^{e(1)}, \; u^{e(2)}]^T$ lying on the straight line described by

$$7u^{e(1)} - 14u^{e(2)} + 4 = 0$$

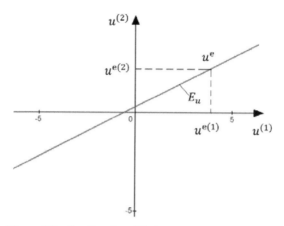

Figure 3.3   Set E_u of equilibrium points in Example 3.1

(see Figure 3.3). In other words, the set $E_u = \{u^e\}$ of the equilibrium points becomes here infinite.

Unfortunately, if rank $B < r$ and $y^0 \notin \Re(B)$, then the equilibrium state may not exist, in principle. To demonstrate this annoying fact, consider the following numerical example.

Example 3.2. Suppose B and B_0^+ are some square matrices given by

$$B = \begin{pmatrix} 2 & 4 \\ 1 & 2 \end{pmatrix}, \quad B_0^+ = \begin{pmatrix} 1 & -2 \\ 2 & -4 \end{pmatrix}$$

and $y^0 \notin \aleph(B_0^+)$, $y^0 \notin \Re(B)$. Since $B_0^+ B = 0_{r \times r}$, in this case, Equation (3.13) has no solutions. This example shows that the closed-loop control system defined by Equations (3.1), (3.6) and (3.12) will have no equilibrium state.

It turns out that the equilibrium state $\{u^e,\ y^e\}$ of the control system above mentioned exists iff

$$\aleph(B_0^+) \cap (y^0 + \Re(B)) \neq \varnothing \tag{3.16}$$

where the symbol "+" denoted the Minkowski sum of two sets of vectors in Euclidean space. Actually, by the definition of set $\Re(B)$ to which the zero vector 0_m belongs, follows that $(-Bu) \in \Re(B)$. Thereby, the vector $y^0 - Bu$ belongs to the set $y^0 + \Re(B)$ representing a linear manifold whose dimension is defined as $\dim \Re(B) = \text{rank } B$ [41, p. 38]. Geometrically, this set is parallel to the linear subspace $\Re(B)$ and does not contain the vector 0_m.

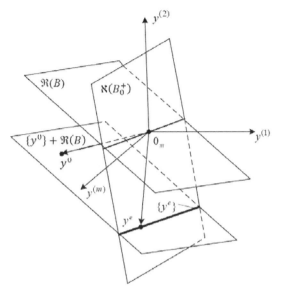

Figure 3.4 Geometric interpretation of the condition

On the other hand, the vector $y^0 - Bu$ should belong to the set $\aleph\left(B_0^+\right)$ to satisfy Equation (3.13). Therefore, the equilibrium point u^e exist if and only if the intersection of $y^0 + \Re\left(B\right)$ and $\aleph\left(B_0^+\right)$ is a non-empty set according to the condition defined by Equation (3.16).

Figure 3.4 gives the geometric interpretation of the condition (3.16).

Unfortunately, the condition (3.16) is not constructive. Another conditions which seems to be constructive are established in the next section.

3.4 Robust Nonadaptive Control

Before going to the design of the robust nonadaptive control of systems whose gain matrices are not full rank, we first establish conditions guaranteeing the existence of the equilibrium state of the closed-loop control systems shown in Figure 2 with rank $B_0 < r$. To do this, as the first we show that

$$\text{rank } P^+ = \text{rank } P \tag{3.17}$$

regardless of the rank of P. Indeed, according to [42, item 6.25] we have rank $P = \dim \Re\left(P\right)$. On the other hand, it is known that $\Re\left(P\right) = \Re\left(P^T\right)$ [41, item 3]. This gives rank $P^+ = \dim \Re\left(P^+\right) = \dim \Re\left(P^T\right) = $ rank $P^T = $ rank P leading to (3.17).

Now, fix a $m \times r$ matrix B_0 whose rank is equal 1. Due to Equation (3.17), we have rank $B_0^+ =$ rank $B_0 = 1$. Such a choice of B_0 gives that this matrix takes the following form:

$$
B_0^+ = \begin{pmatrix}
\beta_0^{(11)} & k_{12}\beta_0^{(11)} & \cdots & k_{m-1,m}\beta_0^{(11)} \\
\beta_0^{(21)} & k_{12}\beta_0^{(21)} & \cdots & k_{m-1,m}\beta_0^{(21)} \\
\vdots & \vdots & \vdots & \vdots \\
\beta_0^{(r1)} & k_{12}\beta_0^{(r1)} & \cdots & k_{m-1,m}\beta_0^{(r1)}
\end{pmatrix}
\tag{3.18}
$$

$$\underbrace{\hspace{8cm}}_{m \text{ columns}}$$

(Its columns are collinear r-dimensional vectors.)

Multiplying B_0 given in the form (3.18) by B of the form (3.2) we obtain the matrix

$$
G = B_0^+ B = \begin{pmatrix}
g^{(11)} & \cdots & g^{(1r)} \\
\vdots & \vdots & \vdots \\
g^{(r1)} & \cdots & g^{(rr)}
\end{pmatrix}
\tag{3.19}
$$

that is the matrix of Equation (3.13). In this expression, $g^{(1k)} = g^{(1k)}(B) = \beta_0^{(11)}d_k, \ldots, g^{(rk)} = g^{(rk)}(B) = \beta_0^{(r1)}d_k$ denote the elements of the kth column of G where

$$
d_k = b^{(1k)} + k_{12}b^{(2k)} + \ldots + k_{m-1,m}b^{(mk)}
\tag{3.20}
$$

is the scalar variable which depends on the elements of its kth column.

Introduce the r-dimensional vector

$$
g^{(k)}(B) = [\beta_0^{(11)}, \ldots, \beta_0^{(r1)}]^T d_k
\tag{3.21}
$$

which is the kth column of $G = G(B)$ depending on B for a fixed B_0. Since Ξ defined by Equation (3.11) is a bounded closed set, the minimum

$$
\underline{d_k}(B) = \min_{B \in \Xi} d_k(B)
$$

and the maximum

$$
\overline{d_k}(B) = \max_{B \in \Xi} d_k(B)
$$

of the variable $d_k(B)$ exist. Taking into account this fact and also Equation (3.20) it can be written

$$\underline{d}_k(B) = \min_{\substack{b^{(ik)} \in [\underline{b}^{(ik)}, \overline{b}^{(ik)}] \\ i = 1, \ldots, m}} \left(b^{(1k)} + k_{12}b^{(2k)} + \cdots + k_{m-1,m}b^{(mk)} \right)$$

$$k = 1, \ldots, r \tag{3.22}$$

$$\overline{d}_k(B) = \max_{\substack{b^{(ik)} \in [\underline{b}^{(ik)}, \overline{b}^{(ik)}] \\ i = 1, \ldots, m}} \left(b^{(1k)} + k_{12}b^{(2k)} + \cdots + k_{m-1,m}b^{(mk)} \right)$$

$$k = 1, \ldots, r) \tag{3.23}$$

Let the condition

$$\underline{d}_k(B)\,\overline{d}_k(B) > 0 \tag{3.24}$$

be satisfied for, at least, one $k \in 1, \ldots, r$.

This condition implies that $\underline{d}_k(B)$ and $\overline{d}_k(B)$ are nonzero and the same sign. Since they are linear continual functions of the m elements of B (according to Equation (3.20)), all components of $g^{(k)}(B)$ are nonzeros for all possible B from Ξ, if the condition (3.24) is satisfied. The transposition of $G = G(B)$ leads to the matrix

$$G^T = \begin{pmatrix} \beta_0^{(11)}d_1 & \cdots & \beta_0^{(r1)}d_1 \\ \vdots & \vdots & \vdots \\ \beta_0^{(11)}d_r & \cdots & \beta_0^{(r1)}d_r \end{pmatrix} \tag{3.25}$$

non-zero elements of which are collinear vectors. By the definition of the null space $\aleph(P) = \{x : \ Px = 0_r\}$ of any $r \times r$ matrix P (see, e.g. ,[42, item 6.24]) and by virtue of the expression (3.25) it can be written

$$\aleph(G^T(B)) = \{u : \beta_0^{(1)T}u = 0_r\} \tag{3.26}$$

where

$$\beta_0^{(1)} = [\beta_0^{(11)}, \ldots, \beta_0^{(r1)}]^T \tag{3.27}$$

Since the dimension of $\aleph(G^T(B))$ is equal to dim $\aleph(G^T(B)) = r - $ rank $G^T(B) = r - 1$ (according to [42, item 6.25]), it is not hard to see that, for each B, the set $\aleph(G^T(B))$ is a hyperplane in the space R^r of some vectors u to which $\beta_0^{(1)}$ given by the expression (3.27) is the normal vector.

Next, consider the right-hand side of Equation (3.13). Substituting the expression (3.18) into this side together with $y^0 = [y^{0(1)}, \ldots, y^{0(m)}]^T$ and taking the expression (3.27) into account, we get

$$B_0^+ y^0 = \beta_0^{(1)^T} L \tag{3.28}$$

where $L = y^{0(1)} + k_{12} y^{0(2)} + \cdots + k_{m-1,m} y^{0(m)}$.

Equation (3.28) shows that when $L \neq 0$, then the right-hand side of Equation (3.13) becomes a nonzero r-dimensional vector orthogonal to the null space of $G^T(B)$ (in accordance with (3.26)):

$$B_0^+ y^0 \perp \aleph(G^T(B)) \tag{3.29}$$

If $L \neq 0$ and the condition (3.29) is satisfied, then Equation (3.13) will be compatible (see [42, item 6.34]). On the other hand, if $L = 0$, then Equation (3.13) becomes homogeneous; it has always a solution.

The fact established above leads to the following basic result.

Lemma 3.1. Fix a matrix B_0 with rank $B_0 = 1$ so that all the elements of B_0^+ are nonzero. If the condition (3.24) with $\underline{d}_k(B)$ and $\overline{d}_k(B)$ defined by Equations (3.22) and (3.23), respectively, are satisfied, at least, for one $k \in 1, \ldots, r$, then the closed-loop control system (3.1), (3.6), (3.12) without any disturbances has an equilibrium state for any B from Ξ and for arbitrary $y^0 \in R^m$.

The geometric interpretation of the property (3.29) written in the form

$$B_0^+ y^0 \perp \aleph(\underbrace{B_0^+ B}_{B \in \Xi})$$

is given in Figure 3.5.

Remark 3.3. It is not hard to see that the computation of $\underline{d}_k(B)$ and $\overline{d}_k(B)$ by Equations (3.22) and (3.23), respectively, leads to solving the standard linear programming problem (LP): find $\min d_k(B)$ and $\max d_k(B)$ $(k = 1, \ldots, r)$ in the presence of constraints $\underline{b}^{(ij)} \leq b^{(ij)} \leq \overline{b}^{(ij)}$ $(i = 1, \ldots, m, \quad j = 1, \ldots, r)$.

With these preliminary results in mind, the asymptotic properties of the closed-loop control system containing the pseudoinverse model-based controller (3.12) are finally established in theorem below.

Theorem 3.1. Subject to Assumptions A1) to A3), if the conditions of Lemma 3.1 added by the condition (3.14) together with the expression (3.15) are

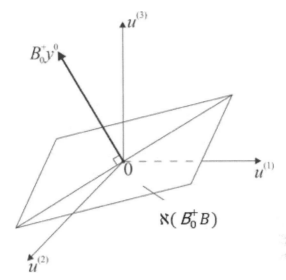

Figure 3.5 Geometric interpretation of the condition (3.29)

satisfied, then the control law (3.12), (3.6) when applied to the plant (3.1) is robust with respect to the uncertainty (3.5), and the ultimate boundedness of the sequences $\{e_n\}$ and $\{u_n\}$ is guaranteed:

$$\limsup_{n \to \infty} ||e_n|| = ||I_m - B_0 B_0^+|| \left[||e_0|| + 2 \sup_{0 <= n < \infty} ||v_n|| \right]$$

$$+2 \sup_{0 = n < \infty} ||v_n||(1 - q)^{-1} < \infty$$

$$\limsup_{n \to \infty} ||u_n - u^e|| = ||I_r - B_0^+ B_0|| \, ||u_0 - u^e||$$

$$+||B_0^+|| \left[(1 - q)^{-1} \sup_{0 <= n < \infty} ||v_n|| \right] < \infty$$

Proof. See [21].   □

Remark 3.4. Note that B_0 may not belong to Ξ to satisfy rank $B_0 = 1$ as illustrated in Figure 3.6.

At first sight, it is hard to verify the condition (3.14) by looking over all possible $\delta^{(ij)}$s from $\left[\underline{\delta}^{(ij)}, \overline{\delta}^{(ij)} \right]$ to compute the value of q (according to the expression (3.15)). In fact, it is not the case. The point is that condition (3.14)

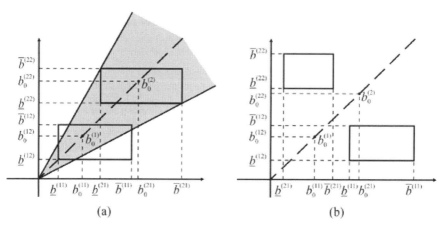

Figure 3.6 Choice of matrix B_0 satisfying the condition rank $B_0 = 1$ (possible cases): (a) there are B_0s belonging to Ξ; (b) there are no B_0s belonging to Ξ (3.29)

holds regardless of how the norm $|| \cdot ||$ is chosen in the expression (3.15). Therefore, the computation difficulty above mentioned can be avoided if we will choose this norm as $||B_0^+ \Delta||_1$, where

$$||P||_1 := \max_{1 < i \leq s} \sum_{j=1}^{s} |p^{(ij)}| \tag{3.30}$$

denotes the so-called 1-norm of a $s \times s$ matrix P. (According to the expression (3.30) this norm represents a piecewise-linear function of $\delta^{(ij)}$). With this in mind, we can consider the maximization problems caused by the expression (3.15) as a LP problems stated as follows.

Find

$$\min \sigma^{(ij)} \qquad \max \sigma^{(ij)} \tag{3.31}$$

under constraints

$$\underline{\delta}^{(ij)} \leq \delta^{(ij)} \leq \overline{\delta}^{(ij)} \quad i = 1, \ldots, m, \; j = 1, \ldots, r \tag{3.32}$$

where

$$\sigma^{(ij)} = \sum_{k=1}^{m} \beta_0^{(ik)} \delta^{(kj)}$$

are the linear forms, depending on the elements of the pseudoinverse matrix $B_0^+ = \left(\beta_0^{(ij)} \right)$, and $\underline{\delta}^{(ij)}$ and $\overline{\delta}^{(ij)}$ are defined by the expression (3.32).

Such a method allows us to reformulate Theorem 3.1 in the terms of LP problems as follows.

Theorem 3.2. Let the conditions of Lemma 3.1 be satisfied. Then if

$$\sum_{j=1}^{r} \max\{ \, |\min \sigma^{(ij)}|, \, |\max \sigma^{(ij)}| \, \} < 1 \ \forall i = 1, \ldots, m$$

with $\min \sigma^{(ij)}$ and $\max \sigma^{(ij)}$ that are the solutions of LP problems given in the expression (3.31), then the controller described in Equations (3.12) together with (3.6) robustly stabilizes the system (3.1) for any B from the set Ξ defined by the expression (3.11).

Proof. Follows immediately from the definition (3.30) of the 1-norm of an arbitrary matrix. $\qquad\qquad\qquad\qquad\qquad\qquad\qquad\qquad\qquad\qquad\qquad\qquad\qquad\quad\square$

Example 3.3. Let $\Xi = \{(b^{(ij)}) : \ 0.4 \leq b^{(11)} \leq 1.4, \, -2 \leq b^{(12)} \leq -1, -1.2 \leq b^{(21)} \leq -0.2, 0.7 \leq b^{(22)} \leq 1.7, -1.7 \leq b^{(31)} \leq -0.7, 1 <= b^{(32)} \leq 2\}$. Next, choose

$$B_0 = \begin{pmatrix} 0.8 & -1.6 \\ -0.7 & 1.4 \\ -1.0 & 2.0 \end{pmatrix}$$

with rank $B_0 = 1$. Such a choice yields

$$B_0^{+} = \begin{pmatrix} 16/213 & -14/213 & -20/213 \\ -32/213 & 28/213 & 40/213 \end{pmatrix}$$

Further, according to (3.18), we get $k_{12} = -7/8, k_{23} = -5/4$. By the formulas (3.22) and (3.23), it can be found $\underline{d}_1(B) = 1.45, \bar{d}_1(B) = 4.575$ showing that $\underline{d}_1(B) \, \bar{d}_1(B) > 0$.

Since the conditions of Lemma 3.1 are here satisfied, we conclude that the closed-loop control system described by Equations (3.1), (3.12) together with (3.6) has the equilibrium state for any $B \in \Xi$ and arbitrary $y^0 \in R^m$.

Let $b^{(11)} = 1, \ b^{(12)} = -1.5, \ b^{(21)} = -0.8, \ b^{(22)} = 1.2, \ b^{(31)} = -1.2, \ b^{(32)} = 1.8$. Choose $y^0 = [0.3, 0.4, 0.3]^T$ to satisfy $y^0 \notin \Re(B)$. Substituting the expressions of B, B_0 and y^0 into Equation (3.13) gives

$$3.2u^{e(1)} - 4.8u^{e(2)} + 0.425 = 0$$

This equation describes the set $E_u = \{u^e\}$ of the vector $u^e = [u^{e(1)}, u^{e(2)}]^T$ which is not one-point set. This set is shown in Figure 3.7.

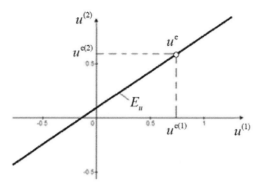

Figure 3.7   Equilibrium states related to the conditions of Example 3.3

Solving the LP problems (3.31) yields $\min\sigma^{(11)} = -194/2130$, $\max\sigma^{(11)} = 306/2130$, $\min\sigma^{(12)} = -106/2130$, $\max\sigma^{(12)} = 394/2130$, $\min\sigma^{(21)} = -612/2130$, $\max\sigma^{(21)} = 388/2130$, $\min\sigma^{(22)} = -788/2130$, $\max s^{(22)} = 216/2130$. This gives

$$\sum_{j=1}^{2}\max\{\,|\min\sigma^{(1j)}\,|,\ |\max\sigma^{(1j)}\,|\} = 70/213 < 1,$$

$$\sum_{j=1}^{2}\max\{\,|\min\sigma^{(2j)}\,|,\ |\max\sigma^{(2j)}\,|\} = 140/213 < 1.$$

Thus, all the conditions of Theorem 3.2 are here satisfied.

Simulation experiment 3.1. To verify how the pseudoinverse controller with fixed B_0^+, before chosen, works, a simulation of the control system with given B, B_0 and y^0 was conducted. The sequences $\{v_n^{(1)}\}$ and $\{v_n^{(2)}\}$ are generated as some i.i.d. random variables belonging to $[-0.07, 0.07]$. Duration of this simulation experiment was equal to 100 steps.

Simulation results are given in Figure 3.8. They demonstrate the boundedness of $\|u_n\|_2$ and $\|y_n\|_2$.

3.5 Robustly-Adaptive Control of Square Systems (Case 1)

In this section, the robustly-adaptive control system in the case when the bounds on the disturbances $v_n^{(i)}$ are known *a priopi* is designed. Let B be

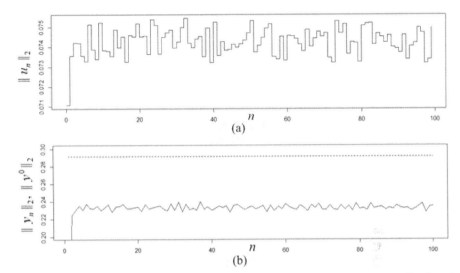

Figure 3.8 Robust pseudoinverse model-based controller: (a) Euclidean norm $||u_n||_2$ of control input vectors; (b) Euclidean norms $||y_n||_2$, $||y^0||_2$ of output vectors (solid line) and desired output vectors (dashed line), respectively

an unknown square singular $r \times r$ matrix, i.e.,

$$\det B = 0 \tag{3.33}$$

Basic idea to deal with a matrix B satisfying the requirement (3.33) is to replace adaptive identification of the true plant having the singular gain matrix B to the adaptive identification of a *fictitious* plant with the nonsingular gain matrix $\tilde{B}$ of the form

$$\tilde{B} = B + \delta_0 I_r \tag{3.34}$$

where I_r denotes the identity $r \times r$ matrix and δ_0 is a fixed quantity [37].

Although $\tilde{B}$ as well as B remain unknown, the requirement

$$\det \tilde{B} \neq 0 \tag{3.35}$$

can always be satisfied by the suitable choice of δ_0 in Equation (3.34). In fact, each ith eigenvalue $\lambda_i(B)$ of B lies in one of the r closed regions of the complex z-plane consisting of all the Gerŝgorin discs [43, p. 146]:

$$\left| z - b^{(ii)} \right| \leq \sum_{\substack{j=1 \\ j \neq i}}^{r} \left| b^{(ij)} \right|, i = 1, \ldots, r \tag{3.36}$$

Since, at least, one of the eigenvalues $\lambda_i(B)$ is equal to zero (due to the singularity of B), by virtue of Equation (3.35) there are the numbers

$$\underline{\beta}^{(i)} := b^{(ii)} - \sum_{\substack{j=1 \\ j \neq i}}^{r} \left| b^{(ij)} \right|, \overline{\beta}^{(i)} := b^{(ii)} + \sum_{\substack{j=1 \\ j \neq i}}^{r} \left| b^{(ij)} \right| \tag{3.37}$$

such that if

$$\left| b^{(i1)} \right| + \cdots + \left| b^{(ir)} \right| \neq 0 \tag{3.38}$$

then either $\underline{\beta}^{(i)} \leq 0$ but $\overline{\beta}^{(i)} > 0$ or $\underline{\beta}^{(i)} < 0$ but $\overline{\beta}^{(i)} \geq 0$. These numbers are defined by the intersection of the ith Gerŝgorin disc with the real axis of the complex z-plane as shown in Figures 3.9 and 3.10, respectively, left. In both cases, $\underline{\beta}^{(i)} \overline{\beta}^{(i)} \leq 0$ if the inequality (3.38) is satisfied because $\underline{\beta}^{(i)}$ and $\overline{\beta}^{(i)}$ cannot have the same sign.

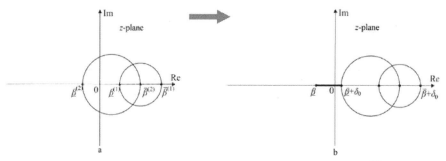

Figure 3.9   The Gerŝgorin discs for $r = 2$ in the case $|\underline{\beta}^{(2)}| < |\overline{\beta}^{(1)}|$

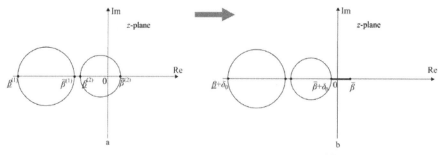

Figure 3.10   The Gerŝgorin discs for $r=2$ in the case $|\overline{\beta}^{(2)}| < |\underline{\beta}^{(1)}|$

Denoting

$$\underline{\beta} := \min\{ \underline{\beta}^{(1)}, \ldots, \underline{\beta}^{(r)} \}, \overline{\beta} := \max\{ \overline{\beta}^{(1)}, \ldots, \overline{\beta}^{(r)} \}, \tag{3.39}$$

consider the following two cases: (a) $|\underline{\beta}| < |\overline{\beta}|$; (b) $|\underline{\beta}| > |\overline{\beta}|$ (The case when $|\underline{\beta}| = |\overline{\beta}|$ can be combined with any two cases.) In order to go to the gain matrix $\tilde{B}$ of the fictitious plant having the form (3.34) in the case (a), it is sufficient to shift the Gerŝgorin disc (3.36) right taking

$$\delta_0 > |\underline{\beta}| \tag{3.40}$$

as shown in Figure 3.9, right. In the case b), the discs (3.36) need to be shifted left according to

$$\delta_0 < -|\overline{\beta}| \tag{3.41}$$

See Figure 3.10, right. In both cases, the nonsingularity of $\tilde{B}$ is guaranteed. Nevertheless, the conditions (3.40) and (3.41) cannot be satisfied, as yet. In fact, the numbers $\underline{\beta}$ and $\overline{\beta}$ given by the expressions (3.39) depend of $\underline{\beta}^{(i)}$ and $\overline{\beta}^{(i)}$s defined by Equations (3.37). But they are unknown because $b^{(ij)}$s are all unknown.

The following operations are proposed to choose a number d_0 satisfying the reqiurement (3.35). Introduce

$$\underline{\beta}^{(i)}_{\min} := \underline{b}^{(ii)} - \sum_{\substack{j=1 \\ j \neq i}}^{r} \max\{ |\underline{b}^{(ij)}|, \ |\overline{b}^{(ij)}| \}$$

$$\overline{\beta}^{(i)}_{\max} := \overline{b}^{(ii)} + \sum_{\substack{j=1 \\ j \neq i}}^{r} \max\{ |\underline{b}^{(ij)}|, \ |\overline{b}^{(ij)}| \} \tag{3.42}$$

minimizing and maximizing in $b^{(ij)} \in [\underline{b}^{(ij)}, \ \overline{b}^{(ij)}]$ the right-hand sides of Equations (3.37) for $\underline{\beta}^{(i)}$ and $\overline{\beta}^{(i)}$, respectively.

Further, the number δ_0 is found to satisfy the conditions

$$\delta_0 > -\underline{\beta}_{\min} \quad \text{if} \ |\underline{\beta}_{\min}| < |\overline{\beta}_{\max}|$$
$$\delta_0 > \overline{\beta}_{\max} \quad \text{if} \ |\underline{\beta}_{\min}| > |\overline{\beta}_{\max}| \tag{3.43}$$

where $\underline{\beta}_{\min}$, $\overline{\beta}_{\max}$ represent some quantities defined as follows:

$$\underline{\beta}_{\min} := \{\underline{\beta}_{\min}^{(1)}, \ldots, \underline{\beta}_{\min}^{(r)}\}$$
$$\overline{\beta}_{\max} := \{\overline{\beta}_{\max}^{(1)}, \ldots, \overline{\beta}_{\max}^{(r)}\} \tag{3.44}$$

It can be clarified that if the conditions (3.43) together with Equations (3.42) and (3.44) will be satisfied then the condition (3.35) will without fail be ensured.

After determining the quantity δ_0 we can proceed to the consideration of the fictitious plant. Since the input variables $u_n^{(1)}, \ldots, u_n^{(r)}$ and the disturbances $v_n^{(1)}, \ldots, v_n^{(N)}$ of both true plant and fictitious plant are the same, this feature makes it possible to describe our fictitious plant by the equation

$$\tilde{y}_n = \tilde{B} u_{n-1} + v_n, \tag{3.45}$$

similar to Equation (3.1), where $\tilde{y}_n = [\tilde{y}_n^{(1)}, \ldots, \tilde{y}_n^{(r)}]^T$ denotes the output vector of the fictitious plant.

It is interesting that the components of $\tilde{y}_n$ can be measured while the components of v_n in Equation (3.45) remain unmeasurable. In fact, substituting the expression (3.34) into Equation (3.45) due to Equation (3.1) we produce

$$\tilde{y}_n = y_n + \delta_0 u_{n-1}. \tag{3.46}$$

It is seen from Equation (3.46) that $\tilde{y}_n$ can always be found indirectly having u_n and y_n to be measured.

Now, our problem reduces to the known problem of adaptive control applicable to the fictitious plant (3.45) with the unknown gain matrix $\tilde{B}$ in the presence of arbitrary bounded disturbances $v_n^{(1)}, \ldots, v_n^{(r)}$. Its solving follows the steps of the section above. Namely, the adaptive control law is designed in the form

$$u_n = u_{n-1} + \tilde{B}_n^{-1} \tilde{e}_n, \tag{3.47}$$

in which, instead of the current estimate B_n of B, another $\tilde{B}_n$ is exploited, and the error vector e_n defined in Equation (3.6) is replaced by

$$\tilde{e}_n = y^0 - \tilde{y}_n \tag{3.48}$$

with $\tilde{y}_n$ given by the expression (3.46).

The adaptive identification algorithm used to determine the estimates $\tilde{B}_n$ may be taken as

$$
\tilde{b}_n^{(i)} =
\begin{cases}
\tilde{b}_{n-1}^{(i)} & \text{if } |\tilde{e}_n^{*(i)}| = \varepsilon_i^0, \\
\tilde{b}_{n-1}^{(i)} + \gamma_n^{(i)} \dfrac{\tilde{e}_n^{*(i)} - \hat{\varepsilon}_i \, \text{sign } \tilde{e}_n^{*(i)}}{\|\nabla u_{n-1}\|_2^2} \nabla u_{n-1} & \text{otherwise}, \quad i = 1, \ldots, r
\end{cases}
$$

(3.49)

which is similar to that in [36]. In this algorithm, ε_i^0 and $\bar{\varepsilon}_i$ are given by

$$\varepsilon_i^0 > \bar{\varepsilon}_i = 2\varepsilon_i, \quad i = 1, \ldots, r \tag{3.50}$$

$$\tilde{e}_n^{*(i)} = \nabla \tilde{y}_n^{(i)} - \tilde{b}_{n-1}^{(i)T} \nabla u_{n-1} \tag{3.51}$$

represent the ith component of the identification error $\tilde{e}_n^*$ given as

$$\tilde{e}_n^* = \nabla \tilde{y}_n - \tilde{B}_{n-1} \nabla u_{n-1} \tag{3.52}$$

where $\nabla \tilde{y}_n^{(i)} := \tilde{y}_n^{(i)} - \tilde{y}_{n-1}^{(i)}$, and the notation $\tilde{b}_n^{(i)T} := \left[\tilde{b}_n^{(i1)}, \ldots, \tilde{b}_n^{(ir)} \right]$ of the ith row of $\tilde{B}_n$ is introduced. The coefficients $\gamma_n^{(i)}$s are chosen to satisfy

$$\det \tilde{B}_n \neq 0$$

The feedback adaptive robust control system described in Equations (3.1), (3.47), (3.49) is designed as depicted in Figure 3.11. In this figure, the notation $\nabla \tilde{y}_n^* := \tilde{B}_{n-1} \nabla u_{n-1}$ is introduced.

The asymptotic properties of the adaptive control system are established in the theorem below.

Theorem 3.3. Determine δ_0 using the formula (3.43) together with Equations (3.42) and (3.44), and choose an arbitrary initial $\tilde{B}_0 = B_0 + \delta_0 I$ with $B_0 = (b_0^{(ij)})$ whose elements satisfy the conditions $\underline{b}^{(ij)} \leq b_0^{(ij)} \leq \bar{b}^{(ij)}$. Subject to assumptions A1–A3, the adaptive controller described in Equations (3.47) and (3.49) together with Equations (3.46) and (3.48) when applied to the plant (3.1) yields Equations (3.7) and (3.8).

Proof. See [37]. □

We now present a simulated example for showing the performance of the adaptive controller.

Simulation experiment 3.2. In this experiment, the elements of B were given as $b^{(11)} = 4$, $b^{(12)} = 2$, $b^{(21)} = 2$, $b^{(22)} = 1$ ($\det B = 0$). The interval

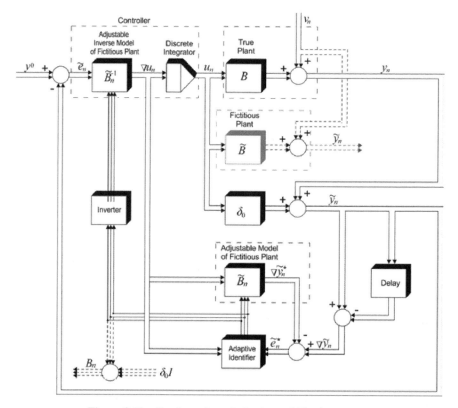

Figure 3.11   Configuration of adaptive stabilization system

estimates of these elements were chosen as follows: $1 \leq b^{(11)} \leq 5$, $0 \leq b^{(12)} \leq 2$, $0 \leq b^{(21)} \leq 2$, $1 \leq b^{(22)} \leq 2$. By the formulas (3.42) and (3.44), it was found: $\underline{\beta}_{\min}^{(1)} = -1$, $\underline{\beta}_{\min}^{(2)} = -1$, $\overline{\beta}_{\max}^{(1)} = 7$, $\overline{\beta}_{\max}^{(2)} = 4$, $\underline{\beta}_{\min} = -1$, $\overline{\beta}_{\max} = 7$. It turned out that $|\underline{\beta}_{\min}| < |\overline{\beta}_{\max}|$. Therefore it is required that $\delta_0 > 1$ to satisfy the inequalities (3.43). Namely, $\delta_0 = 1.1$ was put. From the conditions $b_0^{(11)} \in [1, 5]$, $b_0^{(12)} \in [0, 2]$, $b_0^{(21)} \in [0, 2]$, $b_0^{(22)} \in [1, 2]$ the following initial estimates of the elemetns of B_n were taken: $b_0^{(11)} = 1$, $b_0^{(21)} = 0$, $b_0^{(12)} = 1$, $b_0^{(22)} = 1.9$. Then $\tilde{b}_0^{(11)} = 2.1$, $\tilde{b}_0^{(12)} = 1$, $\tilde{b}_0^{(21)} = 0$, $\tilde{b}_0^{(22)} = 3$.

In this simulation experiment, the sequences $\{v_n^{(i)}\}(i = 1, 2)$ were generated as i.i.d. random variables belonging to $[-1, 1]$. It was put: $y^0 = [1, 3]^T$.

The performance of the adaptive estimation algorithm is shown in Figure 3.12.

Figure 3.13 shows simulation results illustrating a successful behavior of the robustly-adaptive control system when this experiment was conducted.

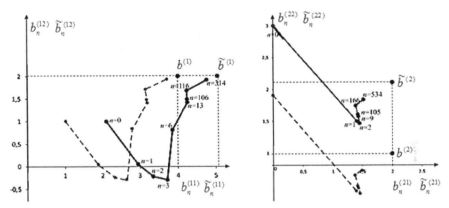

Figure 3.12 Estimated parameters of the fictitious plant (solid line) and the true plant (dashed line)

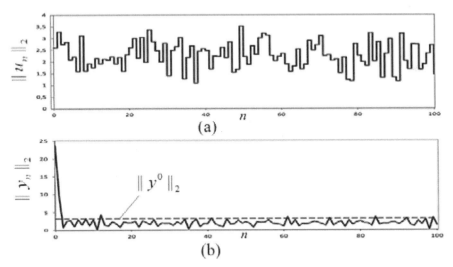

Figure 3.13 Performance of the robustly-adaptive controller in Simulation experiment 3.2: (a) Euclidean norm of input control vector; (b) Euclidean norms of output vector

3.6 Robustly-Adaptive Control of Square Systems (Case 2)

In this case called Case 2 we assume that the bounds $\varepsilon^{(i)}$ on disturbances $v_n^{(i)}$ are unknown *a priori*. Therefore, they need to be adaptively estimated.

Following to [38], the algorithm for the adaptive estimation of the fictitious parameters is chosen as

$$\tilde{b}_n^{(i)} = \tilde{b}_{n-1}^{(i)} - \gamma_n^{(i)} \frac{f(\tilde{e}_n^{*(i)}, \varepsilon_{n-1}^{(i)})}{1 + ||\nabla u_n||^2} \nabla u_n \text{ sign } \tilde{e}_n^{*(i)}, \quad i = 1, \ldots, r \qquad (3.53)$$

where $f(e, \bar{\varepsilon})$ is the dead-zone function of the form

$$f(e, \bar{\varepsilon}) = \begin{cases} 0, & \text{if} |e| \leq \bar{\varepsilon}, \\ |e| - \bar{\varepsilon} & \text{if} |e| > \bar{\varepsilon} \end{cases}$$

which depends on the scalar variable $\tilde{e}_n^{*(i)}$ given by (3.51) and also on the past estimate $\varepsilon_{n-1}^{(i)}$ of unknown $\varepsilon^{(i)}$.

As in [38] the adaptive estimation of the bounds $\varepsilon^{(i)}$ is produced via the recursive algorithm

$$\varepsilon_n^{(i)} = \varepsilon_{n-1}^{(i)} + \gamma_n^{(i)} \frac{f(\tilde{e}_n^{*(i)}, \varepsilon_{n-1}^{(i)})}{1 + ||\nabla u_n||_2^2}, \quad i = 1, \ldots, r \qquad (3.54)$$

The following results concerning the asymptotic properties of the robustly-adaptive control system for Case 2 are valid.

Theorem 3.4. Let the conditions of Theorem 3.3 be satisfied. Then the adaptive controller described in Equations (3.47), (3.48), (3.53) and (3.54) has the properties:

  (a) the estimates caused by Equations (3.53) and (3.54) converge;
  (b) the ultimate boundedness in the forms (3.7) and (3.8) are achieved.

Proof. See [38]. □

Simulation experiment 3.3. The conditions of experiment were taken as in previous experiment except the bounds on the disturbance $v_n^{(1)}$ and $v_n^{(2)}$ given by $-1 \leq v_n^{(1)} \leq 1$ and $-0.5 \leq v_n^{(2)} \leq 0.5$.

Results of this simulation experiment are presented in Figures 3.14–3.17.

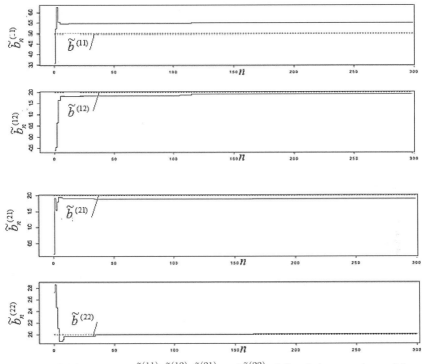

Figure 3.14 Estimates $\tilde{b}_n^{(11)}$, $\tilde{b}_n^{(12)}$, $\tilde{b}_n^{(21)}$ and $\tilde{b}_n^{(22)}$ of Simulation experiment 3.3

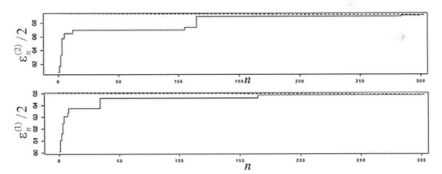

Figure 3.15 Estimates $\varepsilon_n^{(1)}/2$ and $\varepsilon_n^{(2)}/2$ of Simulation experiment 3.3

It is seen from Figures 3.14 and 3.15, that the estimates of unknown system parameters and bounds on disturbances converges as predicted by part (a) of Theorem 3.4. We can also observe that the control inputs and outputs remain bounded.

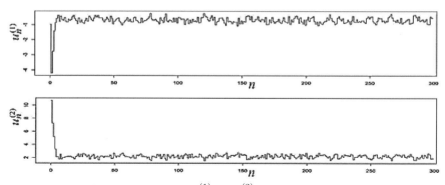

Figure 3.16   Control inputs $u_n^{(1)}$ and $u_n^{(2)}$ of Simulation experiment 3.3

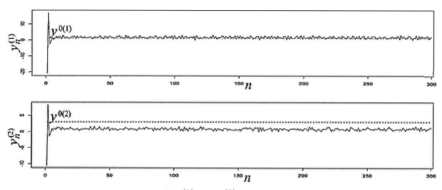

Figure 3.17   Outputs $y_n^{(1)}$ and $y_n^{(2)}$ of Simulation experiment 3.3

3.7 Robustly-Adaptive Control of Nonsquare Systems

Let B be a nonsquare $m \times r$ matrix of the form (3.2) with unknown rank satisfying (3.4). Define the so-called submatrices $B[i_1[k], \ldots, i_r[k]|1, \ldots, r]$ $\in R^{r \times r}$ [43, part I, subsect. 2.2] whose rows represent the rows of B with the numbers $i_1[k], \ldots, i_r[k]$ $(1 \leq i_1[k] < \cdots < i_r[k] \leq m)$. The quantity of these matrices is equal to $N = \begin{pmatrix} m \\ r \end{pmatrix}$. Denoting by $B[k]$ the submatrix which corresponds to a kth subset $\{i_1[k], \ldots, i_r[k]\}$, write the equations of some k plants as:

$$y_n[k] = B[k]u_{n-1} + v_n[k], \qquad k = 1, \ldots, N, \qquad (3.55)$$

where $y_n[k] = [y_n^{(i_1[k])}, \ldots, y_n^{(i_r[k])}]^T \in R^r$ and $v_n[k] = [v_n^{(i_1[k])}, \ldots, v_n^{(i_r[k])}]^T \in R^r$.

In accordance with the approach proposed in the previous section, pass from Equation (3.55) to the equations of the fictitious plants described by

$$\tilde{y}_n[k] = \tilde{B}[k]u_{n-1} + v_n[k], \qquad k = 1, \ldots, N \qquad (3.56)$$

with the same u_{n-1} and $v_{n-1}[k]$. In these equations, $\tilde{y}_n[k]$ denotes the r-dimensional output vector related to the kth fictitious plant whose gain matrix $\tilde{B}[k]$ is defined as follows:

$$\tilde{B}[k] = B[k] + \delta_0[k]I_r, \qquad (3.57)$$

where $\delta_0[k]$ is a fixed quantity depending on k. This quantity is calculated for each $k = 1, \ldots, N$ using the technique described in the previous section. Namely, taking into account the constraints (3.5), $\delta_0[k]$ can always be found to ensure

$$\det \tilde{B}[k] \neq 0 \qquad \forall k = 1, \ldots, N. \qquad (3.58)$$

It follows from Equations (3.55) to (3.57) that

$$\tilde{y}_n[k] = y_n[k] + \delta_0[k]u_{n-1}. \qquad (3.59)$$

This expression shows that although as $\tilde{B}[k]$ as $B[k]$ remain unknown, however, the components of all N the vectors $\tilde{y}_n[k]$ can indirectly be "measured" after measuring the components of y_n and u_{n-1}, and it is essential.

If the conditions (3.58) are satisfied, then the problem of the adaptive stabilization of the true plant (3.1) can be reduced to the problem of simultaneous adaptive stabilization of all N fictitious plants (3.56) with unknown but nonsingular $r \times r$ gain matrices $\tilde{B}[k]$ ($k = 1, \ldots, N$) via forming at each nth time instant a set of N different "potentially" possible controls $u_n[1], \ldots, u_n[N]$ and selecting one of them in accordance with certain choice rule [27] given below.

Following to [39], the adaptive control law to be applicable to any fictitious plant is designed in the form

$$u_n[k] = u_{n-1} + \tilde{B}_n^{-1}[k]\,\tilde{e}_n[k], \qquad k = 1, \ldots, N, \qquad (3.60)$$

where $\tilde{e}_n[k] = y^0[k] - \tilde{y}_n[k]$ with $y^0[k] = [y^{0(i_1[k])}, \ldots, y^{0(i_r[k])}]^T$ defines the output error vector related to the kth fictitious plant at the nth time instant, and $\tilde{B}_n[k] \in R^{r \times r}$ is the current estimate of unknown $r \times r$ matrix $\tilde{B}[k]$ at the same time instant satisfying

$$\det \tilde{B}_n[k] \neq 0 \qquad \forall k = 1, \ldots, N. \qquad (3.61)$$

As the adaptation algorithms, the standard recursive procedures for the adaptive identification of each kth fictitious plant (3.55) described by

$$\tilde{b}_n^{(i)}[k] = \begin{cases} \tilde{b}_{n-1}^{(i)}[k] & \text{if } |\tilde{e}_n^{*(i)}[k]| \le \varepsilon_i^0, \\ \tilde{b}_{n-1}^{(i)}[k] + \gamma_n^{(i)} \dfrac{\tilde{e}_n^{*(i)}[k] - \bar{\varepsilon}_i \, \text{sign } \tilde{e}_n^{*(i)}[k]}{\|\nabla u_{n-1}\|_2^2} \nabla u_{n-1} & \text{otherwise,} \\ i = 1, \ldots, r, \qquad k = 1, \ldots, N \end{cases}$$

(3.62)

are proposed. In these algorithms, $\tilde{b}_n^{(i)}[k]$ denotes the r-dimensional estimate vector obtained by transposing the ith row of $\tilde{B}_n[k]$, and

$$\tilde{e}_n^{*(i)}[k] = \tilde{y}_n^{(i)}[k] - \tilde{y}_{n-1}^{(i)}[k] - \tilde{b}_{n-1}^{(i)T}[k]\nabla u_{n-1} \tag{3.63}$$

represents the scalar variable making sense of the ith component of $\tilde{e}_n^*[k] \in R^r$ that is the identification error vector related to the kth fictitious plant. The coefficients $\gamma_n^{(i)}$s are chosen from the ranges $[\underline{\gamma}^{(i)}, \overline{\gamma}^{(i)}]$ to ensure the requirement (3.61).

Next, add the adaptation algorithms described in the formulas (3.62) together with (3.63) by an algorithm for estimating unknown B defined as follows:

$$b_n^{(i)} = \begin{cases} b_{n-1}^{(i)} & \text{if } |c_n^{*(i)}| \le \varepsilon_i^0, \\ b_{n-1}^{(i)} + \gamma_n^{(i)} \dfrac{e_n^{*(i)} - \bar{\varepsilon}_i \, \text{sign } e_n^{*(i)}}{\|\nabla u_{n-1}\|_2^2} \nabla u_{n-1} & \text{otherwise,} \qquad i = 1, \ldots, m, \end{cases}$$

(3.64)

where $b_n^{(i)T}$ represents the ith row of the estimate matrix B_n, and

$$e_n^{*(i)} = y_n^{(i)} - y_{n-1}^{(i)} - b_{n-1}^{(i)T}\nabla u_{n-1} \tag{3.65}$$

is the ith component of the identification error vector $e_n^* = y_n - y_{n-1} - B_{n-1}\nabla u_{n-1}$ ($\bar{\varepsilon}_i$ and ε_i^0 are given by the conditions (3.50)).

The estimation procedure defined in the algorithm (3.64) together with Equation (3.65) makes it possible to estimate the m predicted output errors $\overrightarrow{e}_{n+1}^{(i)}[k]$ ($i = 1, \ldots, m$) for the each ith output of true plant (3.1) at any n using the formula

$$|\overrightarrow{e}_{n+1}^{(i)}[k]| = |y^{0(i)} - b_n^{(i)T}u_n[k]| + \varepsilon^{(i)}, \qquad i = 1, \ldots, m. \tag{3.66}$$

The synthesis of the adaptive controller is finished by the choice of the control u_n from the set $\{u_n[1], \ldots, u_n[N]\}$ with $u_n[k]$ given by

Equation (3.60). This choice is implemented by the rule giving the minimum of the 1-norm of $\vec{e}_{n+1}[k] = [\vec{e}_{n+1}^{(1)}[k], \ldots, \vec{e}_{n+1}^{(m)}[k]]^T$ according to

$$u_n = \arg \min_{u_n[k]} \sum_{i=1}^{m} |\vec{e}_{n+1}^{(i)}[k]|, \tag{3.67}$$

where $\vec{e}_{n+1}^{(i)}[k]$s are specified by Equation (3.66).

The asymptotic properties of the adaptive controller described in this section are given in theorem below (the main result).

Theorem 3.5. Consider the feedback control system containing the plant (3.1) in which $r < m$, and the adaptive controller defined in Equations (3.62), (3.67) together with Equations (3.59), (3.66) and (3.61). Using the constraints (3.5), determine $\delta_0[1], \ldots, \delta_0[N]$ to satisfy the requirement (3.58). Let assumptions A1–A3 be valid. Then, this controller applied to the plant (3.1) guarantees that the control objectives (3.7) and (3.8) will be achieved.

Proof. See [39]. □

Note that Theorem 3.5 does not guarantee that the ultimate error $\lim_{n\to\infty} \sup \|e_n\|$ will become as in the nonadaptive case when there is no parametric uncertainty and the pseudoinverse model-based controller proposed in [21] can by applied.

Simulation experiment 3.4. A simulation experiment was conducted to illustrate the performance of the proposed adaptive control in the case when $r = 2$, $m = 3$. As the gain matrix,

$$B = \begin{pmatrix} 4 & 2 \\ 2 & 1 \\ 3 & 1.5 \end{pmatrix}$$

of not full rank (rank $B = 1$) was taken. Since $N = 3$, it produces the following three submatrices:

$$B[1] = \begin{pmatrix} 4 & 2 \\ 2 & 1 \end{pmatrix}, B[2] = \begin{pmatrix} 4 & 2 \\ 3 & 1.5 \end{pmatrix} \text{ and } B[3] = \begin{pmatrix} 2 & 1 \\ 3 & 1.5 \end{pmatrix}$$

Further, the three vectors $y_n[1] = [y_n^{(1)}, y_n^{(2)}]^T$, $y_n[2] = [y_n^{(1)}, y_n^{(3)}]^T$ and $y_n[3] = [y_n^{(2)}, y_n^{(3)}]^T$ was introduced to describe the plants (3.55) having the gain matrices $B[1]$, $B[2]$ and $B[3]$, respectively.

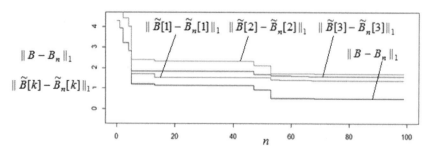

Figure 3.18   Variables describing the adaptive estimation processes

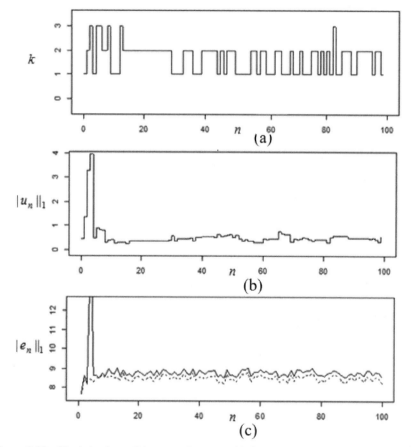

Figure 3.19   The behaviour of the control system: (a) the current number k of control $u_n[k]$ chosen from $\{u_n[1],\ u_n[2],\ u_n[3]\}$ at given n; (b) the 1-norm of control vector; (c) the 1-norm of output vector in adaptive case (solid line) and in nonadaptive optimal case (dashed line)

The quantities $\delta_0[1] = 1.1$, $\delta_0[2] = 1.2$ and $\delta_0[3] = 1.3$ were taken to satisfy the conditions (3.58) guaranteeing $\tilde{B}[k]$ to be nonsingular were derived from Equation (3.5). The initial $\tilde{B}_0[1]$, $\tilde{B}_0[2]$ and $\tilde{B}_0[3]$ were chosen as $\tilde{B}_0[k] = B_0[k] + \delta_0[k]I_r$ with the initial elements of $B_0[k]$ which were selected from B inside the corresponding ranges $[\underline{b}^{(ij)}, \overline{b}^{(ij)}]$ specified as follows: $b^{(11)} \in [1, 5]$, $b^{(12)} \in [0, 2]$, $b^{(21)} \in [0, 2]$, $b^{(22)} \in [1, 2]$, $b^{(31)} \in [1, 4]$, $b^{(32)} \in [0, 5]$. Namely, we set $b_0^{(11)} = 1$, $b_0^{(12)} = 1$ $b_0^{(21)} = 0$, $b_0^{(22)} = 1.9$, $b_0^{(31)} = 2$, $b_0^{(32)} = 2.1$. The desired output vector was given as $y^0 = [1, 3, 7]^T$.

The performance of the simulated adaptive control system with the disturbance sequences $\{v_n^{(i)}\} = v_0^{(i)}, v_1^{(i)}, \ldots$ generated as some i.i.d. random variables taken from $-0.1 \leq v_n^{(1)} \leq 0.1$, $-0.2 \leq v_n^{(2)} \leq 0.2$, $-0.08 \leq v_n^{(3)} \leq 0.08$ is presented in Figures 3.3 and 3.19.

Figures 3.19(a)–(c) demonstrate that the performance of the proposed adaptive controller applied to the static multivariable plant having some nonsquare gain matrix with not full rank is successful enough.

3.8 Conclusion

This chapter shed a light on the problem regarding the robust control of the discrete-time multivariable memoryless systems with the noninvertible gain matrices of the not full rank. It has been established that if certain conditions have been satisfied then the nonadaptive pseudoinverse model-based controller can be robust against plant uncertainty.

Again, the chapter answered the second question: Can robustly-adaptive controller using as adjustable inverse model principle cope with the interval parametric uncertainty of the system to be controlled? Namely, it has been shown that such a possibility exists because these controllers are able to guarantee the ultimate boundedness of output and control input signals. However, this important feature is achieved via an "overparametrization" of the uncertain systems. Nevertheless, the simulation experiments showed their efficiency.

References

[1] V.M. Kuntsevich, et al (Eds). *Control Systems: Theory and Applications. Series in Automation, Control and Robotics*, River Publishers, Gistrup, Delft, 2018.

[2] Y.P. Kondratenko, V.M. Kuntsevich, A.A. Chikrii, V.F. Gubarev, (Eds). Advanced Control Systems: *Theory and Applications. Series in Automation, Control and Robotics*, River Publishers, Gistrup, 2021.

[3] S. Skogestad and I. Postlethwaite, *Multivariable Feedback Control*, Wiley, Chichester, 1996.

[4] P. Albertos and A. Sala, *Multivariable Control Systems: an Engineering Approach*, Springer, London, 2006.

[5] T. Glad and L. Ljung, *Control Theory: Multivariable and Nonlinear Methods*, Taylor & Francis, New York, 2000.

[6] L. Tan, *A Generalized Framework of Linear Multivariable Control*, Elsevier, Oxford, 2017.

[7] V.F. Gubarev, M.D. Mishchenko, and B.M. Snizhko, 'Model predictive control for discrete MIMO linear systems', In: Kondratenko, Y.P., Chikrii, A.A., Gubarev, V.F., Kacprzyk, J. (Eds.). *Advanced Control Techniques in Complex Engineering Systems: Theory and Applications*. Dedicated to Prof. V.M. Kuntsevich. Studies in Systems, Decision and Control, pp. 63-81, vol. 203, Springer Nature Switzerland AG, Cham, 2019. DOI: https://doi.org/10.1007/978-3-030-21927-7_10

[8] E. Davison, 'The output control of linear time-invariant multivariable systems with unmeasurable arbitrary disturbances', *IEEE Trans. Autom. Contr.*, pp. 621–631, no. 5, vol. AC-17, Oct. 1972.

[9] B. Francis and W. Wonham, 'The internal model principle of control theory', *Automatica*, pp. 457–465, no. 5, vol. 12, Sep. 1976.

[10] R. Brockett, 'The invertibility of dynamic systems with application to control', *Ph. D. Dissertation*, Case Inst. of Technology, Cleveland, Ohio, 1963.

[11] M.K. Sain and J.L. Massey, 'Invertibility of linear time-invariant dynamical systems', *IEEE Trans. Autom. Contr*, pp. 141–149, no. 2, vol. AC-14, Apr. 1969.

[12] L.M. Silverman, 'Inversion of multivariable linear systems', *IEEE Trans. Autom. Contr.*, pp. 270–276, no. 3, vol. AC-14, Jun. 1969.

[13] H. Seraji, 'Minimal inversion, command tracking and disturbance decoupling in multivariable systems', *Int. J. Control*, pp. 2093–2191, no. 6, vol. 49, Jun. 1988.

[14] G. Marro, D. Prattichizzo, and E. Zattoni, 'Convolution profiles for right-inversion of multivariable non-minimum phase discrete-time systems', *Automatica*, pp. 1695–1703, no. 10, vol. 38, Oct. 2002.

[15] L.M. Lyubchyk, 'Disturbance rejection in linear discrete Multivariable systems: inverse model approach', In: *Proc. 18th IFAC World Congress*, Milano, Italy, 2011, pp. 7921–7926.

[16] L.M. Lyubchyk, 'Inverse Model Approach to Disturbance Rejection Problem', In: Kondratenko, Y.P., Kuntsevich, V.M., Chikrii, A.A., Gubarev, V.F. (Eds), *Advanced Control Systems:Theory and Applications*, River Publishers, Gistrup, pp. 129–166, 2021.

[17] G.E. Pukhov and K.D. Zhuk, *Synthesis of Interconnected Control Systems via Inverse Operator Method*, Nauk. dumka, Kiev, 1966 (in Russian).

[18] T. Lee, G. Adams, and W. Gaines, *Computer Process Control: Modeling and Optimization*, Wiley, New York, 1968.

[19] V. Lovass-Nagy, J.R. Miller, and L.D. Powers, 'On the application of matrix generalized inversion to the construction of inverse systems', *Int. J. Control*, pp. 733–739, no. 5, vol. 24, Apr. 1976.

[20] L.S. Zhiteckii, V.N. Azarskov, K.Yu. Solovchuk and O.A. Sushchenko, 'Discrete-time robust steady-state control of nonlinear multivariable systems: a unified approach', In: *Proc. 19th IFAC World Congress*, Cape Town, South Africa, 2014, pp. 8140–8145.

[21] L.S. Zhiteckii and K.Yu. Solovchuk, 'Pseudoinversion in the problems of robust stabilizing multivariable discrete-time control systems of linear and nonlinear static objects under bounded disturbances', *Journal of Automation and Information Sciences*, pp. 35–48, no. 5, vol. 49, Jul. 2017.

[22] S.P. Bhattacharyya, H. Chapellat, and L.H. Keel, *Robust Control: the Parametric Approach*. Prentice Hall, Upper Saddle River, NJ, 1995.

[23] M. Green and D.J.N. Limebeer, *Linear Robust Control*. Prentice Hall, Englewood Cliffs, NJ, 1995

[24] R. Sanchess-Pena and M. Sznaier, *Robust Systems: Theory and Application*. Willey, New York, 1998.

[25] B.T. Polyak and P.S. Shcherbakov, *Robust Stability and Control*, Nauka, Moscow, 2002 (in Russian).

[26] V.M. Kuntsevich, *Control under Uncertainty: Guaranteed Results in Control and Identification Problems*, Nauk. dumka, Kiev, 2006 (in Russian).

[27] V.F. Sokolov, *Robust Control with Bounded Disturbances*, Komi Scientific Center, Ural Branch of the Russian Academy of Sciences, Syktyvkar, 2011 (in Russian).

[28] V.M. Kuntsevich, 'Estimation of impact of bounded perturbations on nonlinear discrete systems', In: Kuntsevich, V.M., Gubarev, V.F., Kondratenko, Y.P., Lebedev, D.V., Lysenko, V.P. (Eds), *Control Systems: Theory and Applications. Series in Automation, Control and Robotics*, River Publishers, Gistrup, Delft, pp. 3-15, 2018.

[29] B.I. Kuznetsov, T.B. Nikitina, and I.V. Bovdui, 'Multiobjective synthesis of two degree of freedom nonlinear robust control by discrete continuous plant. Tekhnichna elektrodynamika', *Institute of Electrodynamics National Academy of Science of Ukraine*, pp. 10–14, no. 5, Sep. 2020.

[30] V.N. Fomin, A.L. Fradkov, and V.A. Yakubovich, *Adaptive Control of Dynamic Plants*, Nauka, Moscow, 1981 (in Russian).

[31] G.C. Goodwin and K.S. Sin, *Adaptive Filtering, Prediction and Control*, Prentice-Hall, Engewood Cliffs, 1984.

[32] G. Tao. *Adaptive Control Design and Analysis*. John Wiley and Sons, New York, 2003.

[33] K.S. Narendra and A.M. Annaswamy, *Stable Adaptive Systems*, Dover Publications, New York, 2012.

[34] P. Ioannou and J. Sun, *Robust Adaptive Control*, Dover Publications, New York, 2013.

[35] K.J. Åström and B. Wittenmark, *Adaptive Control: 2nd Edition*, Dover Publications, New York, 2014.

[36] L.S. Zhiteckii and K.Yu. Solovchuk, 'Robust Adaptive Controls for a Class of Nonsquare Memoryless Systems', In: Kondratenko, Y.P., Kuntsevich, V.M., Chikrii, A.A., Gubarev, V.F. (Eds), *Advanced Control Systems:Theory and Applications*, River Publishers, Gistrup, pp. 203–226, 2021.

[37] V.N. Azarskov, L.S. Zhiteckii, and K.Yu. Solovchuk, 'Parametric identification of the inter-connected static closed-loop system: a special case', In: *Proc. 12th All-Russian Control Problems Council (VSPU-2014)*, IPU, Moscow, 2014, pp. 2764-2776.

[38] V.N. Azarskov, L.S. Zhiteckii, and K.Yu. Solovchuk, 'Identification approach to the problem of robust control for multi-connected static plant with nonstochastic uncertainties', In: *Proc. 10th Internation Conference System identification and Control Problems (SICPRO'15)*, IPU, Moscow, 2015, pp. 520-538.

[39] L.S. Zhiteckii, V.N. Azarskov, and K.Yu. Solovchuk, 'Adaptive robust control of inter-connected static plants with nonsquare gain matrixes', In: *Proc. 13th All-Russian Control Problems Council (VSPU-2019)*, IPU, Moscow, 2019, pp. 713–718.

[40] L.S. Zhiteckii and K.Yu. Solovchuk, 'Robustly-adaptive control of linear interconnected memoryless systems with nonsquare gain matrices', In: *Proc. Int. Conf. "Automatics–2020"*, Kyiv, Ukraine, pp. 123–124, 2020.

[41] A. Albert, *Regression and the Moore-Penrose Pseudoinverse*, Academic Press, New York, 1972.

[42] V.V. Voevodin and Yu.A. Kuznetsov, *Matrices and Calculations*, Nauka, Moscow, 1984 (in Russian).

[43] M. Marcus and H. Minc, *A Survey of Matrix Theory and Matrix Inequalities*, Aliyn and Bacon, Boston, 1964.

4

Nonlinear Integral Inequalities and Differential Games of Avoiding Encounter

L.P. Yugay

Dr.Sc.(Math), Professor,
Department of Natural Sciences, Uzbek State University of Physical Culture and Sport, Chirchik City, Republic of Uzbekistan
Tel: +998-90-187-1478(m)
E-mail: yugailp@mail.ru

Abstract

In the paper, a survey of the application of integral inequalities in various theoretical branches of mathematics and their applied fields is represented. New integral inequalities of the Hölder type are introduced, and corresponding estimates are proved. Also, the possibilities of applications of demonstrated results in nonlinear differential games of avoiding encounters are indicated.

Keywords: integral inequality, estimation, trajectory, differential game, evasion, avoiding, trajectory.

4.1 Introduction

4.1.1 On the First Integral Inequalities

Integral inequalities are widely used in the theories of differential equations, nonlinear oscillations, stability, as well as in mechanics and averaging problems. With their help, the unique theorems of solutions of differential equations are proved, and various approximate estimates are obtained for solutions of differential equations that cannot be analytically solved. Let us formulate the first variance of Grönwall–Bellman type inequality.

97

Theorem 1 (Inequality of Grönwall–Bellman)

Let $u(t)$ and $v(t)$ are nonnegative and continuous functions on the length $[a, b]$ and

$$v(t) \leq C + \int_a^b v(s) u(s) \, ds, \, a \leq t \leq b, \, C \geq 0. \tag{4.1}$$

Then

$$v(t) \leq C exp \int_a^b u(s) \, ds. \tag{4.2}$$

The inequality (4.2) was proved by T.H. Grönwall in [1] (1919) and extended by R. Bellman in [2] (1943). We must note that Grönwall–Bellman inequality (4.2) was the beginning point for studying many integral inequalities and their applications. Particularly, in [3], P. Hartman writes that the result (4.2) has as its sours the work of G. Peano [4]. Among the publications on nonlinear Grönwall–Bellman type inequalities, the first papers were published by Bihari [5] (1956) and A. Perov [6] (1957). Therefore integral inequalities of this type (linear and nonlinear) are usually associated with Grönwall, Bellman, Bihari, and others. The other names to be mentioned in connection with the further development of the theory of integral inequalities are R.P. Agarval, S.A. Chaplygin, N.B. Azbelev, D.D. Bainov, P.R. Beesack, S.G. Deo, U.D. Dhongade, V. Lakshmikantham, S. Leela, B.G. Pachpatte, J. Popenda, P.S. Simeonov, Z.B. Tsalyuk, D. Willett, J.S.W. Wong, E.C. Young. Wong, E.C. Young. Integral inequalities are necessary for studying various equations (differential, difference, discrete, etc.). This is why the intensive investigations of different integral inequalities and hundreds of publications on them appear. A part of the results can be found in the monographs of E.F. Beckenbach and R. Bellman (1965), V. Lakshmikantham and S. Leela (1969), W. Walter (1970), P.R. Beesack (1975), A.N. Filatov and L.V. Sharova (1976), A.A. Martynyuk and R. Gutowski (1979), R. Rabczuk (1979), D.Ya. Mamedov, S. Ashirov, S. Atdaev (1980), and J. Schroder (1980).

4.1.2 Development of the Theory of Integral Inequalities (Brief Survey)

The theory of integral inequalities development can be conditionally divided into three stages. At the first stage, linear integral inequalities were developed and were widely applied in differential equations research. One of the first publications in this direction [1–3] began research on linear integral inequalities (4.1-4.2) of the Grönwall–Bellman type and their generalizations to various linear and nonlinear differential equations classes. During this period, integral inequalities constituted the central part of the proofs of existence and uniqueness theorems for solutions of nonlinear differential equations. They were also used in evaluating their solutions (one-sided and two-sided) [5–13]. Using integral inequalities, approximate estimates for solutions of differential equations that do not have analytical solutions are obtained.

Primary applications of integral inequalities are outlined in the actual work [10].

The next stage in the development of research of integral inequalities dates back to the 1970–80s. Integral inequalities began to be used for the solution of problems in the theory of nonlinear oscillations and the theory of stability, as well as in mechanics, averaging problems, etc. A detailed review of nonlinear integral inequalities can be found in [10], and applications in nonlinear oscillations and mechanics are presented in [11–13].

As we can see from the listed sections, there are no applications of integral inequalities in such important applied areas as the theory of optimal control and its extension of the theory of conflict-controlled systems (differential games). Perhaps this is due to the thematic interests of the authors because integral inequalities are widely used in the theory of nonlinear optimal and automatic control problems [14–18] and the theory of differential escape games [19–59].

The theory of integral inequalities is intensively developed (third stage). It covers new areas of applications, in particular, the theory of optimal (automatic) control [14–18] and the theory of nonlinear differential games [19–59].

In the theories of optimal and automatic control, as well as in differential games, integral inequalities are applied to:

- identification (estimation) of trajectories of nonlinear controlled and conflict controlled systems [14–30, 52, 58];
- obtaining lower estimates for the distances from the current position of the trajectory to the terminal (target) set [31–42, 53–56];
- estimates of admissible controls that satisfy "energy" (integral) constraints and integral inequalities [39, 44–47, 58, 59].

Recently, there has been an active extension of new applications of integral inequalities in differential games [48]. Thus, in [48–50], the authors, for the first time, consider new classes of controls for the game participants (the pursuer and the evader) based on integral inequalities of the Grönwall–Bellman type. These controls generalize well-known controls satisfying integral constraints [39] and allow to consider differential games with geometrical and integral constraints from a unified position, In the past geometrical and integral constraints on controls were considered in differential games separately.

Note that the newly introduced classes of controls [48–50] are mainly used in pursuit problems of differential games only.

4.2 Investigation of Nonlinear Hölder Type Integral Inequality

4.2.1 Main Results

Lemma 4.1. Let the function

$$z(\xi) = \xi - \gamma_1 \xi^{\alpha_1} - \gamma_2 \xi^{\alpha_2} - \cdots - \gamma_n \xi^{\alpha_n} - \gamma_{n+1},$$

where $\xi \geq 0$, $\gamma_i > 0$ are constants, $i = 1, 2, \ldots n + 1$; $0 < \alpha_j < 1$, $j = 1, 2, \ldots, n$.

Then the function $z(\xi)$ has a unique minimum point and a unique root $\xi_0 > 0$ and is also strictly convex for all $\xi \geq 0$.. In addition, for all $0 \leq \xi \leq \xi_0$, the inequality

$$z(\xi) = \xi - \gamma_1 \xi^{\alpha_1} - \gamma_2 \xi^{\alpha_2} - \cdots - \gamma_n \xi^{\alpha_n} - \gamma_{n+1} \leq 0 \quad (4.3)$$

is hold.

Proof. Let's first investigate the extremum of $z(\xi)$. To do this, we calculate the derivatives of the function, which are equal:

$$z'(\xi) = 1 - \sum_{i=1}^{n} \gamma_i \alpha_i \xi^{\alpha_i - 1}, \quad z''(\xi) = \sum_{i=1}^{n} \gamma_i \alpha_i (1 - \alpha_i) \xi^{\alpha_i - 1}. \quad (4.4)$$

It's easy to see that

$$z(0) = -\gamma_{n+1} < 0 \quad (4.5)$$

and

$$\lim_{\xi \to +\infty} z(\xi) = \lim_{\xi \to +\infty} \xi \left(1 - \sum_{i=1}^{n} \gamma_i \xi^{\alpha_i - 1} - \xi^{-1} \gamma_{n+1}\right) = +\infty. \quad (4.6)$$

Relations (4.4)–(4.6) show that, due to continuity $z(\xi)$, there exists a positive root $\xi_0 > 0$ of the function $z(\xi)$, that is $z(\xi_0) = 0$.

Next, let's find out the behavior of the derivative. We have, taking into account $0 < \alpha_j < 1, \ j = 1, 2, \ldots, n$, that

$$\lim_{\xi \to +\infty} z'(\xi) = \lim_{\xi \to +\infty} (1 - \sum_{i=1}^{n} \gamma_i \alpha_i \xi^{\alpha_i - 1}) = 1,$$

$$\lim_{\xi \to +0} z'(\xi) = \lim_{\xi \to +0} (1 - \sum_{i=1}^{n} \gamma_i \alpha_i \frac{1}{\xi^{1-\alpha_i}}) = -\infty.$$

From this and the continuity of the function $z'(\xi)$, we obtain that there is a point, such that $z'(\xi_*) = 0$.

Further investigation of the critical point ξ_* shows that at this point $z''(\xi_*) > 0$, therefore, ξ_* is a point of a local minimum. Further, since $z''(\xi) > 0$ for all $\xi > 0$ (including the point $\xi = \xi_*$) then $z(\xi)$ it is a strictly convex function for all $\xi \geq 0$, and such functions, the extremum point $\xi_* > 0$ is the only point of the global minimum of the function $z(\xi)$ в области $\xi \geq 0$. in the domain $\xi \geq 0$. It is easy to verify that $\xi_0 > \xi_* > 0$.

Let $0 \leq \xi \leq \xi_0$, then from the property of the root ξ_0 we have, which proves (4.3). Lemma 1 is proved.

Theorem 4.1 (Main result). Let the function $u(t)$ be continuous for $t_0 \leq t \leq t_0 + 1$, $t_0 \geq 0$, and satisfies the Hölder type integral inequality (4.7):

$$0 \leq u(t) \leq f(t) + \int_{t_0}^{t} \left[a(s) + b(s)u(s) + \sum_{i=1}^{n} c_i(s)u^{\alpha_i}(s) \right] ds \quad (4.7)$$

where all functions $f(t), \ a(t), \ b(t)$ и $c_i(t)$ are given, nonnegative, integrable, and are not equal identically to zero on the interval $[t_0, \ t_0+1]$, $0 < \alpha_i < 1, i = 1, 2, \ldots n$.

Then there is a constant $h_0 \in (t_0; t_0 + 1]$, such that the function $u(t)$ for all $t \in [t_0; t_0 + h_0]$ satisfies the inequality

$$u(t) \leq f(t) + \int_{t_0}^{t} a(s) \, ds + \delta \xi_0 + \sum_{i=1}^{n} \xi_0^{\alpha_i} g_i(t), \quad (4.8)$$

where ξ_0 is the positive root of the equation

$$\xi - \sum_{i=1}^{n} d_i \xi^{\alpha_i} - d_{n+1} = 0, \quad (4.9)$$

$$g_i(t) = \left(\int_{t_0}^{t} c_i^{\frac{1}{1-\alpha_i}}(s)\,ds \right)^{1-\alpha_i},$$

and constants $\delta, h_0, b_1, d_i, i = 1, 2, \ldots, n+1$, are positive and calculated effectively by the functions and the parameters $\alpha_i \in (0; 1)$ given in (4.7).

Proof. Without loss of generality, we can assume in Theorem 1 $t_0 = 0$.

Let's positive. From (4.7), by using Hölder's integral inequality, we obtain that for all $0 \le t \le h \le 1$ the inequalities

$$u(t) \le f(t) + \int_0^t a(s)\,ds + \delta \int_0^t u(s)\,ds + \int_0^t \left(\sum_{i=1}^n c_i(s)u^{\alpha_i}(s) \right)\,ds \le$$

$$f(t) + \int_0^t a(s)\,ds + \delta \int_0^h u(s)\,ds + \sum_{i=1}^n \left(\int_0^h (u^{\alpha_i}(s))^{\frac{1}{\alpha_i}}\,ds \right)^{\alpha_i} \left(\int_0^t c_i^{\frac{1}{1-\alpha_i}}(s)\,ds \right)^{1-\alpha_i} \le$$

$$f(t) + \int_0^t a(s)\,ds + \delta \int_0^h u(s)\,ds + \sum_{i=1}^n \left(\int_0^h u(s)\,ds \right)^{\alpha_i} \left(\int_0^t c_i^{\frac{1}{1-\alpha_i}}(s)\,ds \right)^{1-\alpha_i}$$

are fulfilled, which immediately give us the inequality

$$u(t) \le f(t) + \int_0^t a(s)\,ds + d \int_0^h u(s)\,ds + \sum_{i=1}^n \left(\int_0^h u(s)\,ds \right)^{a_i} g_i(t).$$

$$(4.10)$$

Further, we will estimate the integral

$$\xi = \int_0^h u(s)\,ds$$

in (4.10). To do this, we integrate (10) once again with $0 \le t \le h \le 1$, then we have:

$$\int_0^h u(t)\,dt \le \int_0^h [f(t) + \int_0^t a(s)\,ds]\,dt + \delta h \int_0^h u(s)\,ds$$

$$+ \sum_{i=1}^n \left(\int_0^h u(s)\,ds \right)^{\alpha_i} \int_0^h g_i(t)\,dt. \qquad (4.11)$$

Let introduce the following notations:

$$\overline{\gamma}_i = \int_0^1 g_i(t)\,dt,\, i = 1, 2, \ldots n, \quad \overline{\gamma}_{n+1} = \int_0^1 [f(t) + \int_0^1 a(s)\,ds]dt,$$

then from (4.11), one has

$$\xi \leq \delta h \xi + \sum_{i=1}^n \overline{\gamma}_i\, \xi^{\alpha_i} + \overline{\gamma}_{n+1},$$

or

$$(1 - \delta h)\, \xi \leq \sum_{i=1}^n \overline{\gamma}_i\, \xi^{\alpha_i} + \overline{\gamma}_{n+1}. \tag{4.12}$$

Let us define the following constants:

$$h_0 = \frac{1}{2}\min\{1; \frac{1}{\delta}\}, \gamma_i = \frac{1}{1 - \delta h_0}\overline{\gamma}_i, \gamma_{n+1} = \frac{1}{1 - \delta h_0}\overline{\gamma}_{n+1}.$$

Then for each fixed, $h \in [0; h_0]$ the inequality $1 - \delta h > 0$ is fulfilled, and (11) can be written in the form

$$\xi - \sum_{i=1}^n \gamma_i\, \xi^{\alpha_i} - \gamma_{n+1} \leq 0. \tag{4.13}$$

Note that in (13) $\xi = \xi(h)$, where $h \in [0; h_0]$, γ_i and γ_{n+1} are positive constants, not depending on.

If we put $\xi(h_0) = x$, then from (4.13) we obtain for $h = h_0$ the inequality

$$x - \sum_{i=1}^n \gamma_i\, x^{\alpha_i} - \gamma_{n+1} \leq 0. \tag{4.14}$$

Obviously, in (4.14) x is a variable depending on the function $u(t), t \in [0; h_0]$.

Applying Lemma 1 to (4.14), we have that there is a unique root $\xi_0 > 0$ of the function $z(x) = x - \sum_{i=1}^n \gamma_i\, x^{\alpha_i} - \gamma_{n+1}$, such that inequality (4.14) holds for all $0 \leq x \leq \xi_0$, this implies for all $h \in [0; h_0]$ inequalities

$$\int_0^h u(s)\,ds \leq \int_0^{h_0} u(s)\,ds = x \leq \xi_0. \tag{4.15}$$

Applying (4.15) to (4.10), we get the required inequality (4.8). Theorem 1 is proved.

4.2.2 Applications of Integral Inequalities in Differential Games of Avoiding an Encounter

Let a vector differential equation describe conflict controlled process (differential game)

$$\dot{z} = f(z, u, v), \qquad (4.16)$$

where $z \in R^n, u \in P \subset R^p, v \in Q \subset R^q$, P and Q are non-empty compact sets, containing starting points $0_p \in R^p$ and $0_q \in R^q$, a function $f(z, u, v)$ is continuous over $X = R^n \times P \times$ и Q and satisfies for any $(z, u, v) \in X$ to inequalities

$$< z, \ f(z, u, v) >\leq C\left(1 + |z|^2\right),$$

$$|f(z_2, u, v) - f(z_1, u, v)| \leq L_\alpha |z_2 - z_1|^\alpha, \qquad (4.17)$$

where $z_1 \in R^n$, $z_2 \in R^n$, $0 < \alpha \leq 1$, $C \geq 0$ and $L_\alpha \geq 0$ are constants.

Terminal set M is a linear subspace of R^n.

Parameters u and v in (4.16) are chosen by opposing sides (players) as measurable functions $u = u(t) \in P$, $v = v(t) \in Q$, $t \geq 0$. Player, choosing $v = v(t) \in Q$ (evader), sets itself a task for any acceptable behavior $u(t) \in P$ to avoid encountering the corresponding trajectory $z(t)$ of equation (4.16), starting from any point $z_0 \in R^n \backslash M$ $(z_0 = z(0))$, from M for any $t \geq 0$. Such problem is called global avoiding encounter problem(global evasion problem), which was formulated by L.S. Pontrjagin and E.F. Miščenko in [31]. At every moment $t \geq 0$ the evader by forming value of $v(t) \in Q$ may use values $u(s) \in P$ and z(s) with $s \leq t$, but he does not know its future values at $s > t$.

In addition, according to the conditions of the avoiding problem, the evading player has a particular advantage over the pursuer for the successful solution of the collision avoidance problem. The benefits are set in different ways (geometric, integral, mixed, etc.) and are usually included in a conflict-controlled process's dynamics (4.16).

Using its advantages in dynamics and information, the evader firstly constructs a local special control function $v_c(t) \in Q$, which ensures for him, for any admissible controls of the pursuer, to avoid   on a small-time interval $0 \leq t \leq \theta_0$ ($\theta_0 > 0$ is a constant) the collision M with the corresponding trajectory $z(t)$ starting at point $z_0 = z(0) \in R^n \backslash M$. Then, based on the local deviation, the evader an algorithm for avoiding the trajectory $z(t)$ from M for all $t \geq 0$ is constructed (global avoiding). At the stage of local avoidance, the main task is to prove that the distance $\rho(z(t), M)$ from the moving point $z(t)$ of the trajectory to the terminal set is positive for all $t \geq 0$, then for these t $z(t) \notin M$, and thus, global avoidance is possible. However, it can be tough to prove inequality $\rho(z(t), M) > 0$, especially for nonlinear systems. Therefore, for the distances, they try to obtain lower bounds. Namely, at this stage, it becomes necessary to use integral inequalities. Let us explain this. If the players have chosen controls $u(t) \in P$ and $v_c(t) \in Q, t \in [0, 1]$, then to these controls and a point $z_0 \notin M$ corresponds the solution of system (4.16) with the initial condition z $(0)= z_0$, which has the form [8,9]:

$$z(t) = z_0 + \int_0^t f(z(s), u(s), v(s)) \, ds.$$

Further, based on his advantage and Hölder condition (4.17), the evading player constructs his control in such a way that provides him the inequality

$$|z(t) - z_0| \leq d_1(1 + |z_0|^\alpha)t + \int_0^t (d_2 + L_\alpha |z(s) - z_0|^\alpha) \, ds, \quad (4.18)$$

where constants d_1 and d_2 are positive, $0 < \alpha \leq 1$.

Inequality (4.18) is a nonlinear integral inequality of the Hölder type (4.7). With its help, as well as other conditions of the avoidance problem (4.16–4.17), the main inequality in the theory of differential games of avoiding encounter is obtained [32,39], which in the general case has the form

$$\rho(z(t), M) \geq \gamma t^k, \quad (4.19)$$

where $t \in [0, \theta_0]$, γ, k and θ_0 are positive constants are effectively calculated from the initial data of the game (4.16).

Inequality (4.19) ensures that the trajectory does not hit the terminal set on some "small" time interval. Then, the possibility of global avoidance (evasion) from the point $z_0 \notin M$ is proved for all $t \geq 0$. Note that even in the cases of applying the linear integral Grönwall–Bellman inequality [1, 2, 10], obtaining lower bounds of type (4.19) involves extensive calculations and proofs [32, 39].

4.2.3 Comparison of Results

1. If in (4.7) we assume $n = 1$ and 1) $\alpha_1 = 0$ or $\alpha_1 = 1$; b) $c_1(t) \equiv 0$ for $t \geq t_0$, then each case leads to the consideration of well-known linear integral inequalities and estimates of Grönwall–Bellman type [1, 2].

2. The case when in (4.7) $n = 1$, $a(t) \equiv 0$, $f(t) \equiv f_0 \geq 0$ (f_0 is a constant for all $t \geq t_0$), $\alpha_1 \geq 0$, is studied in [7] (if $\alpha_1 \neq 0$ or $\alpha_1 \neq 1$ we have nonlinear case). It is necessary to note, that the proof in [7] does not extend in the general case for the proof of the inequality (4.7).

3. The case when in (4.7) $n = 1$, $a(t) \equiv 0$, $b(t) \equiv 0$, $0 \leq t \leq 1$, $0 < \alpha_1 < 1$, considered in [7,12,13].

4.2.4 Remarks

Remark 4.1. First, Hölder type inequality was considered in [6] and investigated in [12,13,58].

Remark 4.2. In many papers, as an integrand in (4.7), a more general function is considered that has the property of monotonicity on $u(\cdot)$ of the function being estimated and some other additional conditions. As a result, we obtain general estimates in terms of inverse functions or mappings [10,11], which are not constructive and are not adapted to the study of avoidance problems in the theory of differential games.

Remark 4.3. Inequality (4.8) and the constant h_0 are especially important for obtaining a lower bound for the distance from a moving point of the trajectory (4.17) of a conflict-controlled dynamical system to a given terminal set [32–35,37,39].

Remark 4.4. The problem of approximate calculation of the root ξ_0 of the function $z(\xi)$ is solved in [14, 15] for $n = 1$ only.

4.3 Conclusion

The main aim of this work is to investigate new nonlinear Hölder type integral inequalities, generalized well-known integral inequalities [6, 12, 13].

New integral inequalities of the Hölder type are proved, and corresponding solutions of nonlinear integral inequalities of the Hölder type are obtained, which generalized the well-known [6, 12 ,13]. The results on integral inequalities presented in this chapter allow for solving various control problems [14–18] and are applicable to solving new nonlinear problems in the theory of differential avoiding games. They may also be used for constructing new control functions(strategies) of players in differential games of avoiding encounters [58, 59]. Further, the received results make it possible to apply them to research solutions of nonlinear differential equations (existence, uniqueness, continuity, continuous dependence, etc.).

A short excursion into the historical domain of the development of integral inequalities is represented.

Acknowledgments

The author is grateful to the Academician of the National Academy of Sciences of Ukraine, Professor A.A. Chikrii, for his kind encouragement of this work. He also thanks to the Organizing Committee of AUTOMATION2020 for the invitation to participate in this high-level event.

References

[1] T.N. Grönwall, 'Note on the derivatives with respect to a parameter of the solutions of a system of differential equations,' Ann. of Math. 20: 2 (1919), pp. 292–296. DOI 10.2307/1967124.

[2] R. Bellman. 'The stability of solutions of linear differential equations,' Duke Math. J. 10(1943), pp. 643–647.

[3] P. Hartman, Ordinary Differential Equations, John Wiley & Sons, NewYork- London-Sydney, 1964, 720 p.

[4] G. Peano. Sull' integrabilitá delle equazione differenzialli di primo ordine, Atti. R.Accad.Torino, 21(1885/18886), 667–685 [III.2.4].

[5] I. Bihari, 'A generalization of a lemma of Bellman and its application to uniqueness problems of differential equations,' Acta Math. Acad. Sci. Hungr. 7:1(1956), pp. 71–94.

[6] A.I. Perov, 'On Integral Inequalities,' Proc. of the Seminar on Functional Analysis, Voronej St. Univ. − Issue 5. − 1957.

[7] S.A. Chaplygin, 'A new method of approximate integration of differential equations,' Trudy TsAGI 130 (1932), pp. 5–17.

[8] E. Coddington and N. Levinson, 'Theory of Ordinary Differential Equations', McGraw-Hill Book Company, Inc. New York–Toronto–London,1955, − 474 p. (Russian Transl., Moscow: IL.1958).

[9] L.S. Pontrjagin, 'Ordinary differential equations,' −M.: Nauka, 1970, − 332 p.

[10] D. Bainov, P. Simeonov, ' Integral Inequalities and Applications.' Dordrecht, Springer, Mathematics and its Applications (East European Series, vol. 57), 1992, −256p.

[11] A. N. Filatov, L.V. Sharova, 'Integral inequalities and the theory of nonlinear oscillations,' Nauka, Moscow, 1976.

[12] Sh. Gamidov, 'Some integral inequalities for boundary value problems for differential equations,' Differential Equations, 5: 3 (1969), pp. 463–472.

[13] A.A. Martynyuk, R.Gutovski, 'Integral inequalities and stability of motion,' Kyiv: Nauk?va Dumka, −1979.

[14] V.M. Kuntsevich et al. (Eds), *Control Systems: Theory and Applications.*Series in Automation, Control and Robotics, River Publishers, Gistrup, Delft, 2018.

[15] Y.P. Kondratenko, V.M. Kuntsevich, A.A. Chikrii, V.F. Gubarev (Eds). *Advanced Control Systems: Theory and Applications.* Series in Automation, Control and Robotics, River Publishers, Gistrup, 2021.

[16] Y.P. Kondratenko, A.A. Chikrii, V.F. Gubarev, J. Kacprzyk (Eds). *Advanced Control Techniques in Complex Engineering Systems: Theory and Applications.* Dedicated to Professor Vsevolod M. Kuntsevich. Studies in Systems, Decision and Control, Vol. 203. Cham: Springer Nature Switzerland AG, 2019.

[17] V.M. Kuntsevich, 'Estimation of Impact of Bounded Perturbations on Nonlinear Descrete Systems,' In: Kuntsevich, V.M. et al. (Eds). *Control Systems: Theory and Applications.* Series in Automation, Control and Robotics, River Publishers, Gistrup, Delft, 2018, pp. 3–15.

[18] J. Kacprzyk, et al. 'A Status Quo Biased Multistage Decision Model for Regional Agricultural Socioeconomic Planning Under Fuzzy Information,' In:Kondratenko, Y.P., Chikrii, A.A., Gubarev, V.F., Kacprzyk, J. (Eds) Advanced Control Techniques in Complex Engineering Systems: Theory and Applications. Dedicated to Prof. V.M.Kuntsevich. Studies in Systems, Decision and Control, Vol. 203. Cham: Springer Nature Switzerland AG, 2019, pp. 201–226. DOI: https://doi.org/10.1007/978-3-030-21927-7_10

[19] R.Isaacs, 'Differential Games', John Wiley and Sons, Inc., New York–London–Sydney, 1965, − 480 p.

[20] L.S. Pontrjagin. 'Selected Proceedings'− M.: MAKS Press, 2004, −552 p.

[21] N.N. Krasovskiĭ, A.I. Subbotin, 'Closed-loop differential games,' Springer–Verlag, Berlin and New-York, 1980, −456 p.

[22] A.B. Kurzhanskiĭ, 'Control and Observation under Uncertainty,' −M.: Nauka, 1977, −392 p.

[23] L.A. Petrosjan, 'On a set of differential games of survive in the space R_n' Dokl. Acad.Nauk SSSR, 1965, V. 161, No. 1, –P.52-54.

[24] M.I. Zelikin, N.T. Tynjanskiĭ, 'Determinate differential games', Usp.Mat. Nauk, 1965, V. 20. No. 4. –P.151–157.

[25] Yu.S. Osipov, 'Alternative in differential-difference game,' Dokl. Soviet Acad.Sc., 1971, V. 197, No. 5, – P. 1022–1025.

[26] A.I. Subbotin, A.G. Chentsov, 'Optimization of Guarantee in Control Problems', -M.: Nauka, 1981, – 288p.

[27] B.N. Pshenichnyĭ, V.V. Ostapenko, 'Differential games', Kiiv: Naukova Dumka, 1992, –261p.

[28] A.A. Čhikriĭ, 'Conflict controlled processes', Boston-London-Dordrecht, Kluwer Acad.Publ., 1997, – 424 p.

[29] M. Sh, Mamatov, Kh. Kh. Sobirov, 'To the theory of differential games of pursuit on position,' Results of Science and Technics, Modern math. and its Appl.Subject Review, 2018, v.144, –P. 39–46.

[30] M. Sh. Mamatov, M. Tukhtasinov, 'On a Problem of Pursuit in Distributed Control Systems,' Cybernetics and System Analysis, Kyiv, 2009, No. 2, – P. 153–158.

[31] L.S. Pontrjagin, E.F. Miščenko, 'A problem of evasion of one controlled object from an other,' 1969, Dokl. Acad. Nauk SSSR, V. 189, No. 4, –P. 721–723.

[32] L.S. Pontrjagin, 'A linear differential evasion game', Trudy Mat.Inst.Steklov, 19 71, V. 112, –P. 30–63.

[33] E.F. Miščenko, N, Satimov, 'The evasion problem in differential games with nonlinear controls,' Differencialnie Uravneniya, V. 9, 10, 1973. –P.1792-1797.

[34] E.F. Miščenko, M.S. Nikolskiĭ, N. Satimov, 'The problem of avoiding encounter in N-person differential games', Proc. of the Steklov Institute of Mathematics, 1980, Iss. 1, –P. 111–136.

[35] N.Satimov, 'On a one way of avoiding encounter in differential games,' Matematicheskiy sbornik, 99(141), 1976, –P. 280–293.

[36] F.L. Chernousko, A.A. Melikyan, 'Game Problems of Control and Search', – Nauka, 1978,–270 p.

[37] P.B. Gusyatnikov, L.P.Yugay, 'On an evasion problem in nonlinear differential games with terminal set of compound structure,' Izvestiya Acad. Nauk SSSR, 1977, Techn.kibernetika, No. 2, –P. 8–13.

[38] R.V. Gamkrelidze, G.L. Kharatshvili, 'A differential game of evasion with nonlinear control' SIAM J. Control, May 1974, Vol. 12, No 2, –P. 332–349.

[39] N.Yu. Satimov, B.B. Rikhsiev, 'Methods of solving of avoiding encounter problem in mathematical theory of control,' Fan, Tashkent, 2000, −176 p.

[40]   N.Yu. Satimov, 'To the theory of differential evasion games' Mat. Sbornik,1977, V. 103, No. 3. − P. 430–444.

[41] N. Satimov, 'Composition of differential equation on its solutions and proof of one lemma of L.S. Pontrjagin,' Diff. Equations, 1978, V. 14, No. 7,− P. 1208–1214.

[42] J. Yong, 'Evasion without superiority,' J. of Math. Anal. and Appl., 1988, V. 134, No. 1, −P. 116–124.

[43] V.N. Ushakov, 'Extremal strategies in differential games with integral constraints,' J. Appl. Math.Mech., 1972, Vol. 36, No 1, −P. 12–19. DOI: 10.1016/0021-8928(72)90076-7

[44] N. Satimov, A. Fazilov, A. Khamdamov, 'On pursuit and evasion problems in differential and discrete games of many persons with integral restrictions,' Differential equations, 1984, Vol. 20, No. 8, − P. 1388–1396.

[45] N. Mamadaliev, 'On a one Pursuit Problem with Integral Restrictions on Controls of Players', Siberian Math. Journal, Jan.-Febr., Novosibirsk, 2015,V.56, No. 1, −P. 129–148.

[46]   N. Mamadaliev, 'Linear Differential Games with Integral Restrictions on Controls and Delay,' Mathematical Notes, 2012, No. 5, - P. 750–760.

[47] L.D. Bercovitz, W.H.Fleming, 'On differential games with integral pay-off,' Contributions to the theory of games, Ann. of Math. Studies, No 3, 1957, −P. 413–435.

[48] B.T. Samatov, 'On a Pursuit-Evasion Problem under a Linear Change of the Pursuer Resource,' Siberian Advances in Mathematics, Allerton Press, Inc. Springer, New York, 23(4), 2013, − P. 294–302.

[49] B.T. Samatov, E.T. Umarov, 'Differential Games with Constraints of the Gronwall Type,' Abstracts of the VI International Sc. Conference " Modern Problems of the Applied Mathematics and Information Technology–Al–Khorezmiy 2018", Tashkent, September, 13-15, 2018, −P. 183–184.

[50] B.Samatov. G. Ibragimov, I. Khodjibaeva, 'Pursuit–Evasion Differential Games with Gronwall-Type Constraints on Controls,' Ural Mathematical Journal, Vol. 6, No 2, 2020, -P. 95–107.

[51] A.O. Chikrii, G.T. Chikrii. Matrix resolving functions in dynamic games of approach.Cybernetics and System Analysis, 2014, 50(2), pp. 201–217.

[52] L.P. Yugay, 'Linear differential game without superiority,' Annals of differential equations, China, 1992, No. 2, −P. 158–163.

[53] L.P. Yugay, 'On an optimization of evasion parameters in minmax differential games,' Bulletin of Korean Mathematical Society, Vol. 34, No. 4, November 1997, P. 495–508.

[54] L.P. Yugay, 'Avoiding from discrete terminal set by geometrical restrictions on controls,' J.Problems of Controls and Informatics, Kyiv, Inst. of Cyb. of NAS of Ukraine named after V.M. Glushkov, 1998, No. 1, − P. 4–11.

[55] L.P. Yugay, 'To the Problem of Pendulum's Swinging,' Proc. of 13-th Int'l Conference "Stability and Oscillation of Nonlinear Control Systems" (Pyatnitsky's Conference), Moscow, 1-3 of June 2016, −P. 429–432.

[56] L.P. Yugay, 'L.S. Pontryagin's lemma in differential evasion games', Optimal Control and Differential Games: Proc. of Int'l Conf. dedicated to the 110th- anniversary of L.S. Pontrjagin, Dec. 12–18, 2018, Moscow, Russia, P. 295–296.

[57] L.P. Yugay, 'Differential Evasion Game with Locally Inertial Controls,' Bulletin of L.N. Gumilyov Eurasian National University (Kazakhstan), Mathematics, Computer Science, Mechanics Series, no. 4, vol. 129, 2019, −P. 54–66.

[58] L.P. Yugay, On Nonlinear Integral Inequality in Collision Avoiding Problems.Abstracts of the XXVI-th International Conference AUTOMATION-2020 devoted to the memory of Academicians L.S. Pontryagin, N.N. Krasovsky and B.N. Pshenichny, Kyiv, Ukraine, 13-15.10.2020, –P.142.

[59] L.P. Yugay, 'The Problem of Trajectories Avoiding a Sparse Terminal Set,' Journal: Doklady of Russian Academy of Sciences, Mathematics, Springer, Pleiades Publishing, Ltd., 2020, Vol.102, no. 3, − P. 538–541. DOI: 10.1134/S1064562420060228

5

Principle of Time Dilation in Game Problems of Dynamics

G.Ts. Chikrii

Glushkov Institute of Cybernetics of the National Academy of Sciences
of Ukraine, 40 Glushkov Avenue, 03187, Kyiv-187, Ukraine
E-mail: g.chikrii@gmail.com

Abstract

In this chapter, we consider the linear differential game of approaching a
cylindrical terminal set. The case when classic Pontryagin's condition does
not hold is studied. Instead, a considerably weaker condition, incorporating
the function of time dilation, is introduced. This makes it possible to expand
the range of problems that can be solved analytically, by using information
about the enemy's control in the past.

The study is carried out in the framework of the Method of Resolving
Functions. The gist of the method consists in constructing certain scalar
function associated with the parameters of the conflict-controlled process and
characterizing the gain of the first player at each moment. If the total gain
achieves the predetermined value then this means that the game terminates in
a certain guaranteed time. The pursuer's control, realizing the game goal, is
constructed based on the Filippov–Castaing theorem on measurable choice.

To illustrate the developed scheme, we analyze in detail the problem of
approaching two controlled oscillation systems in geometric coordinates.

Keywords: Differential game, conflict-controlled process, set-valued mapping, modified Pontryagin's condition, function of time dilation, mathematical pendulum.

5.1 Introduction

The mathematical theory of conflict-controlled processes (differential games, dynamic games) has at its disposal a wide range of fundamental methods to study various nature-controlled processes which function in conditions of conflict and uncertainty.

Attempts to construct optimal behavior of the opposing sides when solving game dynamic problems inevitably lead to the ideology of dynamic programming, closely related to the Hamilton–Jacobi—Bellman–Isaacs equation, the main equation in the theory differential games [1]. In a slightly different form, namely in the terms of sets, these ideas are reflected in the Pontryagin–Pshenichnyi backward procedures [2, 3]. These are the method of alternated integral and the method of semi-group operators T_ε.

According to the aforementioned ideology, in the course of the process of counter-acting at each moment, one of the parties maximizes and another minimizes the payoff functional. Thus, optimization of the conflict-controlled process is reduced to calculating continual *maxmin* or *minmax* of the functional, depending on the current information available to each of the parties.

The desire to find optimal solutions for counter-acting parties in the game problems of dynamics encounters great difficulties of mathematical nature. Therefore, a number of effective mathematical methods have been created for deciding dynamic games, which provide a guaranteed result and give sufficient conditions for the goal achievement without worrying about optimality, which is quite justified from the practical point of view.

These are the First Direct Method of L.S. Pontryagin [2], the Rule of Extremal Aiming of N.N. Krasovskii [4, 5], the Method of Programmed Iterations [6], and the Method of Resolving Functions [7]. Adjacent to them are the works [8, 9]. Various applications of game methods to attack applied problems are contained in the books [10, 11].

In the course of a conflict-controlled process, information about the current state of the process, as a rule, delays. One of the reasons is that a certain time is required to process it. The author of this chapter developed the method of reducing the approach problem with delay of information to the equivalent problem with complete information, but slightly changed dynamics [12, 13], followed by application of the known classic methods.

When performing such important tasks as, for example, interception of a moving target, in order to achieve the goal, a certain advantage of the pursuer in terms of control resources and (or) dynamic capabilities is necessary.

Classic Pontryagin's condition [2] plays this role in general problem statements. There exist various forms of this condition [7]. One of them consists in incorporating the solid part of the terminal set into Pontryagin's condition. Another way concerns the suppression of the evader's control resource by the way of multiplying it by a certain matrix function and subsequent subtraction of the used resource from the terminal set. In so doing, Pontryagin's condition falls into two conditions. The latter technique is appropriate, for example, in the case when the opposing objects have different inertia.

However, in many cases of conflict withstanding Pontryagin's condition does not hold, e.g. the problems of soft meeting and the pursuit problems for oscillatory processes.

M.S. Nikolskij and D. Zonnevend [14, 15] suggested weakening Pontryagin's condition by the way of incorporating into it a certain function, later on, called the function of time dilation. The establishment of the close relation of the modified condition with the passage from the original game with complete information to an auxiliary game with special kind information delay [13] gave impetus to the development of an efficient approach the so-called principle of time dilation for solving problems for which Pontryagin's condition does not hold.

Essentially, the transition is made from the original game with complete information to the game with the same dynamics and the terminal set, yet with a special kind of information delay, decreasing as the game trajectory approaches the terminal set and vanishing as it hits the target... An important point is that Pontryagin's condition for the latter game involves the time function of time dilation.

We extend the concept of the function of time dilation to the class of functions of bounded variation. This allows applying the above-mentioned technique to solving a wide range of game problems for the processes, described by the systems of second-order differential equations [16, 17] and for the problems of soft meeting [18].

In this chapter, we apply the principle of time dilation to the game constructions based on the method of resolving functions [7, 19] This method is closely related with the Minkowski' inverse functionals, constructed for special set-valued mappings. An attractive feature of the method of resolving functions is that it provides the full theoretical substantiation of the classic rule of parallel pursuit and the approach along the ray [20]. Moreover, when substantiating the game constructions and obtaining meaningful results on their basis, the modern technique of set-valued mappings is effectively used.

The major contributor to the creation of the mentioned method was a special *minmaximin* function in the theory of avoidance of a group of pursuers [21]. It allowed BN Pshenichnyi to formalize the situation of the environment [22]. Subsequently, the theory of group pursuit was gathering force [7, 23, 24]. The broad possibilities of the method of resolving functions are illustrated on the problem of successive pursuit, the commercial traveler type problem by using Apollonian circles [25], and in the problems with state constraints [7].

The accumulative principle inherent in the method looks quite natural. Its versatility allows, in a unified scheme, to encompass conflict-controlled processes described by functional differential equations [26, 27], partial differential equations [28], equations with fractional derivatives [29], impulse systems [30], non-stationary processes [31, 32], and stochastic systems [33]. Therewith, the approach conditions allow to apply of the quasi- and strobo-scopic strategies [4, 8] in the course of the game, based on the theorems on measurable choice [34].

5.2 Statement of the Game Problem. Classic and Modified Pontryagin's Condition

Let the motion of a conflict-controlled object in the finite-dimensional space R^n be described by the system of differential equations

$$\dot{z} = Az + u - v, \tag{5.1}$$

A is a quadratic matrix of order n.

Parameters of the players' controls u and v are picked up from the domains of players controls U and V, which are compacts from R^n.

Besides, a terminal set M^*, having a cylindrical form, is given:

$$M^* = M_0 + M, \tag{5.2}$$

where M_0 is a linear subspace in R^n, M- a compact from L, the orthogonal complement to M_0 in R^n.

The goal of the first player (the pursuer) is in the shortest time to bring a trajectory of the process (5.1) to the terminal set (5.2) with the help of an appropriate choice of a control parameter

The second player (the evader) strives, by choosing parameter v, to avoid meeting the trajectory of the process (5.1) with the set M^* within a finite time or, if it is impossible, to postpone this meeting as much as possible.

To fully formulate the problem of approach, it is necessary to stipulate what information is available to the players in the process of conflict counteraction. Let's take the side of the first player and find out what result he can guarantee himself.

We assume that the second player chooses as admissible controls arbitrary measurable functions with values from the compact V. Let us denote by Ω_E the set of all admissible controls of the evader.

If at each moment of making a decision t, $t \geq 0$, the first player uses information about the initial state of the process z_0 and the prehistory of control of the evader

$$v_t(\cdot) = \{v(s) : v(s) \in V, s \in [0, t]\},$$

i.e. $u(t) = u(z_0, t, v_t(\cdot))$, then, we say that he applies quasi-strategy.

In the case the pursuer at the current moment t uses only information about the initial state z_0 and instantaneous control of the evader $v(t)$, i.e. $u(t) = u(z_0, t, v(t))$, then they say about counter-control by N.N. Krasovskii [4], prescribed by stroboscopic strategy by O.Hayek [8].

In this chapter, we discuss using delayed information by the first player t with the goal of solvability of the original problem of approach. The case of lack of advantage of the first player in control resources is considered, that is when Pontryagin's condition [2] does not hold. Let us proceed to the scheme of solving the problem (5.1), (5.2) based on the method of resolving functions [7, 19] and formulation of the above-mentioned condition [2].

Let us denote by π the operator of orthogonal projection acting from R^n onto L and take into consideration the set-valued mapping $W(t) = \pi e^{At} u * \pi e^{At} v$, where $*$ is the operation of geometric subtraction by Minkowski [2, 7].

Pontryagin's condition. The set-valued mapping $W(t)$ has non-empty images for $t \in [0, +\infty)$.

Under this condition, the scheme of the method of resolving functions allows obtaining sufficient conditions for the solvability of the game problem of approach for various classes of strategies [7, 19].

However, there are many problems for which Pontryagin's condition does not hold.

In what follows we outline the procedure which makes it feasible to weaken this condition. It is related to the principle of time dilation [16, 17] and allows a widening class of conflict-controlled processes, for which the problem of approach can be solved in a finite time, in their number, in the framework of the method of resolving functions.

Modified Pontryagin's condition. There exists a differentiable monotonically increasing function $I(t), I(t) \geq t, t \geq 0, I(0) = 0$, such that the set-valued mapping

$$W_I(t) = \pi e^{At} U * \dot{I}(t) \pi e^{AI(t)} V, t \geq 0,$$

has non-empty images.

5.3 Method Scheme

Since the mapping $W_I(t)$ is measurable and closed-valued, then, by the theorem on measurable choice [34], there exists a measurable selection $\gamma_I(t), \gamma_I(t) \in W_I(t)$. Let us fix it for use in the sequel and introduce a function

$$N(t, u_0(\cdot)) = \pi e^{At} z_0 + \int_0^{I(t)-t} \pi e^{A(I(t)-s)} u_0(s) ds + \int_0^t \gamma_I(t-s) ds.$$

It depends on the initial state and the control on the initial interval of time $u_0(\tau), u_0(\tau) \in U, \tau \in [0, I(t) - t]$.

Following the ideology of the method of resolving functions [7, 19], we analyze the mapping

$$A_I(t, s, v, u_0(\cdot)) = \{\alpha \geq 0 : [W_I(t-s, v) - \gamma_I(t-s)] \bigcap \alpha$$
$$[M - N(t, u_0(\cdot))] \neq \} \tag{5.3}$$

where

$$W_I(t, v) = \pi e^{At} U - \dot{I}(t) \pi e^{AI(t)} v, t \geq 0, v \in V.$$

Its support function in the direction +1 is called the resolving function [7]:

$$\alpha_I(t, s, v, u_0(\cdot)) = \sup\{\alpha : \alpha \in A_I(t, s, v, u_0(\cdot))\}.$$

From expression (5.3) it follows that, if for some $u_0(\cdot)$ and $t, t \geq 0$, $N(t, u_0(\cdot)) \in M$, then $A_I(t, s, v, u_0(\cdot)) = [0, +\infty)$ and $\alpha_I(t, s, v, u_0(\cdot)) = +\infty$ for all $s \in [0, t], v \in V$.

Let us consider the set

$$T_I(z_0, \gamma_I(\cdot)) = \{t \geq 0 : \sup_{u_0(\cdot) \in U_0^t} \inf_{v(\cdot) \in \Omega_E} \int_0^t \alpha_I(t, s, v(s), u_0(\cdot)) ds \geq 1\},$$

$$\tag{5.4}$$

where
$$U_0^t = \{u_0(s) : u_0(s) \in U, s \in [0, I(t) - t]\}.$$

If the inequality in parentheses (5.4) is not fulfilled for all $t \geq 0$, then we put $T_I(z_0, \gamma_I(\cdot)) = \emptyset$. In what follows, we assume that the exact upper bound $u_0(\cdot)$ is achieved in (5.4). For example, as a rule, in the case, the set U contains 0, for simplicity's sake, control of the first player on the initial time interval is set identically equal to zero that providing fulfillment of the inequality in (5.4).

Before proceeding to the formulation and proof of the main result, let us perform some rearrangements in the representation of the system solution (5.1), taking into account the availability of information of the first player about his adversary control choice.

Let t be an arbitrary moment. Using the Cauchy formula, for the chosen controls of the players, we present o the solution of system (5.1) in projection onto the subspace L:

$$\pi z(t) = \pi e^{At} z_0 + \int_0^t \pi e^{A(t-s)} u(s) ds - \int_0^t \pi e^{A(t-s)} v(s) ds.$$

Taking into account the crucial role of the function of time dilation, we separate the section of initial control of the evader. Then we have

$$\pi z(I(t)) = \pi e^{AI(t)} z_0 + \int_0^{I(t)-t} \pi e^{A(I(t)-s)} u_0(s) ds +$$
$$+ \int_{I(t)-t}^{I(t)} \pi e^{A(I(t)-s)} u(s) ds - \int_0^{I(t)} \pi e^{A(I(t)-s)} v(s) ds. \quad (5.5)$$

In order to equalize the limits of integration, we change the variables in the last two terms of formula (5.5), s by $I(t) - t + s$ and $I(t) - I(t - s)$, respectively.

As a result, we obtain the following representation of the solution

$$\pi z(I(t)) = \pi e^{AI(t)} z_0 + \int_0^{I(t)-t} \pi e^{A(I(t)-s)} u_0(s) ds +$$
$$+ \int_0^t \pi e^{A(t-s)} u(I(t) - t + s) ds$$
$$- \int_0^t \pi e^{AI(t-s)} v(I(t) - I(t - s)) \cdot \dot{I}(t - s) ds. \quad (5.6)$$

5.4 Main Statement

Theorem *Let the modified Pontruagin's condition with some function $I(t)$ be satisfied for the conflict-controlled process (5.1), (5.2) and let the set M be convex.*

Then, if for the initial state z_0 there exists a measurable selection $\gamma_I(t)$ of the set-valued mapping $W_I(t)$, $t \geq 0$, such that

$$T_I(z_0, \gamma_I(\cdot)) \neq \quad and T_* \in T_I(z_0, \gamma_I(\cdot)),$$

then the trajectory of the process (5.1) can be brought to the terminal set at the moment $I(T_*)$.

Proof. Let us study the process (5.1), (5.2) on the interval $[0, I(T_*)]$. On the initial interval $[0, I(T_*) - T_*]$, we choose the initial control $u_0(\cdot)$ from the condition for a *maximum* of the following expression

$$\inf_{v(\cdot) \in \Omega_E} \int_0^{T_*} \alpha_I(T_*, s, v(s), u_0(\cdot)) ds.$$

Then, on the interval $(I(T_*) - T_*, I(T_*)]$ of the length T_*, we count the time from 0 to T_*. Let us consider two cases, namely, $N(T_*, u_0(\cdot)) \notin M$, and $N(T_*, u_0(\cdot)) \in M$, the latter corresponding to the resolving function equal to $+\infty$.

In the first case, under the scheme of the method of resolving function [19], we introduce into consideration the controlling function

$$h_I(t) = 1 - \int_0^t \alpha_I(T_*, s, v(I(T_*) - I(T_* - s)), u_0(\cdot)) ds,$$

$$v(s) = v(I(T_*) - I(T_* - s)) \in V. \tag{5.7}$$

Here $v(s)$ is an arbitrary admissible control of the evader.

By virtue of the continuity of a function $h_I(t)$, from the definition of the time T_* and the theorems of mathematical analysis, it follows that there exists time t_*, such that $h_I(t_*) = 0$.

We name $[0, t_*]$ by the active interval and $(t_*, T_*]$ – by the passive one. To them correspond the intervals $[I(T_*) - T_*, I(T_*) - T_* + t_*]$ and $(I(T_*) - T_* + t_*, I(T_*)]$, respectively. The corresponding representations of solutions at $t = T_*$, not separating the active and passive intervals, are given by formulas (5.5) and (5.6).

Let us choose the pursuer control as follows. In the case $N(T_*, u_0(\cdot)) \notin M$, we consider the set-valued mapping

$$U(I(T_*) - T_* + s, v(I(T_*) - I(T_* - s))) = \{u(I(T_*) - T_* + s) \in U :$$
$$\pi e^{A(T_* - s)} u(I(T_*) - T_* + s) - \pi e^{AI(T_* - s)} v(I(T_*) - I(T_\circ - s)) \cdot \dot{I}(T_* - s)$$
$$- \gamma_I(T_* - s) \in \in \bar{\alpha}_I(T_*, s, v(I(T_*) - I(T_* - s)), u_0(\cdot))$$
$$[M - N(T_*, u_0(\cdot))]\}. \tag{5.8}$$

Here

$$\bar{\alpha}_I(T_*, s, v(I(T_*) - I(T_* - s)), u_0(\cdot)) =$$
$$= \begin{cases} \alpha_I(T_*, s, v(I(T_*) - I(T_* - s)), u_0(\cdot)), s \in [0, t_*], \\ 0, s \in (t_*, T_*]. \end{cases}$$

In view of the measurability of this mapping, there exists its selection, which is a superposition of measurable functions [34]

$$u_I(s) = u(I(T_*) - T_* + s, v(I(T_*) - I(T_* - s))), s \in [0, T_*]. \tag{5.9}$$

We assign just this function to be control of the first player. It is easy to see that the pursuer chooses his control based on delayed information about the evader's control since in view of the above-outlined scheme,

$$I(T_*) - T_* + s \geq I(T_*) - I(T_* - s).$$

If, otherwise, $N(T_*, u_0(\cdot)) \in M$, then control of the first player is chosen according to a similar scheme, but with the resolving function equal to zero on the entire interval $[0, T_*]$. This case [19] is said to correspond to the first direct method of L.S. Pontryagin [2].

Setting $t = T_*$ and separating the active and the passive segments, we finally derive the representation of the solution from formula (5.6):

$$\pi z(I(T_*)) = \pi e^{AI(T_*)} z_0 + \int_0^{I(T_*) - T_*} \pi e^{A(I(T_*) - s)} u_0(s) ds+$$

$$+ \int_0^{t_*} \pi e^{A(T_* - s)} u(I(T_*) - T_* + s) ds + \int_{t_*}^{T_*} \pi e^{A(T_* - s)} u(I(T_*) - T_* + s) ds-$$

$$- \int_0^{T_*} \pi e^{AI(T_* - s)} v(I(T_*) - I(T_* - s)) \cdot \dot{I}(T_* - s) ds.$$

In the case $N(T_*, u_0(\cdot)) \bar{\in} M$, using the above-described rules of control selection on the active and passive parts, we obtain

$$\pi z(I(T_*)) \in N(T_*, u_0(\cdot)) +$$

$$+ \int_0^{T_*} \bar{\alpha}_I(T_*, s, v(I(T_*) - I(T_* - s)), u_0(\cdot))[M - N(T_*, u_0(\cdot))]ds =$$

$$= N(T_*, u_0(\cdot))[1 - \int_0^{T_*} \bar{\alpha}_I(T_*, s, v(I(T_*) - I(T_* - s)), u_0(\cdot))ds] +$$

$$+ \int_0^{T_*} \bar{\alpha}_I(T_*, s, v(I(T_*) - I(T_* - s)), u_0(\cdot))Mds = M. \tag{5.10}$$

Note that in the last equation of (5.10), the convexity of the solid part of the terminal set was taken into account.

If, otherwise, $N(T_*, u_0(\cdot)) \in M$, then, in view of the choice of control (5.8), (5.9) with the zero resolving function, from (5.10) readily follows the inclusion

$$\pi z(I(T_*)) \in M.$$

Remark 5.1. The scheme of the method and main assertion are valid for much more general classes of conflict-controlled processes than described by (5.1), in particular, for generalized functional-differential systems, including fractional, impulse, and time-delay systems.

Remark 5.2. The function of time dilation can be discontinuous, but absolutely continuous on the intervals of its continuity [17].

5.5 Example

To illustrate the proposed approach, let us investigate the game problem of convergence in geometric coordinates of two controlled systems of second-order, describing the dynamics of the mathematical pendulum. Let their dynamics be given by the equations

$$\ddot{x} + a^2 x = \rho u, x \in R^n,$$
$$\ddot{y} + b^2 y = \sigma v, y \in R^n. \tag{5.11}$$

Here x and y are geometric coordinates of the players, the control parameters are u v, $\|u\| \le 1, \|v\| \le 1, a^2$ and b^2 are the stiffness control factors, ρ

and σ – force coefficients, a, b, ρ, σ – positive numbers, $a > b$, therewith a, b are proper frequencies of circle oscillations of corresponding systems [35]. The initial positions and velocities of the players are

$$x(0) = x_0, \dot{x}(0) = \dot{x}_0, y(0) = y_0, \dot{y}(0) = \dot{y}_0.$$

With the help of change of variables

$$z_1 = x, z_2 = \dot{x}, z_3 = y, z_4 = \dot{y},$$

we pass from the system of second-order (5.11) to the system of the first order

$$\dot{z}_1 = z_2,$$
$$\dot{z}_2 = -a^2 z_1 + \rho u, \qquad (5.12)$$
$$\dot{z}_3 = z_4,$$
$$\dot{z}_4 = -b^2 z_3 + \sigma v,$$

$z_0 = (z_1^0, z_2^0, z_3^0, z_4^0), z_1^0 = x_0, z_2^0 = \dot{x}_0, z_3^0 = y_0, z_4^0 = \dot{y}_0.$

The terminal set takes the form

$$M_0 = \{z = (z_1, z_2, z_3, z_4) : z_1 = z_3\}, M = \{0\}.$$

The matrix of orthogonal projecting is

$$\pi = (\ E \quad 0 \quad -E \quad 0\ )^T,$$

where 0 and E are unit and zero matrices of order n, respectively.

The control domains have forms

$$U = \begin{pmatrix} 0 \\ \rho S \\ 0 \\ 0 \end{pmatrix}, V = \begin{pmatrix} 0 \\ 0 \\ 0 \\ \sigma S \end{pmatrix},$$

where S – the unit ball is centered at the origin.

Meeting of the controlled objects at some moment t, $t > 0$, testifies that $\pi z(t) = 0$.

The fundamental matrix of (5.12) has the form [35]

$$e^{At} = \begin{pmatrix} \cos at \cdot E & \frac{1}{a}\sin at \cdot E & 0 & 0 \\ -a\sin at \cdot E & \cos at \cdot E & 0 & 0 \\ 0 & 0 & \cos bt \cdot E & \frac{1}{b}\sin bt \cdot E \\ 0 & 0 & -b\sin bt \cdot E & \cos bt \cdot E \end{pmatrix},$$

where

$$A = \begin{pmatrix} 0 & E & 0 & 0 \\ -a^2 E & 0 & 0 & 0 \\ 0 & 0 & 0 & E \\ 0 & 0 & -b^2 E & 0 \end{pmatrix}.$$

Let us write Pontryagin's condition [2]. In the case in study, it takes the form

$$\bigcap_{\|v\| \le 1} \bigcup_{\|u\| \le 1} [\rho/a |\sin at| \, u - \sigma/b |\sin bt| \, v] = \rho/a |\sin at| \, S * \sigma/b |\sin bt| \, S =$$

$$= (\rho/a |\sin at| - \sigma/b |\sin bt|) S \neq \emptyset \; \forall t \ge 0.$$

One can see that for the fulfillment of Pontryagin's condition it is necessary that the following inequality be satisfied

$$\rho/a |\sin at| - \sigma/b |\sin bt| \ge 0 \forall t \ge 0.$$

But it can hold only under very burdensome conditions [7].

Let us write the modified Pontryagin's condition that incorporates the function of time dilation $I(t)$

$$W_I(t) = (\rho/a |\sin at| - \sigma/b \dot{I}(t) |\sin bI(t)|) S \neq \emptyset \; \forall t \ge 0.$$

This condition is true if the following inequality has a place

$$\rho/a |\sin at| - \sigma/b \dot{I}(t) |\sin bI(t)| \ge 0 \forall t \ge 0. \tag{5.13}$$

Let us find the function of time dilation, which satisfies all requirements mentioned in its definition and provides fulfillment of the inequality (5.13). We set

$$I(t) = \frac{a}{b} t \forall t \ge 0. \tag{5.14}$$

Since $a > b$, then $I(t) > t$, $I(0) > 0$, $\dot{I}(t) = \frac{a}{b}$.

Under additional condition

$$\rho/a^2 \ge \sigma/b^2, \tag{5.15}$$

inequality (5.13) is readily satisfied.

For the function of time dilation (5.14) the set-valued mapping $W_I(t)$ has the form:

$$W_I(t) = a(\rho/a^2 - \sigma/b^2) |\sin at| \cdot S. \tag{5.16}$$

It is easy to see that, under condition (5.15), it has non-empty images.

Let us proceed to implement the scheme of the method of resolving functions to the problem at hand.

The set-valued mapping $W_I(t)$ contains zero for all $t \geq 0$. Therefore, for the simplicity sake of mathematics, we choose for a measurable selection $\gamma_I(t)$ of the mapping $W_I(t)$, the function identically equal to zero, i.e. $\gamma_I(t) \equiv 0$. Moreover, with the same goal for simplicity, since the control domain of the first player is a unit ball centered at the origin, we set his starting control $u_0(\tau)$ equal to zero: $u_0(\tau) \equiv 0, \tau \in [0, I(t) - t]$ for all $t > 0$.

It should be emphasized that the aforementioned point in time $t > 0$ has not yet been found and will be discussed in the sequel. Function $N(t, u_0(\cdot))()$ takes the following form:

$$N(t, 0) = \pi e^{At} z_0 = \cos bI(t) y_0 - \cos aI(t) x_0$$
$$+ {}^1\!/_b \sin bI(t) \dot{y}_0 - {}^1\!/_a \sin bI(t) \dot{x}_0.$$

Using formulas (5.16) and (5.3), we obtain

$$W_I(t, v) = {}^\rho\!/_a |\sin at| S - {}^{\sigma a}\!/_{b^2} |\sin at| v,$$

$$A_I(t, s, v, 0) = \{\alpha \geq 0 : -\alpha N(t, 0) \in W_I(t - s, v)\}.$$

In view of the spherical form of the set-valued mapping $W_I(t, v)$, the resolving function,

$$\alpha_I(t, s, v, 0) = \sup\{\alpha : \alpha \in A(t, s, v, 0)\},$$

is the greatest root of the following quadratic equation for α:

$$\left\| \frac{\sigma a}{b^2} |\sin a(t - s)| v - \alpha N(t, 0) \right\| = {}^\rho\!/_a |\sin a(t - s)|.$$

Because the norm is a square root of the scalar product of vectors, we square both parts of the above equation and obtain the quadratic equation for α. Its greatest root stands for the resolving function and has the form

$$\alpha_I(t, s, v, 0) = \frac{\frac{\sigma a}{b^2} c(t, s)(v, N(t, 0))}{\|N(t, 0)\|^2} +$$

$$+ \frac{[\{\frac{\sigma a}{b^2} c(t, s)\}^2 (v, N(t, 0))^2 + \|N(t, 0)\|^2 (\rho^2\!/_{a^2} c^2(t, s) - \frac{\sigma^2 a^2}{b^4} c^2(t, s) \|v\|^2)]^{1/2}}{\|N(t, 0)\|^2},$$

$$c(t, s) = |\sin a(t - s)|.$$

Minimum in $v, \|v\| \leq 1$, of this function, is achieved on the element $v = -\frac{N(t,0)}{\|N(t,0)\|}$ and is equal to

$$\min_{\|v\| \leq 1} \alpha_1(t, s, v, 0) = \frac{\left(\frac{\rho}{a} - \frac{\sigma a}{b^2}\right) |\sin a(t-s)|}{\|N(t,0)\|}.$$

This can be verified using the Lagrange multiplier method [36].

Thus, by virtue of (5.4), the issue concerning the existence of a finite time for the game termination is reduced to the existence of a positive root of the following equation for t:

$$a \int_0^t \left(\frac{\rho}{a^2} - \frac{\sigma}{b^2}\right) |\sin a(t-s)| \, ds = \|N(t,0)\|. \qquad (5.17)$$

Here

$$N(t,0) = \cos bI(t)y_0 - \cos aI(t)x_0 + \frac{1}{b}\sin bI(t)\dot{y}_0 - \frac{1}{a}\sin bI(t)\dot{x}_0,$$

where the function $I(t)$ is given by the formula (5.14).

Since, $N(0,0) = y_0 - x_0 \neq 0$, then $\|N(t,0)\| > 0$ for $t = 0$ and the left side of (5.17) turns into zero at $t = 0$. Herewith, $N(t,0)$ is bounded at $t > 0$ while the left-hand part of (5.17) is growing without bound as $t \to +\infty$. Since, $\frac{\rho}{a^2} > \frac{\sigma}{b^2}$, then a finite root of equation (5.17) always exists.

Let T_* be a positive root of equation (5.17) and let controls of the first player on the active and passive parts, separated by the zero of the controlling function (5.7), are chosen in compliance with the theorem-proof, that is, from the relationships (5.8), (5.9). Then at the moment T_*, the conflict-controlled process (5.12) hits the terminal set for any admissible behavior of the second player.

Remark 5.3. Similar control problems are analyzed in [37, 38].

5.6 Conclusion

We present a method for solving the game problems of bringing a trajectory to the terminal set in the case when the classical Pontryagin's condition is not satisfied. This became feasible due to the introduction of the function of time dilation and a weaker modified condition based on it

The method of resolving functions is used as a basis for investigation. This method allows obtaining sufficient conditions for the approach in a finite guaranteed time. In so doing, the first player constructs its control based on delayed information on control of the second player, and the choice of control itself is carried out based on the theorem on measurable choice.

The results are illustrated by the example of the game approach in geometric coordinates of two second-order oscillatory systems, describing the dynamics of mathematical pendulums.

References

[1] R.F. Isaacs, 'Differential Games', New York–London–Sydney: Wiley Interscience, 479 p., 1965.

[2] L.S. Pontryagin, 'Selected scientific papers', 2, Moscow: Nauka, 576 p., 1988. (in Russian).

[3] B.N. Pschenitchnyi, 'ε-strategies in Differential Games', Topics in Differential Games, New York, London, Amsterdam: North-Holland Publ. Co., pp. 45–99, 1973.

[4] N.N. Krasovskii, 'Game Problems on the Encounter of Motions, Moscow: Nauka, 420 p., 1970 (in Russian).

[5] B.N. Pschenitchnyi, 'Linear Differential Games', Avtom, Telemekh., p. 65–78, No 1, 1968 (in Russian).

[6] A.I. Subbotin, A.G. Chentsov, 'Optimization of Guaranteed Result in Control Problems', Moscow: Nauka, 286 p., 1981 (in Russian).

[7] A.A. Chikrii, 'Conflict controlled processes', Boston; London, Dordrecht: Springer Science and Business Media, 424 p., 2013.

[8] O. Hajek, 'Pursuit games', New York: Academic Press, 266 p., 1975.

[9] M.S. Nikolskii, 'L.S. Pontryagin's First Direct Method in Differential Games', Izdat. Gos.Univ., Moscow, 65 p., 1984 (in Russian).

[10] G. Siouris, 'Missile Guidance and Control Systems', New York: Springer-Verlag, 666 p., 2004.

[11] G. Siouris, 'Aerospace Avionics Systems: A Modern Synthesis', San Diego: Academic Press, 466 p., 1993.

[12] G.Ts. Chikrii, 'An approach to solution of linear differential games with variable information delay', Journal of Automation and Information Sciences: Begell House Inc., p. 163–170, No 3, 4, vol. 27, 1995.

[13] G.Ts. Chikrii, 'Using the effect of information delay in differential pursuit games', Cybernetics and Systems Analysis, p. 233–245, No. 2, vol. 43, 2007.

[14] M.S. Nikolskij, 'Application of the first direct method in the linear differential games', Izvestia Acad. Nauk SSSR, p. 51–56, vol. 10, 1972 (in Russian).

[15] D. Zonnevend, 'On one method of pursuit', Doklady Akademii Nauk SSSR, p. 1296–1299, vol. 204, No 6, 1972 (in Russian).

[16] G.Ts. Chikrii, 'Principle of time stretching in evolutionary games of approach', Journal of Automation and Information Science, Begell House, Inc., p. 12–26, vol. 48, No. 5, 2016.

[17] G.Ts. Chikrii, 'Principle of time stretching for Motion Control in Condition of Conflict', Advanced Control Systems: Theory and Applications, River Publishers Series in Automation, Control and Robotics, p. 53–82, 2021.

[18] G.Ts. Chikrii, 'One Approach to solution of complex game problems for some quasilinear evolutionary systems', International Journal of Mathematics, Game Theory and Algebra, p. 307–314, vol. 14, 2004.

[19] A.A. Chikrii, 'An analytical method in dynamic pursuit games', Proceedings of the Steklov Institute of Mathematics, p. 69–85, vol. 271, 2010.

[20] Locke S. Arthur, Guidance, D.Van Nostrand Company, Inc. Princeton, 776 p., 1955.

[21] A.A. Chikrii, 'Linear problem of avoiding several pursuers', Engineering Cybernetics, p. 38–42, vol. 14, No 4, 1976.

[22] B.N. Pschenitchnyi, 'Simple Pursuit by Several Objects', Kibernetika, p. 145–146, No 3, 1976 (in Russian).

[23] B.N. Pschenitchnyi, A.A. Chikrii, J.S. Rappoport, 'Group Pursuit in Differential Games', Opt. Invest. Stat., Germany, p. 13–27, 1982 (in Russian).

[24] A.A. Chikrii, 'Differential Games with Many Pursuers', Mathematical Control Theory, Banach Center Publ., PWN, Warsaw, p. 81–107, vol. 14, 1985 (in Russian).

[25] A.A. Chikrii, S.F. Kalashnikova, 'Pursuit of a group of evaders by a single controlled object', Cybernetics, pp. 437–445, No 4, 1987.

[26] A.A. Chikrii, J.S. Rappoport, K.A. Chikrii, 'Multivalued mappings and their selectors in the theory of conflict-controlled processes', Cybernetics and Systems Analysis, p. 719–730, vol. 43, No 5, 2007.

[27] A.A. Chikrii, V.K. Chikrii, 'Image Structure of Multivalued Mappings in Game Problems of Motion Control', Journal of Automation and Information Sciences, pp. 20–35 vol. 48, No. 3, 2016.

[28] L.A. Vlasenko, A.G. Rutkas, A.A. Chikrii, 'On a differential game in an abstract parabolic system', Proceedings of the Steklov Institute of Mathematics, pp. 254–269, vol. 293, Issue 1 Supplement, 2016.

[29] A.A. Chikrii, S.D. Eidelman, 'Game problems for fractional quasilinear systems', Computers and Mathematics with Applications, p. 835–851, vol. 44, No. 7, 2002.

[30] A.G. Nakonechnyi, E.A. Kapustyan, A.A. Chikrii, 'Control of impulse system in conflict situation', Journal of Automation and Information Science, p. 54–63, No. 9, 2019.

[31] L.V. Baranovskaya, A.A. Chikrii, Al.A. Chikrii, 'Inverse Minkowski functional in a nonstationary problem of a group pursuit', Journal of Computer and Systems Sciences International, p. 101–106, vol. 36, No. 1, 1997.

[32] L.V. Baranovskaya, Al.A. Chikrii, 'Game problems for a class of hereditary systems', Journal of Automation and Information Science, p. 87–97, vol. 29, No 2, 1997.

[33] L.A. Vlasenko, A.G. Rutkas, A.A. Chikrii, 'On a differential game in a stochastic system', Proceedings of the Steklov Institute of Mathematics, p. 185–198, vol. 309, suppl.1, 2020.

[34] J.-P. Aubin, H. Frankowska, 'Set-valued analysis', Boston; Basel; Berlin: Birkhauser, 1990.

[35] N.V. Vasilenko, 'Theory of Oscillations, Kiev: Vyshcha Shkola, 430 p., 1992 (in Russian).

[36] A.D. Joffe, V.M. Tikhomirov, 'Theory of Extremal Problems', North-Holland, Amsterdam, 480 p., 1979.

[37] V.M. Kuntsevich et al (Eds). Control Systems: Theory and Applications. Series in Automation, Control and Robotics, River Publishers, Gistrrup, Delft, 2018.

[38] Y.P. Kondratenko, V.M. Kuntsevich, A.A. Chikrii, V.F. Gubarev (Eds). Advanced Control Systems: Theory and Applications, Series in Automation, Control and Robotics, River Publishers, Gistrrup, 2021.

6

Method of Upper and Lower Resolving Functions for Pursuit Differential-difference Games with Pure Delay

Lesia Baranovska

Institute of Applied System Analysis of the National Technical University of Ukraine "Igor Sikorsky Kyiv Polytechnic Institute", 37 Peremohy av., Kyiv, 03056, Ukraine
E-mail: lesia@baranovsky.org

Abstract

A modification of the method of resolving functions for differential-difference games with pure delay is considered. The lower and upper resolving functions are introduced. For the local game of approach, the scheme of a method is constructed and sufficient conditions for the completion of the game are formulated.

Keywords: Differential-difference games, differential games, dynamic games, games of approach, conflict-controlled processes, game theory.

6.1 Introduction

Differential games is a section of mathematical control theory [1–3] studying the manipulation of moving objects operated under conditions of uncertainty and conflict. We consider the differential-difference games of approach. In the theory of differential games, along with Isaacs ideology [4], several methods have been developed that provide a guaranteed result. The most famous methods are the first direct method of L.S. Pontryagin [5, 6], the method of extreme aiming of M.M. Krasovskii [7, 8], and the method of resolving functions of A.O. Chikrii [9–15]. In this chapter, the method of resolving functions will be used. The first direct method of L.S. Pontryagin and the method of resolving functions give a theoretical justification for the classical rule of parallel pursuit. Most processes depend not only on the current state, but they have essential knowledge of the prehistory of the state of the process, which can't be ignored. Today, systems of differential-difference equations are increasingly used in control theory. The method scheme for differential-difference games is developed in works [16–19]. In [20–22], the problem of approach for a group of pursuers and a single evader is considered. The scheme of the method of resolving functions for differential-difference systems of neutral type was developed in [23]. For objects with different inertia in [24], a modification of the Pontryagin condition by introducing the body part of the terminal set is proposed. Differential games in case of failure of control devices are considered in the works [25–27].

In this chapter, we first introduce the upper and lower resolving functions of different types [28] in the method of resolving functions for the differential game of approach with pure delay, which allows developing the method even if the Pontryagin condition is not fulfilled. Sufficient conditions have been obtained to complete the game.

6.2 Statement of the Problem

Denote by 2^{R^n} a set of all subsets of space R^n, by $K(R^n)$ a set of all nonempty compacts in R^n, and by $K(R^n)$ a set of all nonempty convex compacts in R^n. By a set-valued mapping is meant a mapping acting from R^n to 2^{R^n} [9] and transforming each element $x \in R^n$ into a set in R^n.

Consider a controlled object whose dynamics is described by the linear differential-difference system with pure delay in a Euclidean space R^n

$$\dot{z}(t) = Bz(t - \tau) + \varphi(u, v), \quad z \in R^n, \ u \in U, \ v \in V, \qquad (6.1)$$

where B is a square constant matrix of order n; z is the state vector, involving geometric coordinates, velocities, accelerations of the pursuer, and evader; $U, V \in K(R^n); \varphi : U \times V \to R^n$ is jointly continuous in its variables.

The initial condition

$$z(t) = z^0(t), \quad -\tau \le t \le 0, \qquad (6.2)$$

is absolutely continuous on $[-\tau, 0]$.

The controls $u(s)$ and $v(s)$ that are allowed are the Lebesgue measurable functions.

Denote

$$\Omega_U = \{u(s) : \ u(s) \in U, \ s \in [0, +\infty)\},$$

$$\Omega_V = \{v(s) : \ v(s) \in V, \ s \in [0, +\infty)\}.$$

The function $u(\cdot) \in \Omega_U \quad (v(\cdot) \in \Omega_V)$, will be called an open-loop pursuer's (evader's) control. This control is chosen by the pursuer (evader) based on knowledge of the initial condition.

The function $v_t(\cdot) = \{v(s) : \ s \in [0, t], \ v(\cdot) \in \Omega_V\}$, will be called prehistory of the control of evader at a time t, $\quad t \geq 0$ [9].

We define the pursuer's "quasistrategy" [29] as a mapping $U\left(t, z^0(\cdot), v_t(\cdot)\right)$. It assigns a Lebesgue measurable function to each moment $t \geq 0$, initial condition (6.2) and arbitrary prehistory $v_t(\cdot)$ of the control of evader, taking its values in control domain U:

$$u(t) = U\left(t, z^0(\cdot), v_t(\cdot)\right), \quad t \geq 0, \tag{6.3}$$

Strobostrophic strategies [30] are a special case of 'quasistrategies'. The counter control of the pursuer is based on information on its initial state (6.2) of the process (6.1) and the instantaneous value of the control of the evader

$$u(t) = U\left(z^0(\cdot), v(t)\right), \quad t \geq 0. \tag{6.4}$$

The terminal set has a cylindrical form [9]:

$$M^* = M_0 + M, \tag{6.5}$$

where M_0 is a linear subspace in R^n and M is a compact set from the orthogonal complement (L) of M_0 in R^n.

The goal of the pursuer (u) is in the shortest time to bring a trajectory of the process (6.1), (6.2) to a terminal set (6.5). The goal of the evader (v) is to avoid a trajectory of this process from meeting with the set (6.5) on a whole semi-infinite interval of time or if it is impossible to delay the moment of the meeting as much as possible. The game is evolving in the time interval $[0, T]$, where T is a moment when a trajectory of the process brings to a set (6.5) $M^*; T > 0$ such that $z(T) \in M^*$ or $\pi z(T) \in M, \pi$ is the orthogonal project from R^n onto the subspace L [9].

In this game, we take the side of the pursuer. We will find sufficient conditions on the parameters of the problem (6.1), (6.2), (6.5), ensuring the game termination for a certain guaranteed time.

6.3 Scheme of the Method

Let $\Delta = \{(t,s): 0 \leq s \leq t < +\infty\}$ be the flat cone; $\varphi(U,v) = \{\varphi(u,v) = u - v: u \in U\}$.

Consider the set-valued mappings $W(t,s,v) = \pi K(t,s)\varphi(U,v)$ on the set $\Delta \times V$, where $K(t,s)$ is a matrix-valued function [31, 32].

For the practical finding of the fundamental matrix $K(t,s)$ of systems of differential-difference equations, it is more convenient to decompose this function into a series.

__Definition 6.1__ [33–36]. For each $s=1,2,\ldots$ the time-delay exponential is defined as follows

$$exp_\tau\{B,t\} = \begin{cases} \Theta, & -\infty < t < -\tau; \\ I, & -\tau \leq t < 0; \\ I + B\frac{t}{1!} + B^2\frac{(t-\tau)^2}{2!} + \ldots + B^k\frac{(t-(s-1)\tau)^s}{s!}, \\ \quad (s-1)\tau \leq t \leq s\tau. \end{cases}$$

__Lemma 6.1__ [35]. Let $z(T)$ be a continuous solution to the system (6.1) with pure time-delay under the initial condition (6.2). Then,

$$z(T) = exp_\tau\{B,t\}z^0(-\tau) + \int_{-\tau}^0 exp_\tau\{B, t-\tau-s\}\dot{z}^0(s)ds + \\ + \int_0^t exp_\tau\{B, t-\tau-s\}\varphi(u(s),v(s))ds.$$

Consider the set-valued mapping

$$W(t,s,v) = \pi exp_\tau(B,t)\varphi(U,v),$$

$$W(t,s) = \bigcap_{v \in V} W(t,s,v), \quad 0 \leq s \leq t < +\infty.$$

Remark 6.1. For the linear process (6.1) ($\varphi(u,v) = u - v$) we have

$$W(t,s) = \pi exp_\tau\{B,t\}U \underset{*}{\_} \pi exp_\tau\{B,t\}V,$$

$\underset{*}{\_}$ is a geometric subtraction of the sets (Minkowski's difference) [38].

Let [28, 37]

$$domW = \{(t,s) \in \Delta: W(t,s) \neq \emptyset\}.$$

__Pontryagin Condition.__ $domW = \Delta$.

It is further assumed that the Pontryagin condition is not satisfied.

Condition 6.1. The mapping $W(t, s, v)$ is closed-valued on the direct product of the cone Δ and compact V.

Let's call the function $\gamma(t, s)$, $\gamma : \Delta \to L$, a shift function. It's almost everywhere limited and measurable in t, integrated into s, $s \in [0, t]$ for all $t \in [0, +\infty)$.

We put

$$\xi(t) = \xi(t, z^0(t), \gamma(t, \cdot)) = \pi \, exp_\tau \{B, t\} z^0(-\tau) + \\ + \int_{-\tau}^0 \pi \, exp_\tau \{B, t - \tau - s\} \dot{z}^0(s) ds + \int_0^t \gamma(t, s) ds.$$

Denote

$$R(t, s, v) = \{\alpha \geq 0 : \; [\pi \, exp_\tau \varphi(u, v) - \gamma(t, s)] \cap \alpha \, [M - \xi(t)] \neq \emptyset\},$$
(6.6)

$$R : \; \Delta \times V \to 2^{R_+}, \quad R(t, s) = \bigcap_{v \in V} R(t, s, v), \quad (t, s) \in \Delta.$$

Denote the lower and upper resolving functions [28] first type

$$\alpha^*(t, s, v) = \sup \{\alpha : \; \alpha \in R(t, s, v)\},$$

$$\alpha_*(t, s, v) = \inf \{\alpha : \; \alpha \in R(t, s, v)\}.$$

Condition 6.2. The set-valued mapping $R(t, s)$ has non-empty images on the cone Δ.

Finally, denote the lower and upper resolving functions [28] second type:

$$\alpha^*(t, s) = \sup \{\alpha : \; \alpha \in R(t, s)\},$$

$$\alpha_*(t, s) = \inf \{\alpha : \; \alpha \in R(t, s)\},$$

and numerical functions

$$\alpha^*(t) = \int_0^t \alpha^*(t, s) ds, \quad \alpha_*(t) = \int_0^t \alpha_*(t, s) ds.$$

To the upper resolving function, which meaningfully means the maximum gain of the pursuer at the moment s in the game on the interval $[0, t]$ when counteracting v, let's put by the set

$$T\left(z^0(\cdot), \gamma(\cdot, \cdot)\right) = \left\{t \geq 0 : \; \inf_{v(\cdot) \in V(\cdot)} \int_0^t \alpha^*(t, s, v(s)) \, ds \geq 1\right\} \quad (6.7)$$

and its element

$$t\left(z^0(\cdot),\gamma(\cdot,\cdot)\right)=\inf\left\{t:\ t\in T\left(z^0(\cdot),\gamma(\cdot,\cdot)\right)\right\}.$$

If for some t, $t > 0$, $\alpha^*(t,s,v) \equiv +\infty$, $s \in [0,t]$, $v \in V$, then in this case the value of the integral in relation (6.7) is naturally put equal $+\infty$, and the corresponding inequality is performed automatically, and $t \in T\left(z^0(\cdot),\gamma(\cdot,\cdot)\right)$. In the case when inequality in (6.6) does not hold for all $t > 0$, we assume $T\left(z^0(\cdot),\gamma(\cdot,\cdot)\right) = \emptyset$, then, $t\left(z^0(\cdot),\gamma(\cdot,\cdot)\right) = +\infty$, respectively.

Theorem 6.1. Let the conflict-controlled process (6.1), (6.2), (6.5) satisfy Conditions 6.1 and 6.2, $M = coM$, and for a given function $z^0(\cdot)$ and a shift function $\gamma(\cdot,\cdot)$ $T = T\left(z^0(\cdot),\gamma(\cdot,\cdot)\right) \neq \emptyset$.

Then, the trajectory of the process (6.1), (6.2) can be under condition $\alpha_*(T) < 1$, brought by the pursuer (u) from $z^0(\cdot)$ to the terminal set (6.3) at the moment T by the control in the form of (6.4); and under the condition $\alpha^*(T) > 1$—in the counter-control class, with any permissible counteractions of the evader.

Proof. Let $v(\cdot)$ be an arbitrary measurable function taking values from the control domain V.

Suppose that $\alpha^*(T,s,v) \neq +\infty$, $s \in [0,T]$, $v \in V$. We introduce the controlling function

$$h(t) = 1 - \int_0^t \alpha^*(T,s,v(s))ds - \int_0^T \alpha_*(T,s)ds, \quad s \in [0,T].$$

If $h(0) = 1 - \int_0^T \alpha_*(T,s)ds = 1 - \alpha_*(T) > 0$, we have

$$h(T) = 1 - \int_0^T \alpha^*(T,s,v(s))ds \leq 0.$$

Hence there exists a switching time t_*, $t_*(v(\cdot))$, $0 < t_* \leq T$, such that $h(t_*) = 0$.

The whole process of approach is divided into time sections, such as active $[0,t_*)$ and passive $[t_*,T]$, where t_* is the moment of switching from one law of choosing the counter-control of the pursuer to another, depending on the prehistory of running away [9].

By the foregoing, we define the following law of choice of the control of pursuer.

We consider the set-valued mapping

$$U_1(s,v) = \{u \in U : \pi\, exp_\tau\{B, T - \tau - s\}\varphi(u,v) -$$
$$- \gamma(T - \tau - s) \in \alpha^*(T, s, v)\,[M - \xi(T)]\},$$
$$U_2(s,v) = \{u \in U : \pi\, exp_\tau\{B, T - \tau - s\}\varphi(u,v) -$$
$$- \gamma(T - \tau - s) = 0\}, \quad s \in [t_*, T]. \tag{6.8}$$

It follows from the construction of the mappings $U_1(s,v)$ and $U_2(s,v)$has non-empty images. As stated in [28, 39–42] there are exist measurable selections $u_1(s,v)$ and $u_2(s,v)$.

The pursuer's control on the active interval $[0, t_*)$is constructed in the form

$$u_1(s) = u_1(s, v(s)).$$

On the passive interval $[t_*, T]$ we set the control of pursuit in the following form

$$u_2(s) = u_2(s, v(s)).$$

We will show that if the pursuer chooses control as described above, then the trajectory of the process (6.1), (6.2) will reach the terminal set (6.5) at the time T under arbitrary admissible controls of the evader.

By Lemma 6.1, the Cauchy formula for the system (6.1) under the condition (6.2) implies the representation

$$\pi z(T) = \pi\, exp_\tau\{B, T\}z^0(-\tau) + \int_{-\tau}^{0} \pi\, exp_\tau\{B, T - \tau - s\}\dot{z}^0(s)ds+$$

$$+ \int_0^T \pi\, exp_\tau\{B, T - \tau - s\}\varphi(u(s), v(s))ds. \tag{6.9}$$

We add and subtract the value $\int_0^T \gamma(T, s)ds$, applies (6.8), $h(t_*) = 0$, and get the following:

$$\pi z(T) = \pi\, exp_\tau\{B, T\}z^0(-\tau) + \int_{-\tau}^{0} \pi\, exp_\tau\{B, T - \tau - s\}\dot{z}^0(s)ds+$$

$$+ \int_0^{t_*} \pi\, exp_\tau\{B, T - \tau - s\}\varphi(u_1(s), v(s))ds$$

$$+ \int_{t_*}^{T} \pi\, exp_\tau\{B, T - \tau - s\}\varphi(u_2(s), v(s))ds\pm$$

$$\pm \int_0^T \gamma(T, s)ds \in \pi\, exp_\tau\{B, T\}z^0(-\tau)$$

$$+ \int_0^{t_*} \alpha^*(T, s, v(s))[M - \xi(T)]ds +$$

$$+ \int_{t_*}^T \alpha^*(T, s)[M - \xi(T)]ds + \int_0^T \gamma(T, s)ds =$$

$$= \xi(T) \left(1 - \int_0^{t_*} \alpha^*(T, s, v(s))ds - \int_{t_*}^T \alpha^*(T, s)ds \right) +$$

$$+ \int_0^{t_*} \alpha^*(T, s, v(s))Mds + \int_{t_*}^T \alpha^*(T, s)Mds =$$

$$= \left(\int_{t_*}^T \alpha^*(T, s)ds + \int_{t_*}^T \alpha^*(T, s)ds \right) M = M.$$

From (6.6) we have that the case $\alpha^*(T, s, v) = +\infty$ for some $s \in [0, T]$, $v \in V$, is possible only if $0 \in M - \xi(T)$, $0 \in exp_\tau\{B, T\}\varphi(U, v) - \gamma(T, s)$, and in this case it is obvious that $R(T, s, v) = [0, +\infty)$, $s \in [0, T]$, $v \in V$, and $\alpha^*(T, s) = 0$, $s \in [0, T]$.

This makes it possible to choose as the resolving functions at those points $s \in [0, T]$ where $\alpha^*(T, s, v(s)) = +\infty$, an arbitrary finite measurable function that takes values on the interval $[0, +\infty)$ with only one condition that the final function on the interval $[0, T]$ provides relation $h(t_*) = 0$ for a certain switching moment t_*, $t_* \in [0, T]$.

Thus, the construction of control in the active and the passive time sections is reduced to the previous case.

6.4 Conclusion

The concept of upper and lower resolving functions of two types for differential-difference games of approach with pure time delay is introduced at the first time. Sufficient conditions are obtained for the trajectory of the process (6.1), (6.2) to converge with a given cylindrical terminal set (6.5) in the case when the Pontryagin condition does not hold.

In the future, it is planned to expand the proposed scheme for processes that are described by systems of differential-difference equations with several delays based on the use of the Cauchy formulas [43–45].

It is planned to apply the proposed scheme to specific examples and to realize the visualization of the trajectory of the objects on the plane [46, 47].

We also plan to shift the proposed scheme of the method of revolving functions for group pursuit problems [48, 49] and nonstationary pursuit [50–53].

References

[1] Kuntsevich, V.M., et al (Eds). *Control Systems: Theory and Applications.* Series in Automation, Control, and Robotics, River Publishers, Gistrup, Delft, 2018.

[2] Kondratenko, Y.P., Kuntsevich, V.M., Chikrii, A.A., Gubarev, V.F. (Eds). *Advanced Control Systems: Theory and Applications.* Series in Automation, Control, and Robotics, River Publishers, Gistrup, 2021.

[3] Kondratenko, Y.P., Chikrii, A.A., Gubarev, V.F., Kacprzyk, J. (Eds). *Advanced Control Techniques in Complex Engineering Systems: Theory and Applications.* Dedicated to Professor Vsevolod M. Kuntsevich. Studies in Systems, Decision, and Control, Vol. 203. Cham: Springer Nature Switzerland AG, 2019.

[4] R. Isaacs. Differential Games: *A Mathematical Theory with Applications to Warfare and Pursuit, Control and Optimization.* John Wiley & Sons Inc, New York, 1965.

[5] L.S. Pontryagin. *Selected scientific works.* Nauka, Moscow, 2, 1988.

[6] M.S. Nikol'skii. *Pontryagin's First Direct Method in Differential Games.* Mosk. Gos. Univ., Moscow, 1984 (in Russian).

[7] N.N. Krasovskii, A.I. Subbotin. *Game-theoretical control problems.* Springer-Verlag, New York, 1988.

[8] V.N. Ushakov. *Krasovskii's Unification Method and the Stability Defect of Sets in a Game Problem of Approach on a Finite Time Interval.* In: Kondratenko Y., Chikrii A., Gubarev V., Kacprzyk J. (eds) Advanced Control Techniques in Complex Engineering Systems: Theory and Applications. Studies in Systems, Decision, and Control, vol 203. Springer, Cham, 2019. https://doi.org/10.1007/978-3-030-21927-7_5.

[9] A. Chikrii. *Conflict-Controlled Processes.* Springer Science &Business Media, 2013.

[10] A.A. Chikrii, I.S Rappoport, K.A. Chikrii. Multivalued Mappings and their Selectors in the Theory of Conflict-Controlled Processes. *Cybernetics and Systems Analysis,* 43(5):719–730, 2007.

[11] A.A. Chikrii, R. Petryshyn, I. Cherevko., Y. Bigun. Method of Resolving Functions in the Theory of Conflict—Controlled Processes. In: Kondratenko Y., Chikrii A., Gubarev V., Kacprzyk J. (eds) *Advanced Control*

Techniques in Complex Engineering Systems: Theory and Applications. Studies in Systems, Decision, and Control, vol 203. Springer, Cham, 2019. https://doi.org/10.1007/978-3-030-21927-7_1.

[12] J. Albus, A. Meystel, A.A. Chikrij, A.A. Belousov, A.I. Kozlov. Analytical Method for Solution of the Game Problem of Self Landing for Moving Objects. *Cybernetics and Systems Analysis*, 37(1):75–91, 2001.

[13] A.A. Chikrii, S.F. Kalashnikova. Pursuit of a Group of Evaders by a Single Controlled Object. *Cybernetics*, 23(4):437–445, 1987.

[14] A.A. Chikrii, I.S. Rappoport. Systems Analysis Method of Resolving Functions in the Theory of Conflict-Controlled Processes. *Cybernetics and Systems Analysis*, 48(4):512–531, 2012.

[15] A.A. Chikrii, G.T. Chikrii. Matrix Resolving Functions in Game Problems of Dynamics. *Proceedings of the Steklov Institute of Mathematics* 268(SUPPL. 1), 2010.

[16] L.V. Baranovska. Quasi-Linear Differential-Difference Game of Approach. *Understanding Complex Systems*, 2019, pp. 505–524. https://doi.org/10.1007/978-3-319-96755-4_26.

[17] L.V. Baranovska. On Quasilinear Differential-Difference Games of Approach. *Journal of Automation and Information Sciences*, 49(8):53–67, 2017. https://doi.org/10.1615/JAutomatInfScien.v49.i8.40.

[18] Lesia V. Baranovska. Pursuit differential-difference games with pure time-lag. *Discrete and Continuous Dynamical Systems* – Series B 24(3):1021–1031, 2019. https://doi.org/10.3934/dcdsb.2019004.

[19] L.V. Baranovska. Group Pursuit Differential Games with Pure Time-Lag. In: Sadovnichiy, V., Zgurovsky, M. (eds.) *Contemporary Approaches and Methods in Fundamental Mathematics and Mechanics. Understanding Complex Systems*, pp. 475–488. Springer, Cham, 2021. https://doi.org/10.1007/978-3-030-50302-4_23.

[20] G.G. Baranovskaya, L.V. Baranovskaya. Group Pursuit in Quasilinear Differential-Difference Games. *Journal of Automation and Information Sciences*, 29(1):55–62, 1997. https://doi.org/10.1615/JAutomatInfScien .v29.i1.70.

[21] L.V. Baranovskaya, A.A. Chikrij, Al.A. Chikrij. Inverse Minkowski functionals in a nonstationary problem of group. *Izvestiya Akademii Nauk. Teoriya i Sistemy Upravleniya*, (1):109–114, 1997.

[22] L.V. Baranovskaya, A.A. Chikrij, Al.A. Chikrij. Inverse Minkowski functionals in a nonstationary problem of group. *Journal of Computer and Systems Sciences International*, 36(1):101–106, 1997.

[23] L.V. Baranovskaya. A method of resolving functions for one class of pursuit problems. *Eastern-European Journal of Enterprise Technologies*, vol.2, 4(74):4–8, 2015. https://doi.org/10.15587/1729-4061.2015.39355.

[24] L.V. Baranovska. Method of resolving functions for the differential-difference pursuit game for different-inertia objects. *Studies in Systems, Decision and Control*, 69:159–176, 2016. https://doi.org/10.1007/978-3-319-40673-2_7.

[25] A.A. Chikrij, L.V. Baranovskaya, Al.A. Chikrij. The game problem of approach under the condition of failure of controlling devices. *Problemy Upravleniya i Informatiki (Avtomatika)*, (4):5–13, 1997.

[26] L.V. Baranovskaya, Al.A. Chikrii. Game Problems for a Class of Hereditary Systems. *Journal of Automation and Information Sciences*, 29(2):87–97, 1997. https://doi.org/10.1615/JAutomatInfScien.v29.i2-3.120.

[27] A.A. Chikriy, L.V. Baranovskaya, Al.A. Chikriy. An Approach Game Problem under the Failure of Controlling Devices. *Journal of Automation and Information Sciences*, 35(5):1–8, 2000. https://doi.org/10.161 5/JAutomatInfScien.v32.i5.10.

[28] A.A. Chikrii, V.K. Chikriy. Image Structure of Multivalued Mappings in Game Problems of Motion Control. *Journal of Automation and Information Science*, 48(3):20–35, 2016. https://doi.org/10.1615/jautomatin fscien.v48.i3.30.

[29] A.A. Chikrii. An analytical method in dynamic pursuit games, *Proceedings of the Steklov Institute of Mathematics*, vol. 271, no. 1, 2010, pp. 85.

[30] O. Hayek. *Pursuit Games* (Academic, New York, 1975), Math. Sci. Eng. 120.

[31] L.E. Elsgolts, S.B. Notkin. *Differential Equations with Deviating Argument*. Nauka, Moscow, 1971.

[32] R. Bellman, K. Cooke. *Differential-Difference Equations*, Academic, Cambridge, 1963.

[33] D.Ya. Khusainov, D.D. Benditkis, J. Diblik. Weak delay in system with an aftereffect. *Funct. Differ. Equ.* 9(3-4):385-404, 2002.

[34] J. Diblik, B. Morvkov, D. Khusainov, A. Kukharenko. Delayed exponential functions and their application to representations of solutions of linear equations with constant coefficients and with single delay. In: *Proceedings of 2nd International Conference on Mathematical Models for Engineering Science*, 2011.

[35] D.Ya. Khusainov, J. Diblik, M. Ruzhichkova. *Linear dynamical systems with aftereffect. In: Representation of Decisions, Stability, Control, Stabilization.* GP Inform-Analytics Agency, Kyiv, 2015.

[36] J. Diblik, D. Khusainov. Representation of solutions of linear discrete systems with constant coefficients and pure delay. *Advances in Difference Equations,* 2006. https://doi.org/10.1155/ADE/2006/80825.

[37] R.T. Rockafellar. *Convex Analysis.* Princeton University Press, Princeton, 1970.

[38] B.N. Phenichnyi. Simple Pursuit by Several Objects. *Kibernetika,* 3:145-146, 1976.

[39] J.-P. Aubin, I. Ekeland. *Applied Nonlinear Analysis,* Wiley, New York, 1984.

[40] J.-P. Aubin, He. Francowska. *Set-Valued Analysis.* Birkhause, Boston, 1990.

[41] R.J. Aumann. Integrals of set-valued functions. *J. Math. Anal. Appl.,* 12:1–12, 1965.

[42] A.A. Chikrii. Multivalued mappings and their selections in game control problems. *Journal of Automation and Information Science,* 27(1):27–38, 1995.

[43] M. Medved', M. Pospíŝil. Sufficient conditions for the asymptotic stability of nonlinear multidelay differential equations with linear parts defined by pairwise permutable matrices. *Nonlinear Anal.,* 75:3348–3363, 2012.

[44] M. Pospíŝil. Representation of solutions of systems of linear differential equations with multiple delays and nonpermutable variable coefficients. *Mathematical modeling and analysis,* 25(2):303–322, 2020.

[45] M. Pospíŝil, F. Jaroŝ. On the representation of solutions of delayed differential equations via Laplace transform. *Electronic Journal of Qualitative Theory of Differential Equations,* 117:1–13, 2016.

[46] L. Baranovska, D. Hyriavets, K. Dovzhanytsia, V. Mukhin. Visualization of pursuit differential game on a plane. *Selected Papers of the XX International Scientific and Practical Conference "Information Technologies and Security" (ITS 2020),* 2020, pp. 99–113.

[47] A.A. Chikrii, L.A. Sobolenko, S.F. Kalashnikova. A numerical method for the solution of the pursuit-and-evasion problem. *Cybernetics,* 24(1):53–59, 1988.

[48] M. Pittsyk, A.A. Chikrii. On group pursuit problem. Journal of Applied Mathematics and Mechanics, 46(5):584–589, 1982.

[49] A.A. Chikrii, P.V. Prokopovich. Simple pursuit of one evader by a group. *Cybernetics and Systems Analysis*, 28(3):438–444, 1992.

[50] Al.A. Chikrii. On nonstationary game problem of motion control. *Journal of Automation and Information Sciences,* 47(11):74-83, 2015. https://doi.org/10.1615/JAutomatInfScien.v47.i11.60.

[51] I.I. Kryvonos, Al.A. Chikrii, K.A. Chikrii. On an approach scheme in nonstationary game problems. *Journal of Automation and Information Sciences,* 45(80):32–40, 2013. https://doi.org/10.1615/JAutomatInfScien.v45.i8.40.

[52] V.A. Pepelyaev, Al.A. Chikrii. On the game dynamics problems for nonstationary controlled processes. *Journal of Automation and Information Sciences,* 49(3):13–23, 2017. https://doi.org/10.1615/JAutomatInfScien.v49.i3.30.

[53] Y.N. Onopchuk, Al.A. Chikrii. The analytical method to solve nonstationary differential games of pursuit. *Cybernetics and Systems Analysis,* 49(4):603–615, 2013. https://doi.org/10.1007/s10559-013-9547-7.

7

Adaptive Method for the Variational Inequality Problem Over the Set of Solutions of the Equilibrium Problem

Y. Vedel, S. Denisov, and V. Semenov

Faculty of Computer Science and Cybernetics, Taras Shevchenko National University of Kyiv, Kyiv, Ukraine
E-mail: {yana.vedel, denisov.univ, semenov.volodya}@gmail.com

Abstract

In this chapter we consider a two-level variational problem: a variational inequality problem over the set of solutions of the equilibrium problem. An example of such a problem is the search for the normal Nash equilibrium. To solve this problem, we propose a new iterative proximal algorithm that combines the ideas of a two-stage proximal point method, adaptability, and scheme of iterative regularization. In contrast to the previously applied rules for choosing the step size, the proposed algorithm does not calculate any values of the bifunction at additional points; it does not require knowledge of information about the Lipschitz constants of the bifunction, the Lipschitz constant and strong monotonicity constant of the operator. For monotone Lipschitzian bifunctions and strongly monotone Lipschitz continuous operators, a theorem on the strong convergence of the algorithm is proved. It is shown that the proposed algorithm is applicable to monotone two-level variational inequalities in real Hilbert spaces.

Keywords: Two-level problem, variational inequality problem, equilibrium problem, strong convergence, two-stage proximal method, scheme of iterative regularization.

7.1 Introduction

In numerical optimization, and the theory of ill-posed (non-correct) problems, the following technique for solving problems with a non-unique solution is widespread [1–4]. The problem is associated with a family of perturbed auxiliary problems that are uniquely and correctly (well-posed) solvable. A particular solution to the original problem is obtained as the limit of solutions to perturbed auxiliary problems with a decrease in perturbations. The solutions found in this way satisfy certain additional conditions. For example, the minimality of the norm of the normal solution to an optimization problem obtained by the Tikhonov regularization algorithm. On the other hand, in the operation research, optimization problems arise according to sequentially specified criteria (lexicographic, sequential, or multi-stage optimization) [5, 6]. Note that the ideas of the Tikhonov regularization method permeate many studies in the field of automatic control [7–10].

Two-level variational inequalities arose as a natural generalization of well studied lexicographic optimization problems with two criteria, as well as in the analysis of ordinary optimization problems with functional constraints in the form of a monotone variational inequality. As an independent mathematical object, two-level variational inequalities began to be considered in [11]. The papers [12, 13] are devoted to the solvability of more general n-level variational inequalities. By now, approximate methods of one-stage for solution of all these problems are known, which are ideologically close to the methods of penalty and regularization [6, 14].

One of the popular directions of contemporary applied nonlinear analysis is the study of equilibrium problems (equilibrium programming problems) [15–23], to which mathematical programming problems, variational inequalities and many game theory problems can be reduced. Recently, algorithms for solving variational inequalities and equilibrium problems have become popular in the ML community. On their basis it is possible to design effective learning methods for GANs [24–26].

The Chapter deals with a two-level problem: variational inequality problem on the set of solutions to the equilibrium programming problem. A similar problem was considered in [22], where a strongly converging algorithm was proposed that used the operation of calculating the value of the resolvent of a bifunction. The latter significantly increased the complexity of the algorithm. Below, to solve this problem, an adaptive version of the previously studied algorithm [23] based on the two-stage proximal point method [20, 21] and the idea of scheme of iterative regularization [1, 27] will be proposed. The proposed algorithm does not calculate the values of the bifunction at any additional points; it does not require knowledge of information about the Lipschitzian constants of the bifunction, the constants of Lipschitz and strong monotonicity of the operator.

For monotone bifunctions of Lipschitzian type and strongly monotone Lipschitz continuous operators, a theorem on the strong convergence of the algorithm is proved. It is shown that the proposed algorithm is applicable to monotone two-level variational inequalities in real Hilbert spaces.

The results obtained are useful in the theory of optimal control. Indeed, the problem of optimal control of the system can be formulated as a variational inequality problem. In the case of non-uniqueness of the optimal control, an additional criterion can be introduced for the regularization of the problem and a two-level variational in equality can be obtained. And to solve the obtained problem, we use the algorithm proposed and studied in this Chapter.

7.2 Problem Formulation

Everywhere below H is real infinite-dimensional Hilbert space with inner product $(\cdot, \cdot)$ and induced norm $\|\cdot\|$. For an operator $T : H \to H$, set $D \subseteq H$, and bifunction $G : H \times H \to R$ we denote $VI(T, D)$ and $EP(G, D)$ sets

$$\{\xi \in D : \ (T\xi, \eta - \xi) \geq 0 \ \forall \eta \in D\},$$

$$\{\xi \in D : \ G(\xi, \eta) \geq 0 \ \forall \eta \in D\},$$

respectively.

For closed convex set $C \subseteq H$ consider two-level problem:

$$\text{find } x \in VI(F, EP(G, C)). \tag{7.1}$$

Assume the following conditions hold [23]:

C1) $G(x, x) = 0 \ \forall x \in C$;
C2) $G(x, y) + G(y, x) \leq 0 \ \forall x, y \in C$(monotonicity);
C3) $\forall x \in C$ functional $G(x, \cdot)$ is closed and convex on C;
C4) $\forall y \in C$ functional $-G(\cdot, y)$ is weakly closed on C;
C5) for all $x, y, z \in C$ holds

$$G(x, y) \leq G(x, z) + G(z, y) + a \|x - z\|^2 + b \|z - y\|^2,$$

where a, b are positive constants (Lipschitz continuity);

C6) $EP(G, C) \neq \varnothing$;

C7) $F : C \to H$–μ-strongly monotone and L-Lipschitz continuous operator.

Remark 7.1. Condition C5 is provided by Giandomenico Mastroeni in [16]. For example, bifunction

$$G(x, y) = (Tx, y - x)$$

with Lipschitz continuous operator $T : C \to H$ satisfies C5 with $a = \frac{L\varepsilon}{2}$ and $b = \frac{L}{2\varepsilon}$ where $L > 0$ is Lipschitzian constant of operator T, $\varepsilon > 0$.

With given conditions the set $EP(G, C)$ is closed and convex and problem (7.1) has unique solution $\bar{x} \in C$ [17].

Remark 7.2. The most known case of (7.1) is search of normal solution of variational inequality (with $Fx = 2x$, $G(x, y) = (Tx, y - x)$ where $T : C \to H$):

$$\|x\|^2 \to \min, x \in C : (Tx, y - x) \geq 0 \ \forall y \in C.$$

Let us remind one important definition. Let $\varphi : H \to R \cup \{+\infty\}$ be proper convex closed functional. Proximal operator associated with functional φ is operator

$$\text{prox}_\varphi x = \arg \min_{\eta \in \text{dom} \, \varphi} \left(\varphi(\eta) + \frac{1}{2} \|\eta - x\|^2 \right).$$

This operator is firmly non-expansive and

$$\varphi(\eta) - \varphi(z) + (z - x, \eta - z) \geq 0$$

for all $\eta \in \text{dom} \, \varphi$ if and only if $z = \text{prox}_\varphi x$ [28].

Below we provide strongly converging iterative algorithm for solving the two-level variational problem (7.1) that does not require knowledge of constants from conditions C5, C7. First, we approximate the two-level problem (7.1) by one-level and more regular problem of equilibrium programming.

7.3 Tikhonov–Browder Approximation

Consider auxiliary equilibrium problem depending on small parameter $\varepsilon > 0$:

$$\text{find} \quad x \in C : G(x, y) + \varepsilon (Fx, y - x) \geq 0 \ \forall y \in C. \tag{7.2}$$

Following Bakushinskii [1], we call the equilibrium problem (7.2) the Tikhonov–Browder approximation of two-level problem (7.1).

Remark 7.3. For solving incorrect extremum problems that type of approximation was proposed by Tikhonov for building regularization methods. Later Felix Browder [2, 3] applied similar regularization technique.

From results of [17] follows the existence and uniqueness of solution $x_\varepsilon \in C$ of problem (7.2) for all $\varepsilon > 0$. Auxiliary elements $x_\varepsilon \in C$ have several properties important for subsequent constructions.

Lemma 7.1 ([23]). The following inequalities take place:

1) $\|x_\varepsilon\| \leq \frac{1}{\mu}\left(1 + \frac{1}{\mu}L\right)\|F\bar{x}\| \, \forall \varepsilon > 0;$

2) $\|x_\varepsilon - x_\delta\| \leq |\varepsilon - \delta|\frac{1}{\varepsilon\mu}\left(1 + L\frac{1}{\mu}\right)\|F\bar{x}\| \, \forall \varepsilon, \delta > 0.$

Proof. Let $\varepsilon > 0$. For $x_\varepsilon \in C$ and an arbitrary element $w \in EP(G, C)$, we have

$$G(x_\varepsilon, w) + \varepsilon(Fx_\varepsilon, w - x_\varepsilon) \geq 0 \text{ and } G(w, x_\varepsilon) \geq 0.$$

Adding the inequalities and using the monotonicity of the bifunction G, we obtain

$$(Fx_\varepsilon, w - x_\varepsilon) \geq 0,$$

that is,

$$(Fw, w - x_\varepsilon) \geq (Fx_\varepsilon - Fw, x_\varepsilon - w).$$

The strong monotonicity of the operator F and the Cauchy–Schwarz inequality give

$$\mu \|w - x_\varepsilon\| \leq \|Fx_\varepsilon\|,$$

whence 1) follows.

Let us prove 2). Let $x_\varepsilon \in C$ and $x_\delta \in C$ be solutions of problem (7.2) with $\varepsilon > 0$ and $\delta > 0$, respectively. We have

$$G(x_\varepsilon, x_\delta) + \varepsilon(Fx_\varepsilon, x_\delta - x_\varepsilon) \geq 0 \text{ and } G(x_\delta, x_\varepsilon) + \delta(Fx_\delta, x_\varepsilon - x_\delta) \geq 0.$$

Adding the inequalities and using the monotonicity of the bifunction G, we obtain

$$\varepsilon(Fx_\varepsilon, x_\delta - x_\varepsilon) + \delta(Fx_\delta, x_\varepsilon - x_\delta) \geq 0.$$

Let us rewrite the last inequality us

$$(\delta - \varepsilon)(Fx_\delta, x_\varepsilon - x_\delta) \geq \varepsilon(Fx_\varepsilon - Fx_\delta, x_\varepsilon - x_\delta).$$

Using the strong monotonicity of the operator F, we obtain

$$|\delta - \varepsilon| \, \|Fx_\delta\| \, \|x_\varepsilon - x_\delta\| \geq \varepsilon\mu \, \|x_\varepsilon - x_\delta\|^2 \,,$$

that is,

$$\|x_\varepsilon - x_\delta\| \leq \frac{|\delta - \varepsilon|}{\varepsilon\mu} \|Fx_\delta\| \,.$$

Let us estimate from above the norm of Fx_δ

$$\|Fx_\delta\| \leq \|F\bar{x}\| + \|Fx_\delta - F\bar{x}\| \leq \|F\bar{x}\| + L \, \|x_\delta - \bar{x}\| \leq \left(1 + \frac{L}{\mu}\right) \|F\bar{x}\| \,.$$

Using the last inequality in the estimate for $\|x_\varepsilon - x_\delta\|$, we obtain 2).   ■

Lemma 7.2 ([23]). Assume that conditions C1–C4 and C6, C7 are satisfied. Then

$$\lim_{\varepsilon \to 0} \|x_\varepsilon - \bar{x}\| = 0.$$

Proof. By first inequality of Lemma 7.1, from $\{x_\varepsilon\}_{\varepsilon > 0}$ we can extract a sequence (x_{ε_n}) $(\varepsilon_n \to 0)$ that weakly converges to an element $v \in C$. Using the weak upper semicontinuity of the functional $G(\cdot, y)$, we pass to the limit in the inequality

$$G(x_{\varepsilon_n}, y) + \varepsilon_n (Fx_{\varepsilon_n}, y - x_{\varepsilon_n}) \geq 0 \ \forall y \in C.$$

We get

$$G(v, y) \geq 0 \ \forall y \in C,$$

that is, $v \in EP(G, C)$. And passing to the limit in inequality

$$(Fw, w - x_{\varepsilon_n}) \geq (Fx_{\varepsilon_n}, w - x_{\varepsilon_n}) \geq 0 \ \forall w \in EP(G, C),$$

we obtain

$$(Fw, w - v) \geq 0 \ \forall w \in EP(G, C),$$

that is, $v = \bar{x}$. Let us show that

$$\lim_{\varepsilon \to 0} \|x_\varepsilon - \bar{x}\| = 0.$$

This follows from inequality

$$(F\bar{x}, \bar{x} - x_{\varepsilon_n}) \geq (Fx_{\varepsilon_n} - F\bar{x}, x_{\varepsilon_n} - \bar{x}) \geq \mu \, \|x_{\varepsilon_n} - \bar{x}\|^2 \,.$$

From the uniqueness of the element $\bar{x} \in H$ we obtain $\lim_{\varepsilon \to 0} \|x_\varepsilon - \bar{x}\| = 0.$■

7.4 Algorithm

Following the well-known scheme of two-step proximal point method [20, 21] for solving problem (7.1) in recent work [23] the following method was introduced:

$$
\begin{cases}
z_n = x_n - \alpha_n \lambda_n F x_n, \\
y_n = \mathrm{prox}_{\lambda_n G(y_{n-1}, \cdot)} z_n, \\
x_{n+1} = \mathrm{prox}_{\lambda_n G(y_n, \cdot)} z_n,
\end{cases}
\tag{7.3}
$$

where λ_n were defined considering $\lambda_n \in \left[\underline{\lambda}, \overline{\lambda}\right] \subseteq \left(0, \frac{1}{2(2a+b)}\right)$ and positive sequence (α_n) was such that

$$
\lim_{n \to \infty} \alpha_n = 0, \ \sum_{n=1}^{\infty} \alpha_n = +\infty, \ \lim_{n \to \infty} \frac{\alpha_{n+1} - \alpha_n}{\alpha_n^2} = 0.
$$

For getting rid of explicit usage of information about values of constants a and b with defined limits for λ_n in (7.3) we consider the following algorithm with adaptiveness in choice of parameters λ_n.

Algorithm 7.1.
Initialization. Choose elements $x_1, y_0 \in C$, and $\tau \in \left(0, \frac{1}{3}\right)$, $\lambda_1 \in (0, +\infty)$. Let $n = 1$.
Step 1. Calculate

$$
z_n = x_n - \alpha_n \lambda_n F x_n.
$$

Step 2. Calculate

$$
y_n = \mathrm{prox}_{\lambda_n G(y_{n-1}, \cdot)} z_n.
$$

Step 3. Calculate

$$
x_{n+1} = \mathrm{prox}_{\lambda_n G(y_n, \cdot)} z_n.
$$

Step 4. Calculate

$$
\lambda_{n+1} =
\begin{cases}
\lambda_n, & \text{if} \quad G\left(y_{n-1}, x_{n+1}\right) - G\left(y_{n-1}, y_n\right) - G\left(y_n, x_{n+1}\right) \leq 0, \\
\min\left\{\lambda_n, \frac{\tau}{2} \frac{\|y_{n-1} - y_n\|^2 + \|y_n - x_{n+1}\|^2}{(G(y_{n-1}, x_{n+1}) - G(y_{n-1}, y_n) - G(y_n, x_{n+1}))}\right\}, & \text{otherwise.}
\end{cases}
$$

Let $n := n + 1$ and go to step 1.

In the proposed iterative algorithm, parameter λ_{n+1} depends on locations of points y_{n-1}, y_n, x_{n+1} and values of $G\left(y_{n-1}, x_{n+1}\right)$, $G\left(y_{n-1}, y_n\right)$, and $G\left(y_n, x_{n+1}\right)$. We do not use any information about constants a and b from the inequality. Obviously, the sequence (λ_n) is decreasing. Also, it is lower bounded by $\min \left\{ \lambda_1, \tau 2^{-1} \left(\max \{a, b\} \right)^{-1} \right\}$. We have

$$G\left(y_{n-1}, x_{n+1}\right) - G\left(y_{n-1}, y_n\right) - G\left(y_n, x_{n+1}\right)$$

$$\leq \max\{a, b\} \left(\|y_{n-1} - y_n\|^2 + \|y_n - x_{n+1}\|^2 \right).$$

Regarding positive parameters α_n we assume that the next conditions are satisfied:

D1) $\lim\limits_{n \to \infty} \alpha_n = 0$;

D2) $\sum_{n=1}^{\infty} \alpha_n = +\infty$;

D3) $\lim\limits_{n \to \infty} \frac{\alpha_{n+1} - \alpha_n}{\alpha_n^2} = 0$.

Remark 7.4. For (α_n) we can use $\alpha_n = \frac{1}{n^p}, p \in (0, 1)$.

7.5 Proof of Algorithm Convergence

We provide proof of convergence with the following simple scheme. Let x_{α_n} be a solution of auxiliary problem (7.2) for $\varepsilon = \alpha_n$. Since

$$\|x_n - \bar{x}\| \leq \|x_n - x_{\alpha_n}\| + \|x_{\alpha_n} - \bar{x}\|,$$

$$\lim\limits_{n \to \infty} \|x_{\alpha_n} - \bar{x}\| = 0$$

it is enough to show that the generated by Algorithm 7.1 sequence (x_n) has the next property

$$\lim\limits_{n \to \infty} \|x_n - x_{\alpha_n}\| = 0.$$

Let us remind the well-known elementary lemma about recurrent numerical inequalities.

Lemma 7.3 ([1]). Let the non-negative numerical sequence (ψ_n) satisfies the recurrent inequality

$$\psi_{n+1} \leq (1 - \alpha_n) \psi_n + \beta_n \text{ for all } n \in N$$

where numerical sequences (α_n) and (β_n) have the following conditions $\alpha_n \in (0, 1), \sum \alpha_n = +\infty, \varlimsup\limits_{n \to \infty} \frac{\beta_n}{\alpha_n} \leq 0$. Then $\lim\limits_{n \to \infty} \psi_n = 0$.

Let us prove an important inequality for generated sequences (x_n), (y_n) and auxiliary elements x_{α_n}.

Lemma 7.4. For sequences generated by Algorithm 7.1 (x_n), (y_n) and auxiliary elements x_{α_n} the inequality holds

$$\|x_{n+1} - x_{\alpha_n}\|^2 \leq (1 - \alpha_n \lambda_n \mu) \|x_n - x_{\alpha_n}\|^2 - \left(1 - \tau \frac{\lambda_n}{\lambda_{n+1}}\right) \|x_{n+1} - y_n\|^2$$

$$- \left(1 - 2\tau \frac{\lambda_n}{\lambda_{n+1}} - \alpha_n \lambda_n L^2 \mu^{-1}\right) \|y_n - x_n\|^2$$

$$+ 2\tau \frac{\lambda_n}{\lambda_{n+1}} \|x_n - y_{n-1}\|^2. \tag{7.4}$$

Proof. We have

$$\|x_{n+1} - x_{\alpha_n}\|^2 = \|x_n - x_{\alpha_n}\|^2 - \|x_n - x_{n+1}\|^2$$
$$+ 2 (x_{n+1} - x_n, x_{n+1} - x_{\alpha_n}) = \|x_n - x_{\alpha_n}\|^2 - \|x_n - y_n\|^2$$
$$- \|y_n - x_{n+1}\|^2 - 2 (x_n - y_n, y_n - x_{n+1})$$
$$+ 2 (x_{n+1} - x_n, x_{n+1} - x_{\alpha_n}). \tag{7.5}$$

From the definition of points x_{n+1} and y_n it follows that

$$\lambda_n G (y_n, x_{\alpha_n}) - \lambda_n G (y_n, x_{n+1})$$
$$\geq (x_{n+1} - x_n + \alpha_n \lambda_n F x_n, x_{n+1} - x_{\alpha_n}), \tag{7.6}$$
$$\lambda_n G (y_{n-1}, x_{n+1}) - \lambda_n G (y_{n-1}, y_n)$$
$$\geq - (x_n - \alpha_n \lambda_n F x_n - y_n, y_n - x_{n+1}). \tag{7.7}$$

Using inequalities (7.6), (7.7) for estimating inner products in (7.5), we get

$$\|x_{n+1} - x_{\alpha_n}\|^2 \leq \|x_n - x_{\alpha_n}\|^2 - \|x_n - y_n\|^2 - \|y_n - x_{n+1}\|^2$$
$$+ 2\lambda_n (G (y_n, x_{\alpha_n}) - G (y_n, x_{n+1}) + G (y_{n-1}, x_{n+1}) - G (y_{n-1}, y_n))$$
$$+ 2\alpha_n \lambda_n (F x_n, x_{\alpha_n} - y_n). \tag{7.8}$$

From the rule of calculation of λ_{n+1} it follows the inequality

$$G (y_{n-1}, x_{n+1}) - G (y_{n-1}, y_n) - G (y_n, x_{n+1})$$
$$\leq \tau \frac{1}{2\lambda_{n+1}} \left(\|y_{n-1} - y_n\|^2 + \|x_{n+1} - y_n\|^2\right). \tag{7.9}$$

For estimating an expression $G (y_{n-1}, x_{n+1}) - G (y_{n-1}, y_n) - G (y_n, x_{n+1})$ in (7.8) we use (7.9). We obtain

$$\|x_{n+1} - x_{\alpha_n}\|^2 \leq \|x_n - x_{\alpha_n}\|^2 - \|x_n - y_n\|^2 - \|y_n - x_{n+1}\|^2$$

$$+\tau\frac{\lambda_n}{\lambda_{n+1}}\left\|y_{n-1}-y_n\right\|^2+\tau\frac{\lambda_n}{\lambda_{n+1}}\left\|x_{n+1}-y_n\right\|^2$$

$$+2\lambda_n G\left(y_n,x_{\alpha_n}\right)+2\alpha_n\lambda_n\left(Fx_n,x_{\alpha_n}-y_n\right).$$

We estimate $\|y_{n-1}-y_n\|^2$ in the following way

$$\left\|y_{n-1}-y_n\right\|^2\le 2\left\|y_{n-1}-x_n\right\|^2+2\left\|x_n-y_n\right\|^2.$$

We have

$$\left\|x_{n+1}-x_{\alpha_n}\right\|^2\le\left\|x_n-x_{\alpha_n}\right\|^2-\left\|x_n-y_n\right\|^2-\left\|y_n-x_{n+1}\right\|^2$$

$$+2\tau\frac{\lambda_n}{\lambda_{n+1}}\left\|y_{n-1}-x_n\right\|^2+2\tau\frac{\lambda_n}{\lambda_{n+1}}\left\|x_n-y_n\right\|^2+\tau\frac{\lambda_n}{\lambda_{n+1}}\left\|x_{n+1}-y_n\right\|^2$$

$$+2\lambda_n G\left(y_n,x_{\alpha_n}\right)+2\alpha_n\lambda_n\left(Fx_n,x_{\alpha_n}-y_n\right). \tag{7.10}$$

From monotonicity of bifunction G it follows $G\left(y_n,\ x_{\alpha_n}\right)\le G\left(x_{\alpha_n},y_n\right)$ hence

$$G\left(y_n,\ x_{\alpha_n}\right)-\alpha_n\left(Fx_{\alpha_n},y_n-x_{\alpha_n}\right)\le -G\left(x_{\alpha_n},y_n\right)-\alpha_n\left(Fx_{\alpha_n},y_n-x_{\alpha_n}\right).$$

Since

$$G\left(x_{\alpha_n},y_n\right)+\alpha_n\left(Fx_{\alpha_n},y_n-x_{\alpha_n}\right)\ge 0$$

then

$$G\left(y_n,\ x_{\alpha_n}\right)\le\alpha_n\left(Fx_{\alpha_n},y_n-x_{\alpha_n}\right).$$

Taking into account the last estimation in (7.10) we obtain

$$\left\|x_{n+1}-x_{\alpha_n}\right\|^2\le\left\|x_n-x_{\alpha_n}\right\|^2-\left\|x_n-y_n\right\|^2-\left\|y_n-x_{n+1}\right\|^2$$

$$+2\tau\frac{\lambda_n}{\lambda_{n+1}}\left\|y_{n-1}-x_n\right\|^2+2\tau\frac{\lambda_n}{\lambda_{n+1}}\left\|x_n-y_n\right\|^2+\tau\frac{\lambda_n}{\lambda_{n+1}}\left\|x_{n+1}-y_n\right\|^2$$

$$+2\alpha_n\lambda_n\left(Fx_n-Fx_{\alpha_n},x_{\alpha_n}-y_n\right). \tag{7.11}$$

Let us estimate the upper bound for $\left(Fx_n-Fx_{\alpha_n},x_{\alpha_n}-y_n\right)$. We have

$$\left(Fx_n-Fx_{\alpha_n},x_{\alpha_n}-y_n\right)=\left(Fx_n-Fx_{\alpha_n},x_{\alpha_n}-x_n\right)$$

$$+\left(Fx_n-Fx_{\alpha_n},x_n-y_n\right)\le -\mu\left\|x_{\alpha_n}-x_n\right\|^2$$

$$+L\left\|x_n-x_{\alpha_n}\right\|\left\|x_n-y_n\right\|\le -\mu\left\|x_{\alpha_n}-x_n\right\|^2+\frac{\mu}{2}\left\|x_n-x_{\alpha_n}\right\|^2$$

$$+\frac{L^2}{2\mu}\left\|x_n-y_n\right\|^2=-\frac{\mu}{2}\left\|x_n-x_{\alpha_n}\right\|^2+\frac{L^2}{2\mu}\left\|x_n-y_n\right\|^2. \tag{7.12}$$

From inequalities (7.11) and (7.12) we obtain

$$\left\| x_{n+1} - x_{\alpha_n} \right\|^2 \leq \left(1 - \alpha_n \lambda_n \mu \right) \left\| x_n - x_{\alpha_n} \right\|^2 - \left(1 - \tau \lambda_n \lambda_{n+1}^{-1} \right)$$
$$\times \left\| x_{n+1} - y_n \right\|^2 - \left(1 - 2\tau \lambda_n \lambda_{n+1}^{-1} - \alpha_n \lambda_n L^2 \mu^{-1} \right)$$
$$\times \left\| y_n - x_n \right\|^2 + 2\tau \lambda_n \lambda_{n+1}^{-1} \left\| x_n - y_{n-1} \right\|^2 ,$$

Q.E.D. ∎

Let us prove the estimation from which it follows the convergence to 0 of sequences $\left(\left\| x_n - x_{\alpha_n} \right\| \right)$ and $\left(\left\| x_n - y_{n-1} \right\| \right)$.

Lemma 7.5. For sequences (x_n), (y_n) generated by Algorithm 7.1 and auxiliary elements x_{α_n} for large n the inequality is hold

$$\left\| x_{n+1} - x_{\alpha_{n+1}} \right\|^2 + \frac{2\tau \lambda_{n+1} \lambda_{n+2}^{-1}}{1 - \alpha_{n+1} \lambda_{n+1} \mu} \left\| x_{n+1} - y_n \right\|^2$$
$$\leq \left(1 - \frac{\alpha_n \lambda_n \mu}{2} \right) \left(\left\| x_n - x_{\alpha_n} \right\|^2 + \frac{2\tau \lambda_n \lambda_{n+1}^{-1}}{1 - \alpha_n \lambda_n \mu} \left\| x_n - y_{n-1} \right\|^2 \right)$$
$$+ \frac{2M}{\lambda_n \mu} \frac{\left(\alpha_{n+1} - \alpha_n \right)^2}{\alpha_n^3} , \tag{7.13}$$

where $M = \mu^{-1} \left(1 + L\mu^{-1} \right) \left\| F\bar{x} \right\|$.

Proof. We have

$$\left\| x_{n+1} - x_{\alpha_n} \right\|^2 = \left\| x_{n+1} - x_{\alpha_{n+1}} \right\|^2 + \left\| x_{\alpha_{n+1}} - x_{\alpha_n} \right\|^2$$
$$+ 2 \left(x_{n+1} - x_{\alpha_{n+1}}, x_{\alpha_{n+1}} - x_{\alpha_n} \right) \geq \left\| x_{n+1} - x_{\alpha_{n+1}} \right\|^2 + \left\| x_{\alpha_{n+1}} - x_{\alpha_n} \right\|^2$$
$$- 2 \left\| x_{n+1} - x_{\alpha_{n+1}} \right\| \left\| x_{\alpha_{n+1}} - x_{\alpha_n} \right\| \geq (1 - \varepsilon) \left\| x_{n+1} - x_{\alpha_{n+1}} \right\|^2$$
$$+ \frac{\varepsilon - 1}{\varepsilon} \left\| x_{\alpha_{n+1}} - x_{\alpha_n} \right\|^2 , \tag{7.14}$$

where $\varepsilon > 0$. Let in (7.14) $\varepsilon = \frac{1}{2} \alpha_n \lambda_n \mu$. We get

$$\left\| x_{n+1} - x_{\alpha_n} \right\|^2 \geq \frac{2 - \alpha_n \lambda_n \mu}{2} \left\| x_{n+1} - x_{\alpha_{n+1}} \right\|^2$$
$$- \frac{2 - \alpha_n \lambda_n \mu}{\alpha_n \lambda_n \mu} \left\| x_{\alpha_{n+1}} - x_{\alpha_n} \right\|^2 . \tag{7.15}$$

With rules for values of algorithm's parameters α_n, λ_n for large numbers n we have $1 - \alpha_n\lambda_n\mu > 0$. Taking into consideration the second inequality of Lemma 7.1 from (7.15) we obtain

$$\|x_{n+1} - x_{\alpha_n}\|^2 \geq \left(1 - \frac{\alpha_n\lambda_n\mu}{2}\right)\|x_{n+1} - x_{\alpha_{n+1}}\|^2$$
$$- \frac{(2 - \alpha_n\lambda_n\mu)}{\alpha_n\lambda_n\mu}(\alpha_{n+1} - \alpha_n)^2\,\alpha_n^{-2}\mu^{-1}\left(1 + L\mu^{-1}\right)\|F\bar{x}\|,$$

$$(7.16)$$

for all numbers $n \geq n_0$. Using (7.16) in (7.4) we get (for numbers $n \geq n_0$)

$$\frac{2 - \alpha_n\lambda_n\mu}{2}\|x_{n+1} - x_{\alpha_{n+1}}\|^2 \leq (1 - \alpha_n\lambda_n\mu)\|x_n - x_{\alpha_n}\|^2$$
$$- \left(1 - \tau\lambda_n\lambda_{n+1}^{-1}\right)\|x_{n+1} - y_n\|^2 - \left(1 - 2\tau\lambda_n\lambda_{n+1}^{-1} - \alpha_n\lambda_n L^2\mu^{-1}\right)$$
$$\times \|y_n - x_n\|^2 + 2\tau\lambda_n\lambda_{n+1}^{-1}\|x_n - y_{n-1}\|^2 + \frac{(2 - \alpha_n\lambda_n\mu)}{\alpha_n\lambda_n\mu}\frac{(\alpha_{n+1} - \alpha_n)^2}{\alpha_n^2}M,$$

where $M = \mu^{-1}\left(1 + L\mu^{-1}\right)\|F\bar{x}\|$. Hence it follows

$$\|x_{n+1} - x_{\alpha_{n+1}}\|^2 \leq \frac{2 - 2\alpha_n\lambda_n\mu}{2 - \alpha_n\lambda_n\mu}\|x_n - x_{\alpha_n}\|^2 - \frac{2\left(1 - \tau\lambda_n\lambda_{n+1}^{-1}\right)}{2 - \alpha_n\lambda_n\mu}$$
$$\times \|x_{n+1} - y_n\|^2 - \frac{2\left(1 - 2\tau\lambda_n\lambda_{n+1}^{-1} - \alpha_n\lambda_n L^2\mu^{-1}\right)}{2 - \alpha_n\lambda_n\mu}\|y_n - x_n\|^2$$
$$+ \frac{4\tau\lambda_n\lambda_{n+1}^{-1}}{2 - \alpha_n\lambda_n\mu}\|x_n - y_{n-1}\|^2 + \frac{2M}{\lambda_n\mu}\frac{(\alpha_{n+1} - \alpha_n)^2}{\alpha_n^3}.$$

$$(7.17)$$

Regrouping the variables in (7.17) we obtain

$$\|x_{n+1} - x_{\alpha_{n+1}}\|^2 + 2\tau\frac{\lambda_{n+1}}{\lambda_{n+2}}\frac{1}{1 - \alpha_{n+1}\lambda_{n+1}\mu}\|x_{n+1} - y_n\|^2$$
$$\leq \frac{2 - 2\alpha_n\lambda_n\mu}{2 - \alpha_n\lambda_n\mu}\left(\|x_n - x_{\alpha_n}\|^2 + 2\tau\frac{\lambda_n}{\lambda_{n+1}}\frac{1}{1 - \alpha_n\lambda_n\mu}\|x_n - y_{n-1}\|^2\right)$$
$$- \frac{2\left(1 - 2\tau\lambda_n\lambda_{n+1}^{-1} - \alpha_n\lambda_n L^2\mu^{-1}\right)}{2 - \alpha_n\lambda_n\mu}\|y_n - x_n\|^2$$
$$- \left(\frac{1 - \tau\lambda_n\lambda_{n+1}^{-1}}{1 - \frac{1}{2}\alpha_n\lambda_n\mu} - \frac{2\tau\lambda_{n+1}\lambda_{n+2}^{-1}}{1 - \alpha_{n+1}\lambda_{n+1}\mu}\right)\|x_{n+1} - y_n\|^2.$$

$$+ \frac{2M}{\lambda_n \mu} \frac{(\alpha_{n+1} - \alpha_n)^2}{\alpha_n^3}. \tag{7.18}$$

Since $\tau \in \left(0, \frac{1}{3}\right)$ there exists $\lim\limits_{n \to \infty} \lambda_n > 0$ and $\lim\limits_{n \to \infty} \alpha_n = 0$, then starting from some number n_1 the inequalities

$$\frac{1 - 2\tau \frac{\lambda_n}{\lambda_{n+1}} - \alpha_n \lambda_n L^2 \mu^{-1}}{2 - \alpha_n \lambda_n \mu} > 0,$$

$$\frac{1 - \tau \frac{\lambda_n}{\lambda_{n+1}}}{1 - \frac{1}{2}\alpha_n \lambda_n \mu} - \frac{2\tau \frac{\lambda_{n+1}}{\lambda_{n+2}}}{1 - \alpha_{n+1} \lambda_{n+1} \mu} > 0,$$

$$\frac{2 - 2\alpha_n \lambda_n \mu}{2 - \alpha_n \lambda_n \mu} < 1 - \frac{\alpha_n \lambda_n \mu}{2},$$

will be satisfied. So for all $n \geq N = \max\{n_0, n_1\}$ from (7.18) it follows

$$\|x_{n+1} - x_{\alpha_{n+1}}\|^2 + \frac{2\tau \lambda_{n+1} \lambda_{n+2}^{-1}}{1 - \alpha_{n+1} \lambda_{n+1} \mu} \|x_{n+1} - y_n\|^2$$

$$\leq \left(1 - \frac{\alpha_n \lambda_n \mu}{2}\right) \left(\|x_n - x_{\alpha_n}\|^2 + \frac{2\tau \lambda_n \lambda_{n+1}^{-1}}{1 - \alpha_n \lambda_n \mu} \|x_n - y_{n-1}\|^2\right)$$

$$+ \frac{2M}{\lambda_n \mu} \frac{(\alpha_{n+1} - \alpha_n)^2}{\alpha_n^3},$$

Q.E.D. ∎

Let us formulate the main result of Chapter.

Theorem 7.1. Let the conditions C1–C7 and D1–D3 are satisfied. Then for two sequences (x_n), (y_n) generated by iterative Algorithm 7.1 takes place

$$\lim_{n \to \infty} \|x_n - \bar{x}\| = \lim_{n \to \infty} \|y_n - \bar{x}\| = 0, \tag{7.19}$$

where $\bar{x} \in H$ is the unique solution of problem (7.1).

Proof. From Lemma 7.3 and inequality (7.13) we have

$$\|x_n - x_{\alpha_n}\|^2 + 2\tau \frac{\lambda_n}{\lambda_{n+1}} (1 - \alpha_n \lambda_n \mu)^{-1} \|x_n - y_{n-1}\|^2 \to 0, n \to \infty.$$

Hence

$$\lim_{n \to \infty} \|x_n - x_{\alpha_n}\| = \lim_{n \to \infty} \|y_{n-1} - x_n\| = 0. \tag{7.20}$$

From inequality

$$\|x_n - \bar{x}\| \le \|x_n - x_{\alpha_n}\| + \|x_{\alpha_n} - \bar{x}\|,$$

Lemma 7.2 and (7.20) we obtain (7.19). ∎

Remark 7.5. Easy to see that sequence (z_n) $(z_n = x_n - \alpha_n \lambda_n F x_n)$ also converges strongly to point $\bar{x} \in C$.

7.6 Algorithm for Two-Level Variational Inequalities

Now we consider a particular case of two-level variational problem (7.1): two-level variational inequality problem in real Hilbert space H:

$$\text{find } x \in VI\left(F_2, VI\left(F_1, C\right)\right) \tag{7.21}$$

We assume that the following conditions are satisfied: set $C \subseteq H$ is closed and convex; operator $F_1 : C \to H$ is Lipschitz continuous and monotone; set $VI\left(F_1, C\right)$ is non-empty; operator $F_2 : C \to H$ is strongly monotone and Lipschitz continuous. Let P_C be a metric projection operator on closed convex set C i.e. $P_C x$ is an unique element of C with property

$$\|P_C x - x\| = \min_{z \in C} \|z - x\|.$$

For problem (7.21) Algorithm 7.1 takes the following form.

Algorithm 7.2.

Initialization. Choose elements $x_1, y_0 \in C$, and numbers $\tau \in \left(0, \frac{1}{3}\right)$, $\lambda_1 \in (0, +\infty)$. Let $n = 1$.
Step 1. Calculate

$$z_n = x_n - \alpha_n \lambda_n F_2 x_n.$$

Step 2. Calculate

$$y_n = P_C\left(z_n - \lambda_n F_1 y_{n-1}\right).$$

Step 3. Calculate

$$x_{n+1} = P_C\left(z_n - \lambda_n F_1 y_n\right).$$

Step 4. Calculate

$$\lambda_{n+1} = \begin{cases} \lambda_n, & \text{if } \left(F_1 y_{n-1} - F_1 y_n, x_{n+1} - y_n\right) \le 0, \\ \min\left\{\lambda_n, \dfrac{\tau}{2} \dfrac{\|y_{n-1} - y_n\|^2 + \|x_{n+1} - y_n\|^2}{\left(F_1 y_{n-1} - F_1 y_n, x_{n+1} - y_n\right)}\right\}, & \text{otherwise.} \end{cases}$$

Let $n := n + 1$ and go to step 1.

From Theorem 7.1 the next result follows.

Theorem 7.2. Let H be a real Hilbert space, $C \subseteq H$ is non-empty closed convex set; operator $F_1 : C \to H$ is Lipschitz continuous and monotone; set $VI(F_1, C)$ is non-empty; operator $F_2 : C \to H$ is strongly monotone and Lipschitz continuous; conditions D1–D3 are satisfied. Then two sequences (x_n) and (y_n) generated by Algorithm 7.2 strongly converge to the unique solution of problem (7.21).

Remark 7.6. Like theorem 7.2 results takes place for next modification of Algorithm 7.2 changing the rule of calculation λ_n on

$$\lambda_{n+1} = \begin{cases} \lambda_n, & \text{if} \quad F_1 y_{n-1} - F_1 y_n \neq 0, \\ \min\left\{\lambda_n, \tau \dfrac{\|y_{n-1} - y_n\|}{\|F_1 y_{n-1} - F_1 y_n\|}\right\}, & \text{otherwise,} \end{cases}$$

where $\tau \in \left(0, \frac{1}{3}\right)$.

Remark 7.7. We can build an economic modification of Algorithm 7.2. We change step 3, letting [19]

$$x_{n+1} = P_{C_n}\left(z_n - \lambda_n F_1 y_n\right),$$

where

$$C_n = \left\{z \in H : \ (z_n - \lambda_n F_1 y_{n-1} - y_n, z - y_n) \leq 0\right\}.$$

7.7 Conclusion

In this chapter we consider a two-level variational problem: a variational inequality problem over the set of solutions of the equilibrium problem [22, 23]. To solve this two-level problem, an iterative proximal method is proposed that combines the ideas of a two-stage proximal point method [20, 21], adaptation and scheme of iterative regularization [1, 27]. The algorithm has the structure:

$$\begin{cases} z_n = x_n - \alpha_n \lambda_n F x_n, \\ y_n = \arg\min_{y \in C}\left(\lambda_n G\left(y_{n-1}, y\right) + \frac{1}{2}\|y - z_n\|^2\right), \\ x_{n+1} = \arg\min_{y \in C}\left(\lambda_n G\left(y_n, y\right) + \frac{1}{2}\|y - z_n\|^2\right), \end{cases}$$

where we choose $\lambda_n > 0$ adaptively, and positive sequence (α_n) is such that $\lim\limits_{n\to\infty} \alpha_n = 0$, $\sum_{n=1}^{\infty} \alpha_n = +\infty$ and $\lim\limits_{n\to\infty} \frac{\alpha_{n+1}-\alpha_n}{\alpha_n^2} = 0$.

In contrast to the rules for choosing the step size used earlier [23], the proposed proximal algorithm does not calculate the values of the bifunction at any additional points and does not require knowledge of information about the Lipschitzian constants of the bifunction.

For monotone bifunctions of Lipschitzian type and strongly monotone Lipschitz continuous operators, a theorem on the strong convergence of the algorithm is proved. The proof is based on the results and techniques of [20, 21, 23, 27, 29].

In the future, it is planned to study the possibility of replacing the proximal steps with constructive instructions such as the steps of the subgradient method. It is also interesting to relax the rather constraining condition of monotonicity of the bifunction.

The results obtained are useful in the theory of optimal control. Indeed, the problem of optimal control of the system can be formulated as a variational inequality. In the case of non-uniqueness of the optimal control, an additional criterion can be introduced for the regularization of the problem and a two-level variational inequality can be obtained. And to solve the obtained problem, we use the algorithm proposed and studied in this chapter. In future chapters, we will show in detail how this approach is applied to problems close to those considered in [10].

Acknowlegdments

This work was supported by Ministry of Education and Science of Ukraine (project "Mathematical modelling and optimization of dynamical systems for defense, medicine and ecology", 0119U100337) and the National Academy of Sciences of Ukraine (project "New methods of research of correctness and solution search for discrete optimization problems, variational inequalities and their applications", 0119U101608).

References

[1] A.B. Bakushinskii, A.V. Goncharskii, Iterative Methods for Solving Ill-Posed Problems, Moscow: Nauka, 1989.

[2] F. Browder, "Existence and approximation of solutions of nonlinear variational inequalities," Proc. Nat. Acad. Sci. USA, vol. 56, no. 4, pp. 1080–1086, 1966.

[3] F. Browder, "Convergence of approximants of fixed points of nonexpansive non-linear mappings in Banach spaces," Arch. Rational Mech. Anal., vol. 24, pp. 82–90, 1967.

[4] H. Attouch, "Viscosity Solutions of Minimization Problems," SIAM J. Optim., vol. 6, pp. 769–806, 1996.

[5] V.V. Podinovskii, V.M. Gavrilov, Optimization with respect to successively applied criteria, Moscow: Sovetskoe Radio, 1975.

[6] M. Solodov, "An explicit descent method for bilevel convex optimization," Journal of Convex Analysis, vol. 14, pp. 227–238, 2007.

[7] Kuntsevich, V.M., et al (Eds). Control Systems: Theory and Applications. Series in Automation, Control and Robotics, River Publishers, Gistrup, Delft, 2018.

[8] Kondratenko, Y.P., Kuntsevich, V.M., Chikrii, A.A., Gubarev, V.F. (Eds). Advanced Control Systems: Theory and Applications. Series in Automation, Control and Robotics, River Publishers, Gistrup, 2021.

[9] Kondratenko, Y.P., Chikrii, A.A., Gubarev, V.F., Kacprzyk, J. (Eds). Advanced Control Techniques in Complex Engineering Systems: Theory and Applications. Dedicated to Professor Vsevolod M. Kuntsevich. Studies in Systems, Decision and Control, Vol. 203. Cham: Springer Nature Switzerland AG, 2019.

[10] S.I. Lyashko, D.A. Klyushin, D.A. Nomirovsky, V.V. Semenov, "Identification of age-structured contamination sources in ground water," in: Boucekkine R., Hritonenko N., and Yatsenko Y. (eds.) Optimal Control of Age-Structured Populations in Economy, Demography, and the Environment, Routledge, London–New York, pp. 277–292, 2013.

[11] V.V. Kalashnikov, N.I. Kalashnikova, "Solution of two-level variational inequality," Cybernetics and Systems Analysis, vol. 30, pp. 623–625, 1994.

[12] I.V. Konnov, "On systems of variational inequalities," Izvestiya Vuzov, Matematika, no. 12, pp. 79–88, 1997.

[13] L.D. Popov, "Lexicographic variational inequalities and some applications," Math. Programming. Regularization and Approximation. A Collection of Papers, Tr. IMM, vol. 8, no. 1, pp. 103–115, 2002.

[14] L.D. Popov, "A one-stage method of solving lexicographic variational inequalities," Izvestiya Vuzov, Matematika, no. 12, pp. 71–81, 1998.

[15] G. Kassay, V.D. Radulescu, Equilibrium Problems and Applications, London: Academic Press, 2019.

[16] G. Mastroeni, "On auxiliary principle for equilibrium problems," in Equilibrium Problems and Variational Models, Daniele, P. et al., Eds. Dordrecht: Kluwer Academic Publishers, pp. 289–298, 2003.

[17] P.L. Combettes, S.A. Hirstoaga, "Equilibrium Programming in Hilbert Spaces," J. Nonlinear Convex Anal., vol. 6, pp. 117–136, 2005.

[18] T.D. Quoc, L.D. Muu, N.V. Hien, "Extragradient algorithms extended to equilibrium problems," Optimization, vol. 57, pp. 749–776, 2008.

[19] V.V. Semenov, "Modified Extragradient Method with Bregman Divergence for Variational Inequalities," Journal of Automation and Information Sciences, vol. 50, issue 8, pp. 26–37, 2018.

[20] L. Chabak, V. Semenov, Y. Vedel, "A New Non-Euclidean Proximal Method for Equilibrium Problems," in: Chertov O., Mylovanov T., Kondratenko Y., Kacprzyk J., Kreinovich V., Stefanuk V. (eds.) Recent Developments in Data Science and Intelligent Analysis of Information. ICDSIAI 2018. Advances in Intelligent Systems and Computing, vol. 836, Springer, Cham, pp. 50–58, 2019.

[21] D.A. Nomirovskii, B.V. Rublyov, V.V. Semenov, "Convergence of Two-Stage Method with Bregman Divergence for Solving Variational Inequalities," Cybernetics and Systems Analysis, vol. 55, issue 3, pp. 359–368, 2019.

[22] V.V. Semenov, "Strongly Convergent Algorithms for Variational Inequality Problem Over the Set of Solutions the Equilibrium Problems," in Continuous and Distributed Systems. Solid Mechanics and Its Applications, vol. 211, M.Z. Zgurovsky, V.A. Sadovnichiy (eds.) Springer International Publishing Switzerland, pp. 131–146, 2014.

[23] Ya.I. Vedel, S.V. Denisov, V.V. Semenov, "Algorithm for variational inequality problem over the set of solutions the equilibrium problems," Journal of Numerical and Applied Mathematics, no. 1 (133), pp. 5–17, 2020.

[24] G. Gidel, H. Berard, P. Vincent, S. Lacoste-Julien, "A Variational Inequality Perspective on Generative Adversarial Networks," arXiv:1802.10551, 2018.

[25] A. Bohm, M. Sedlmayer, E.R. Csetnek, R.I. Bot, "Two steps at a time – taking GAN training in stride with Tseng's method," arXiv:2006.09033, 2020.

[26] R.I. Bot, M. Sedlmayer, P.T. Vuong, "A relaxed inertial forward-backward-forward algorithm for solving monotone inclusions with application to gans," arXiv:2003.07886, 2020.

[27] L.D. Popov, "On schemes for the formation of a master sequence in a regularized extragradient method for solving variational inequalities," Russian Mathematics, vol. 48, iss. 1, pp. 67–76, 2004.

[28] H.H. Bauschke, P.L. Combettes, Convex Analysis and Monotone Operator Theory in Hilbert Spaces. Berlin, Heidelberg, New York: Springer, 2011.

[29] Y.I. Vedel, S.V. Denisov, V.V. Semenov, "An adaptive algorithm for the variational inequality over the set of solutions of the equilibrium problem," Cybernetics and Systems Analysis, vol. 57, issue 1, pp. 91–100, 2021.

Part II
Advances in Control Systems Application

8

Identification of Complex Systems in the Class of Linear Regression Models

V. Gubarev, S. Melnychuk*, and N. Salnikov

Space Research Institute, National Academy of Sciences of Ukraine
and State Space Agency of Ukraine,
40, b. 4/1 Glushkov av., Kyiv, 03187, Ukraine
E-mail: melnychuk89s@gmail.com; sergvik@ukr.net
*Corresponding Author:

Abstract

Most methods of analysis and synthesis of control assume the known mathematical model of the control object. To obtain the desired control result, the model must accurately describe the real processes [1]. It is often not possible to obtain quantitative relations between the parameters and variables of the system under consideration on the basis of theoretical laws. For many causal systems, these relations can only be assessed experimentally, during identification. The most developed approach is the so-called stochastic identification, which consists in obtaining unbiased estimates of the parameters for certain model. However, the possibilities of such identification are significantly limited in the case of complex systems. This chapter discusses a more general approach to identification within the regression model class, which searches for approximate solutions that are consistent in accuracy with errors in the data. Since identification problems in complex systems are often incorrectly posed, the proposed method includes regularization that provides practically suitable solutions.

Keywords: Identification, complex discrete system, linear regression, bounded errors, regularization, SVD, regularized dimension.

8.1 Introduction

System identification, i.e. the problem of constructing its mathematical model from experimental data has a long history. It is conducted by scientists who study the behavior of various systems, as well as control specialists who need mathematical models to solve problems of analysis and synthesis. Many results on this problem within the framework of a discrete description have been published in well-known books [2, 3], including monographs [4, 5].

The most interesting problems from a practical point of view are associated with uncertainty in the output and input variables. The corresponding type of identification problem is called "errors in variables" (EIV) and is considered by many scientists [6–9]. Recent publications can be found in [10–13]. The main focus there was on obtaining unbiased parameter estimates and the issue of consistency.

Different approaches were considered within EIV identification, for example, for describing an uncertain dynamic system it was proposed to use a graph metric introduced in [14] with the representation of model uncertainties by the graph subspace consisting of the input and output spaces of the system. Moreover, it was assumed everywhere that the errors are independent and identically distributed (i.i.d.) measurement noises. Recently graph subspace approach was proposed and developed in [15] for EIV identification problem and a new estimation algorithm was derived that is more general than total last square algorithm used in preceding works. It is aimed at estimation of the model parameters based on finite measurement samples of the graph subspace of a given system.

This chapter explores a different approach to address the EIV identification problem with a different interpretation of the uncertainty in the data. We assume that the errors are randomly distributed but limited in magnitude. As in [15], the data series can be large, but always finite. This initial information about the system is more realistic than in the stochastic case.

The approach proposed here is focused on the identification of controlled systems and allows the use of a fairly arbitrary input signal. This allows data collection in active experiments with specially designed informative input signals. The problems of identifying systems of large dimension, including infinite ones, based on inaccurate data, are usually ill-posed. Obtaining unbiased estimates of the model parameters in these cases becomes impossible. Moreover, a regularized solution in ill-posed case can give an approximate model of a lower dimension than that of the original system. Thus, the proposed identification method involves determining the size of the model that guarantees stability, followed by parameter estimation.

The resulting models were tested when simulating the identification process. This chapter is actually developing an approach and results obtained earlier in [16, 17].

8.2 Active Experiments and Informative Data

This chapter focuses on systems of large or infinite dimension, for which obtaining unbiased estimates is meaningless. We assume that the identified model may have a smaller dimension than the original system.

The use of a reduced dimension is most often dictated by the very formulation of the EIV identification. The presence of errors in the initial data of the identification problem, in many cases, turns it into an ill posed one. As a result, the identification of any method becomes very sensitive to random errors in the input data. Re-identification with other errors leads to significantly different parameter estimates, so such solutions are practically useless.

Therefore, in identification, it is proposed to use a regularization that ensures the stability of the obtained solutions. We find and use only those models whose parameters are less sensitive to errors. For this it is enough to find approximate solutions that are consistent in accuracy with the errors in the data.

Naturally, in this case, the accuracy of the resulting models will be limited. Under such inevitable circumstances, or rather, the fundamental properties of identification, it is proposed to take the dimension of the sought model as a regularization parameter. The largest dimension of the model, admissible according to the stability condition, will correspond to the regularized approximate solution. It should be noted that despite the discreteness of the regularization parameter, there is no clear boundary between stable and unstable solutions.

In principle, it is possible to complicate the problem by finding the optimal or quasi-optimal dimension, taking into account the corresponding quality criteria of the model. However, this will greatly complicate the method of its solution, which is undesirable for engineering practice.

Here we consider fairly simple methods for choosing the dimension of the model, which are analogs of the residual principle and quasi-optimal regularization used in methods for solving ill-posed problems. The rejection of unbiased estimates of the stochastic approach allows instead to consider other non-stochastic formulations of the problem and the corresponding solution methods, which, nevertheless, retain the basic techniques of stochastic identification.

Instead of interpreting the error in the data as i.i.d, we consider them arbitrary, but belonging to sets limited to a known value. We restrict consideration to the scalar case, when the object under study has single input $u(t)$ and single output $y(t)$ (t–discrete moments of time). In cases of identification of multidimensional systems, various approaches based on the scalar case can be used. They are not considered in this chapter.

So, instead of exact values of signals $u(t)$ and $y(t)$ we have their measurements

$$\tilde{u}(t) = u(t) + \xi_u(t), \tilde{y}(t) = y(t) + \xi_y(t), \tag{8.1}$$

where $\xi_u(t)$ and $\xi_y(t)$ are errors in data. According to the non-stochastic approach, we assume that $\xi_u(t)$ and $\xi_y(t)$ are unknown arbitrary random sequences satisfying condition

$$|\xi_u(t)| \le \varepsilon_u, |\xi_y(t)| \le \varepsilon_y \tag{8.2}$$

where ε_u and ε_y are sufficiently small compared to the magnitude of $u(t)$ and $y(t)$.

The kind of input signal affects the information contained in the observations. The most important characteristic of an input is its ability to excite all modes of the system. Insufficiently excited modes cannot be identified.

For stochastic identification, the persistently exiting input sequence of a given order [18] is used. An ergodic sequence $u(t)$ is persistently exiting of order l if and only if

$$rank \left(\lim_{N \to \infty} \frac{1}{N} U_{lN} \cdot U_{lN}^T \right) = l, \tag{8.3}$$

where

$$U_{lN} = \begin{bmatrix} u(t) & u(t+1) & \dots & u(t+N-1) \\ u(t+1) & u(t+2) & \dots & u(t+N) \\ \vdots & \vdots & \ddots & \vdots \\ u(t+l-1) & u(t+l) & \dots & u(t+N+l-2) \end{bmatrix}.$$

When $u(t)$ is a zero-mean discrete white noise sequence it has the statistical property

$$\frac{1}{N} U_{lN} \cdot U_{lN}' = \sigma_u^2 \cdot I_l + \varepsilon_N^2 \cdot E_N, \tag{8.4}$$

where " $\prime$ " is matrix transpose, σ_u–dispersion, I_l is l–dimensional unit matrix, ε_N is a sequence such that $\lim\limits_{N\to\infty} \varepsilon_N = 0$ and E_N is a matrix of appropriate dimensions with $\|E_N\| \le 1$.

In the case of non-stochastic identification with bounded uncertainty (8.1), (8.2), we plan an active experiment so that each mode of the system at a certain moment of observation makes the most significant contribution to the output signal, as much as possible.

The chapter examines the identification of only asymptotically stable systems. An initial informative data for identification can be obtained in two ways. The first one is to collect data from many separate experiments, consisting of the finite intervals of excitation by $u(t) \ne 0$ and relaxation with $u(t) = 0$ of the same length l, where l should to be larger then duration of the transient process. An alternative option is a single continuous experiment, which is compound from previous mentioned, where l should guarantee the process is completely damped or the steady-state regime is reached.

To select an informative input excitation, consider a discrete linear stationary system. Such systems can be represented by models in the classes of linear regression or systems of difference equations in the state space, which are known to be equivalent.

If we take a state-space model

$$\vec{x}(k+1) = A\vec{x}(k) + bu(k)$$
$$y(k) = c'\vec{x}(k) + du(k)$$

with a Jordan block realization $A = diag(A_p)$, $\; b = col\,(b_p)$, $\; c = col\,(c_p)$, where

$$A_p = \begin{pmatrix} \alpha_p & -\beta_p \\ \beta_p & \alpha_p \end{pmatrix}, \; b_p = \begin{pmatrix} b_p^c \\ b_p^s \end{pmatrix}, \; c_p = \begin{pmatrix} c_p^c \\ c_p^s \end{pmatrix}, \; \lambda_p = \alpha_p \pm i\beta_p,$$
$$A_p = (\alpha_p), \qquad b_p = (b_p^c), \quad c_p = (c_p^c), \qquad \lambda_p = \alpha_p,$$

$$(8.5)$$

and λ_p are eigenvalues of matrix A, then for controlled and observable systems we can write the input-output ratio in which the parameters are invariants. In the case $y(t) = 0$ and $u(t) = 0$ at $t \le 0$ and $u(0) \ne 0$ at $t > 0$, according to [19], this relation is

$$y(t) = \sum_{j=0}^{t-1} \sum_{p=1}^{P} h_{0p}(t-j)u(j), \qquad (8.6)$$

where $h_{0p}(k) = \rho_p^k [f_p^c \cos \omega_p(k) + f_p^s \sin \omega_p(k)]$, and

$$\rho_p = |\lambda_p|, \omega_p = \arg(\lambda_p), f_p^c = c_p^c b_p^c + c_p^s b_p^s, f_p^s = c_p^c b_p^s - c_p^s b_p^c.$$

Parameters ρ_p, ω_p, f_p^c and f_p^s are invariants characterizing the dynamic properties of the system. It was assumed in the expression that there are no multiple roots, and the case of real roots is a special case of (8.5) corresponding to $\omega_p \equiv 0$, $\beta_p = 0$, $c_p^s = b_p^s = 0$.

When identifying an EIV, the informativeness of the input action can be judged by the signal-to-noise ratio (SNR). The higher SNR for the output of each mode p (8.5), the more informative is the input action. Therefore, we will shape input at the excitation intervals so that at its end, one of the modes has the largest output signal. Different $u(t)$ across experiments should effectively excite all modes of the system.

We will use 2 types of input signals. The first type is a rectangular pulse

$$u(t) = 1, \quad t = [1, \tau], \tag{8.7}$$

where τ is duration, variable across experiments in the range: $\tau = [1, ..., \tau_{\max}]$. Short ones provide the greatest excitation of fast real modes. Increasing the duration will add significant signals from slower modes, regardless of the value of $c_p^c b_p^c$. Therefore, rectangular pulses of different duration should form a set of initial conditions with different contributions of separate modes.

The second type of input is a single harmonic

$$u(t) = \sin(\omega t), \quad t = [1, \tau], \tag{8.8}$$

where frequency ω changes across experiments, limited by discreteness of measurements to the range $(0, \pi/2]$. For complex modes in (8.5), we aim to use the resonant excitation when ω happens to be close to some natural frequency ω_p. When ω are varied with a small step, it is possible to provide resonant excitation of all oscillating modes of the system. In this case, the duration τ is chosen to be sufficient to establish steady-state forced oscillations.

Let us denote the step of the rectangular pulse duration by Δ_0, and the step of the harmonic signal frequency ω by Δ_1. With an appropriate choice of parameters Δ_0, $\tau_{\max}$, Δ_1, and l, it is possible to obtain informative data with an acceptable SNR for identification with a fairly general understanding of the system under study.

8.3 Method of Identification

In this chapter, we will identify a system in the class of linear autoregressive models. When considering complex systems that in this class have a very large or infinite dimension, we come to the need for structural-parametric identification.

When identifying EIV, an approximate model must be found that is consistent with data errors and computational accuracy. Then, in the general case, the dimension of the model may be less than the order of the true system. Therefore, the identification method should include a procedure for finding an acceptable dimension of the model with subsequent parametric identification. In the class of autoregressive models

$$y(t) = -a_1 y(t-1) - a_2 y(t-2) - \ldots - a_n y(t-n) + \\ + b_0 u(t) + b_1 u(t-1) + \ldots + b_m u(t-m) \tag{8.9}$$

the above-described design of active experiment with excitation and relaxation allows splitting the identification problem. Primarily, we estimate dimension n and autoregressive coefficients $\vec{a} = (-a_1, -a_2, \ldots, -a_n)'$ from measured output during relaxation, where $u(t) \equiv 0$. By subtracting the calculated free motion from the measured output, we obtain the forced motion data. Finally we estimate dimension m and control coefficients $\vec{b} = (b_0, b_1, \ldots, b_m)'$ from input and corresponding forced motion.

8.3.1 Model Order Selection

At the first stage of the method, the dimension of the model n is determined. For this, from the measured output data on the relaxation intervals, we form the following matrix

$$\tilde{Y}_{relax} = \begin{bmatrix} \tilde{y}_1^{pulse}(\tau_1^{pulse}+1) & \cdots & \tilde{y}_1^{pulse}(\tau_1^{pulse}+l) \\ \vdots & \ddots & \vdots \\ \tilde{y}_{k_1}^{pulse}(\tau_{k_1}^{pulse}+1) & \cdots & \tilde{y}_{k_1}^{pulse}(\tau_{k_1}^{pulse}+l) \\ \tilde{y}_1^{harm}(\tau_1^{harm}+1) & \cdots & \tilde{y}_1^{harm}(\tau_1^{harm}+l) \\ \vdots & \ddots & \vdots \\ \tilde{y}_{k_2}^{harm}(\tau_{k_2}^{harm}+1) & \cdots & \tilde{y}_{k_2}^{harm}(\tau_{k_2}^{harm}+l) \end{bmatrix}, \tag{8.10}$$

where $\tilde{y}_i^{pulse}(t)$, $i = \overrightarrow{1, k_1}$–output from ith rectangular pulse, τ_i^{pulse}– duration of ith rectangular pulse, k_1 - total number of rectangular pulses, $\tilde{y}_j^{harm}(t)$, $j = \overrightarrow{1, k_2}$–output from jth harmonic input, τ_j^{harm}–duration of jth harmonic input, k_2–total number of input harmonics. We denote the total number of rows by $R = k_1 + k_2$.

By its structure, the matrix $\tilde{Y}_{relax}$ is not Hankel one, but in a certain sense it can be considered as its generalization. We use the SVD decomposition

$$\tilde{Y}_{relax} = U\Sigma V^T, \tag{8.11}$$

where U, Σ, V are matrices of dimensions $R \times R$, $R \times l$ and $l \times l$ respectively. For a system with a finite dimension n, noiseless measurements $y(t)$, $R < n$ and $l < n$, the matrix Y_{relax} will be of incomplete rank, and only the first n singular numbers $\sigma_1, \ldots, \sigma_n$ will be non-zero. Thus, the dimension of the system can be precisely determined by the number of nonzero diagonal elements of Σ.

Since measurements $\tilde{y}(t)$ contain uncertainty, the matrix $\tilde{Y}_{relax}$ with probability almost one will be of full rank and all singular numbers will be positive. However, with small errors bound ε_y, it makes sense to divide it into "signal" and "noise" part. To obtain a model of a given dimension $\hat{n}$, we use the low-rank approximation [20] of $\tilde{Y}_{relax}$. We split matrices (8.11) into blocks

$$U\Sigma V' = \begin{bmatrix} U^S & U^N \end{bmatrix} \begin{bmatrix} \Sigma^S & 0 \\ 0 & \Sigma^N \end{bmatrix} \begin{bmatrix} V^S & V^N \end{bmatrix}', \tag{8.12}$$

so matrix Σ^S includes first $\hat{n}$ singular values. Here indices "S" and "N" denote "signal" and "noise" parts. As a result, we get decomposition

$$\tilde{Y}_{relax} = Y_{relax}^S(\hat{n}) + Y_{relax}^N(\hat{n}), \tag{8.13}$$

where $Y_{relax}^S(\hat{n}) = U^S\Sigma^S V^{S\prime}$ and $Y_{relax}^N(\hat{n}) = U^N\Sigma^N V^{N\prime}$. Expansion (8.13) allows, proceeding from the ε–rank property [20] to find the matrix of the given rank closest to full-rank ones in Frobenius norm:

$$rank\left(Y_{relax}^S(\hat{n})\right) = \hat{n}, \left\|Y_{relax}^N(\hat{n})\right\|_F = \sigma_{\hat{n}+1}, \tag{8.14}$$

where $\| \cdot \|_F$ means Frobenius norm. The value $\sigma_{\hat{n}+1}$ can be used to judge the proximity of the matrix $\tilde{Y}_{relax}$ to the matrix of rank $\hat{n}$. Sometimes, the true dimension n can be determined by the rate of decrease of the singular values. This can be done if there is a pronounced gap between the values of σ_i and σ_{i+1}. However, there is often no clear gap. It all depends on the signal-to-noise ratio of individual modes (8.5). Modes with a low SNR becomes indistinguishable from background noise, making impossible in the non-stochastic case to establish the exact dimension of the system. In such cases, we can find a model of reduced dimension. For this, the principle of regularization is used. The required approximate solution must be consistent in accuracy with the uncertainty of the data.

Consider how the quantities $\sigma_{\hat{n}+1}$ and ε_y are related. Let's represent the matrix $\tilde{Y}_{relax}$ in the form of the exact and noise part, omitting here the "relax" index

$$\tilde{Y} = Y + \Delta Y. \tag{8.15}$$

Inequality (8.2) gives $\|\Delta Y\|_{max} < \varepsilon_Y$. Let us estimate from above the quantity $\sigma_{\hat{n}+1}$. From (8.14) we have $\sigma_{\hat{n}+1} = \left\|Y^N(\hat{n})\right\|_F = \left\|\tilde{Y} - Y^S(\hat{n})\right\|_F$. Substituting (8.15) and using the triangle inequality we derive

$$\sigma_{\hat{n}+1} \leq \left\|Y - Y^S(\hat{n})\right\|_F + \|\Delta Y\|_F .$$

Since the quantity $\left\|Y - Y^S(\hat{n})\right\|_F$ can not be larger then $\left\|Y - \tilde{Y}\right\|_F$, we have

$$\sigma_{\hat{n}+1} \leq 2 \|\Delta Y\|_F .$$

From norm equivalence $\|A\|_F = \sqrt{\sum_i^N \sum_j^M a_{i,j}^2}$, $\|A\|_{max} = \max_{\substack{i = 1, M \\ j = 1, N}} \left|a_{i,j}\right|$, we have estimation

$$\sigma_{\hat{n}+1} \leq 2\sqrt{Rl} \cdot \varepsilon_Y. \tag{8.16}$$

Relation (8.16) means that the modes of the system, starting with such $\hat{n}$, cannot be guaranteed to be identified. The possibility of their identification is determined only by the case of noise realization. This property will be used to determine the dimension of the model, namely the dimension of the vector $\vec{a}$.

In certain cases, by reducing Δ_0 and Δ_1, and, consequently, the magnitude of R, it is possible to achieve an approximation to the exact dimension. Everything will depend on the dynamic properties of the system, determined by its expansion invariants (8.5). Low SNR values appear for modes with close ω_p or modes with small f_p^c and f_p^s. There exist those ε_y, starting from which their signals are not distinguishable against the background of noise. The presence of such modes in the system will lead to the reduction of the model order in the proposed approach.

8.3.2 SVD Parametric Identification

After establishing the dimension n of the autoregressive part of the model, the problem of determining the parameters $\vec{a}$ is being solved. The data of matrix $Y^S(n)$ is used to form an overdetermined system of linear equations.

$$Y^A(n)\vec{a} = \bar{y}^A(n). \tag{8.17}$$

The matrix $Y^A(n)$ in (8.17) is formed from all equations of type

$$(y^r(t-1), \ y^r(t-2), \ \ldots, \ y^r(t-n)) \cdot \vec{a} = -y^r(t), \tag{8.18}$$

where index r denotes row from matrix $Y^S(n)$, $r = \overrightarrow{1, R}$ and t goes through all possible values from $n+1$ up to l. When forming the system, the equations are filtered. If the absolute values of y are less than a certain threshold level

$$K_{NS} \cdot \varepsilon_Y,$$

then such an equation is discarded. This is done in order to ignore noisy data.

The vector a is found from (8.17) by the usual or generalized least squares method (LS) [20]. However, it should first be sure that the matrix is not ill-conditioned. An estimate of the condition number in the norm $|| \cdot ||_2$ can be established from the ratio of the singular values of the matrix Y^A

$$\mathfrak{x}_2(n) = \frac{\sigma_1}{\sigma_n}. \tag{8.19}$$

If the condition number (8.14) is acceptable, then the solution is found by standard, for example, generalized LS. Otherwise, it is necessary to use the regularized LS, which is available in the libraries of standard programs.

8.3.3 Total Model Reconstruction

The described procedure for finding the dimension and coefficients of a vector $\vec{a}$ is fully applicable for the vector $\vec{b}$. As a result, we will have a complete solution to the identification problem. It should be noted here that the dimension of the vector $\vec{b}$ cannot exceed $n + 1$, when the observation equation has the form $y(t) = c'x(t) + du(t)$ and $d \neq 0$, and the value n when $d = 0$. In the latter case, the dimension of $\vec{b}$ may be smaller than n. The task of finding $\vec{b}$ is preceded by a procedure for extracting a purely forced motion from the sequence $\{y(t)\}$. The estimated output signal of purely forced motion $\tilde{\tilde{y}}(t)$ is determined by the relation

$$\tilde{\tilde{y}}(t) = \tilde{y}(t) + a_1\tilde{y}(t-1) + a_2\tilde{y}(t-2) + \ldots + a_n\tilde{y}(t-n).$$

Then the basic equation for finding the vector $\vec{b}$ will be the following equation

$$\begin{aligned} \tilde{y}(t) &= b_0 u(t) + b_1 u(t-1) + \ldots + b_m u(t), & d \neq 0 \\ \tilde{y}(t) &= b_1 u(t-1) + \ldots + b_m u(t), & d = 0 \end{aligned}, \quad (8.20)$$

for any moment of time t where we have nonzero excitation signals. From the data on the intervals of excitation of a suitable duration, we form an overdetermined system

$$W \cdot b = \tilde{y}, \quad (8.21)$$

where $\vec{b}$ is a vector whose components are b_0, b_1, $\ldots$, b_m or b_1, b_2, $\ldots$, b_m, the vector $\tilde{y}$ is composed of $\tilde{y}(t)$ for the time moments corresponding to (8.20), and W is a matrix consisting of the input signals for the appropriate time moments. In (8.21), both the matrix W and vector $\tilde{y}(t)$ are given approximately. Taking into account the specified errors, we will find an approximate solution of (8.21), using the generalized LS. In this case, you should make sure that the problem being solved is correctly posed. If it is ill-conditioned, a regularized solution should be found in accordance with the well-known methods for solving incorrect linear systems. In this case, it is possible to obtain stable approximate solutions, like with a vector $\vec{a}$, by lowering the dimension m of the vector $\vec{b}$.

Bad conditionality of the matrix W indicates that the dimension of the vector $\vec{b}$ is less than n. In such cases, the further actions are similar to those described in Section 2.1, when determining the dimension of the vector $\vec{a}$. To do this, we carry out the SVD decomposition of the matrix W, and check for singular values. After establishing the dimension of the vector $\vec{b}$ its value is found from the redefined system (8.21).

Solving the system of equations (8.21) by generalized LS completes the system identification, resulting in an approximate model consistent with the EIV.

8.3.4 Quasioptimal Model Dimensions

The dimensions of vectors $\vec{a}$ and $\vec{b}$ defined by cited above can be considered as using the residual principle for this purpose [21]. In addition to it, it is permissible to find the dimension of the model based on a method that allows calculating the quasi-optimal value of the regularization parameter, in our case the dimension of the model [21].

The criterion (8.16) for determining the dimension does not guarantee that the coefficients of the found model will be determined consistently, even if the exact dimension is determined.

Then the quasi-optimal dimension should be found. For the varied n in decreasing order the problem of parametric identification is solved and the variation of the coefficients are estimated

$$\left\| \frac{1}{n}[\vec{a}(n+1) - \hat{\vec{a}}(n)] \right\|, \left\| \frac{1}{n}[\vec{b}(n+1) - \hat{\vec{b}}(n)] \right\|, \qquad (8.22)$$

where $\vec{a}(n+1)$ and $\vec{b}(n+1)$ is the solution to the problem of parametric identification of vectors $\vec{a}$ and $\vec{b}$ for dimension $n+1$, and $\hat{\vec{a}}(n)$ and $\hat{\vec{b}}(n)$ are extended vectors formed from solutions for dimension n according to the following rule

$$\hat{\vec{a}}(n) = (-a_1 \ -a_2 \ \ldots \ -a_n \ 0)' = (\vec{a}'(n) \ 0)',$$

$$\hat{\vec{b}}(n+1) = (b_1 \ b_2 \ \ldots \ b_n \ 0)' = (\vec{b}'(n) \ 0)',$$

where $\vec{a}(n)$ and $\vec{b}(n)$ are the values of the corresponding vectors for the dimension n. Various behaviors of the discrete function defined by (8.22) are possible. At first, it decreases, and then begins to grow, while local minima are possible. Then the last local minimum determines the quasi-optimal dimension. It is possible when this function grows continuously, starting from the smallest n. It is also possible a noticeable scatter in the behavior of this function, starting from some point. Then the value n before the scatter determines the desired dimension.

The parameters $\vec{a}$ and $\vec{b}$ corresponding to the quasi-optimal n determine a regularized approximate solution to the identification problem.

8.4 Simulation Results

The purpose of the model experiments was to demonstrate the efficiency of the proposed approach to identification and the method described, as well as to evaluate its effectiveness by examining the quality of the resulting models and the features that arise when solving specific problems. Since the result of identification depends on the invariants of the systems, the capabilities of the method have been demonstrated on the most suitable systems for identification. This allows one to assess the limiting capabilities of the method.

The most suitable form for defining a system with specified dynamic properties is a representation in the form (8.5). Using it we can choose invariants that provide the best conditions for identification. So, we set eigenvalues

λ_p for all modes, ensuring stability, slow decay and sufficient distance from each other. We also set invariants of $|f_p^c|$ and $|f_p^s|$ to be equal for all modes, to provide better observability. The definition of the system in the form (8.5) in canonical observable representation ($c_p^c = 1$, $c_p^s = 0$) makes it easy to pass to equivalent description in a normal observable realization in the state space

$$A = \begin{bmatrix} 0 & 0 & \cdots & 0 & -\bar{a}_n \\ 1 & 0 & \cdots & 0 & -\bar{a}_{n-1} \\ 0 & 1 & \cdots & 0 & -\bar{a}_{n-2} \\ \vdots & \vdots & \ddots & \vdots & \vdots \\ 0 & 0 & \cdots & 1 & -\bar{a}_1 \end{bmatrix}, \bar{b} = \begin{bmatrix} \bar{b}_n \\ \bar{b}_{n-1} \\ \vdots \\ \bar{b}_1 \end{bmatrix}, \bar{c} = \begin{bmatrix} 0 \\ 0 \\ \vdots \\ 1 \end{bmatrix}.$$

and then to an equivalent representation in the form of autoregression (8.9):
$a_i = \bar{a}_i$, $b_i = da_i + \bar{b}_i$, $i = \overline{1,n}$, $b_0 = d$, if m=n.

Thus, we have the opportunity to investigate the identification method for systems with different dynamic properties. It is also important to establish the impact on the quality of models of errors in the data within the constraints defined by (8.1), (8.2). We assume that the values of ε_u and ε_y are also quite small compared to deterministic signals. To correctly compare and evaluate simulation results, we will consider the relative levels of errors

$$\frac{\|\xi_u\|_{\max}}{\|u\|_{\max}} = \frac{\|\xi_y\|_{\max}}{\|y\|_{\max}} = \varepsilon,$$

where ε is a variable value. Taking into account constraints (8.2), it leads to the relations

$$\varepsilon_u = \|u\|_{\max} \cdot \varepsilon, \varepsilon_y = \|y\|_{\max} \cdot \varepsilon. \tag{8.23}$$

Then ε_u and ε_y characterize the relative width of the intervals of membership of errors, and $\|u\|_\infty$ and $\|y\|_\infty$ characterize the same for input and output signals. Using (8.22), the dependence of the quality of the resulting models on ε was investigated.

8.4.1 Algorithm of identification

1. A series of active experiments with the excitation and relaxation:

$$u_i^{pulse}(t) = 1, \ t = [1, \tau_i']$$
$$u_i^{pulse}(t) = 0, \ t = [\tau_i' + 1, \tau_i' + l]' \ \tau_i' = i, \ i = \overrightarrow{1, k_1},$$

$$u_i^{harm}(t) = \sin(\omega_i t), \ t = \left[1, \tau_i''\right]' \ \omega_i = 0.01i$$

$$u_i^{harm}(t) = 0, \ t = \left[\tau_i'' + 1, \tau_i'' + l\right]' \ \tau_i'' = 10 \cdot \min(10, \pi/\omega_i),$$

$$i = \overrightarrow{1, k_2},$$

where $k_1 = 100$, $k_2 = 157$, $l = 1000$. The noisy outputs are stored line by line in the matrix $\tilde{Y}$, the inputs are stored line by line in the matrix $\tilde{U}$.

2. From $\tilde{Y}$ on the relaxation, a matrix $\tilde{Y}_{relax} \in R^{R \times l}$ is formed. $R = k_1 + k_2$.

3. Using the SVD, dimension n is determined.

4. Matrix $\tilde{Y}_{relax}$ is splitted by n into signal $Y_{relax}^S(n)$ and noise $Y_{relax}^N(n)$ parts.

5. A linear system for the regression coefficients $\vec{a}$ is formed from $Y_{relax}^S(n)$. Equation included if its coefficients exceed the SNR threshold $K_{NS} \cdot \varepsilon \cdot \left\|\tilde{Y}\right\|_{\max}$.

6. The regression coefficients of free motion are found.

7. We define forced movement $\tilde{Y}_{forced}$ by subtracting free movement from $\tilde{Y}$.

8. (optional). We build compound matrix $\left[\tilde{Y}_{forced}|\tilde{U}\right]$ and apply SVD. If it is not possible to determine dimension m by singular values, then we accept $m = n$. Extract signal part of $\left[\tilde{Y}_{forced}|\tilde{U}\right] = \left[\tilde{Y}_{forced}|\tilde{U}\right]^S(\hat{m}) + \left[\tilde{Y}_{forced}|\tilde{U}\right]^N(\hat{m})$, and then break it back: $\left[\tilde{Y}_{forced}|\tilde{U}\right]^S = \left[\tilde{Y}_{forced}^S|\tilde{U}^S\right]$.

9. From the matrices $\tilde{Y}_{forced}$ and $\tilde{U}$ (or $\tilde{Y}_{forced}^S$ and $\tilde{U}^S$) we form a system for the control coefficients $\vec{b}$ of regression. The system includes all equations whose coefficients exceed the threshold value, similarly to item 5.

10. By singular values we check whether the chosen dimension $\hat{m}$ is true. If not, we rebuild the system for the specified dimension.

11. We solve the system and find the regression coefficients for control.

Then we check regression model:

1. Check if the resulting model is stable.

2. Calculate the deviation of the model output from the real system when testing with a rectangular pulse followed by relaxation.

8.4.2 Excitation of system modes

Consider the system as a set of individual modes (8.5), each of which corresponds to either a real eigenvalue or a pair of complex conjugate eigenvalues. The mode can be identified only when its contribution to the output is distinguishable against the noise. When $u = 0$ it is defined as $y_p = x_p A_p^c$ for real λ_p, and $y_p = x_p^c A_p^c + x_p^s A_p^s$ for complex λ_p, where p is a mode number, x_p and $\vec{x}_p = \left(x_p^c, x_p^s\right)^T$ are states of real and complex modes.

For real λ_p, the magnitude of the mode output is proportional to the $|x_p|$. Therefore, the best conditions for identifiability correspond to maximizing $|x_p|$. For the complex mode, the magnitude of y_p is determined by the dot product of $\vec{x}_p$ and the vector of coefficients $\vec{A}_p = \left(A_p^c, A_p^s\right)^T$. The best conditions for identifiability correspond to a minimizing the angle between $\vec{x}_p$ and $\vec{A}_p$ and maximizing $\|\vec{x}_p\|_2$. Vector $\vec{A}_p$ is unknown in advance. Free movement of a separate complex mode for $\lambda = \alpha \pm i\beta$ is described by the equation

$$\begin{pmatrix} x_{k+1}^c \\ x_{k+1}^s \end{pmatrix} = \begin{pmatrix} \alpha & -\beta \\ \beta & \alpha \end{pmatrix} \begin{pmatrix} x_k^c \\ x_k^s \end{pmatrix},$$

where the subscript indicates the quantization step number. We represent the eigenvalue in form $\alpha + i\beta = re^{i\varphi}$, where $\alpha = r\cos\varphi$, $\beta = r\sin\varphi$, $r = \sqrt{\alpha^2 + \beta^2} < 1$. Then the transformation matrix of the state vector

$$\begin{pmatrix} \alpha & -\beta \\ \beta & \alpha \end{pmatrix} = r \begin{pmatrix} \cos\varphi & -\sin\varphi \\ \sin\varphi & \cos\varphi \end{pmatrix}$$

is a combination of rotation by an angle φ and scaling with a factor r. The end of the vector $\vec{x}_p$ in free motion describes a contracting spiral on the $Ox^c x^s$ plane, and the angle between $\vec{x}_p$ and $\vec{A}_p$ at each turn approaches to 0 and π closer than $\varphi/2$. The maximum mode output will be reached at the first of these points and is estimated as

$$\|\vec{x}_0\|_2 \cdot r^{\left\lceil \pi/\varphi \right\rceil + 1} \cdot \|\vec{c}\|_2 \cdot \cos\left(\varphi/2\right) \leq \|y\|_{\max},$$

where $\vec{x}_0$ is the state vector at the first relaxation step. Therefore, the best conditions for identifying the mode correspond to maximization $\|\vec{x}\|_2$ for any implementation of $\vec{A}$.

Conclusion: the magnitude of the mode excitation will be defined as the norm of the state vector.

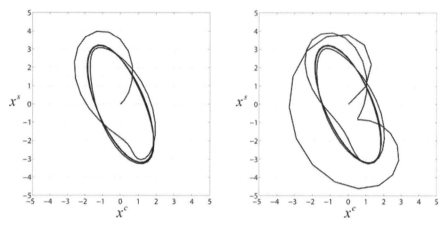

Figure 8.1   State of a complex mode under harmonic excitation with different initial phases

Let us consider the behavior of the state vector of a complex mode upon excitation by a harmonic signal with a different initial phase. Figure 8.1 shows trajectory $(x^c, x^s)'$ of mode $0.95e^{0.5i}$ for excitations $u(t) = \sin(0.2t)$ (left), and $u(t) = \cos(0.2t)$ (right). In both cases, the limiting trajectories coincide.

Conclusion: the steady-state forced motion of the complex mode does not depend on the initial phase of the harmonic signal.

Consider the excitation of a complex mode by harmonic signals of different frequencies. Figure 8.2 shows trajectory $(x^c, x^s)'$ of mode $0.95e^{0.5i}$ for excitations $u(t) = \sin(0.5t)$ (left), and $u(t) = \cos(0.8t)$ (right). The first case is resonant.

The steady-state forced motion of the complex mode under harmonic excitation has trajectory of the form of an ellipse and depends on the ratio of the natural frequency and the input frequency. In the resonant case, the trajectory is close to circular.

Let us consider the trajectories of free motion after the excitation of a mode by a harmonic signal. We use harmonic excitation and turn off control at times corresponding to different points of the elliptical trajectory. Figure 8.3 shows trajectory $(x^c, x^s)'$ of mode $0.95e^{0.5i}$ for input $u(t) =$
$$\begin{cases} \sin(0.2t), & t = \overrightarrow{1, \tau} \\ 0, & t > \tau \end{cases} \text{ for } \tau = 196 \text{ (solid) and } \tau = 204 \text{ (dashed).}$$

Conclusion: in the case of nonresonant harmonic excitation of the complex mode, the value of the relaxation signal can significantly depend on the initial moment.

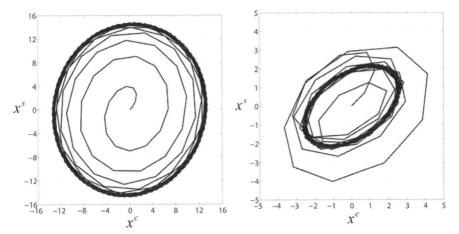

Figure 8.2 State of a complex mode under harmonic excitation of different frequencies

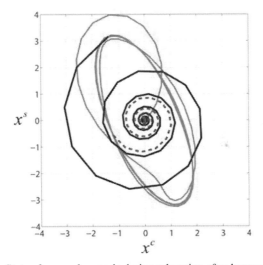

Figure 8.3 State of a complex mode during relaxation after harmonic excitation

Let us now consider the excitation of a complex mode by a constant signal. Figure 8.4 shows trajectory $(x^c, x^s)'$ of mode $0.95e^{0.5i}$ for $u(t) = 1, \ t \le \tau$ (solid) and $u(t) = 0, \ t > \tau$ (dashed).

Conclusion: constant excitation moves the state of the complex mode along a contracting spiral to a certain limiting value. When the control is turned off, it spirals back to zero.

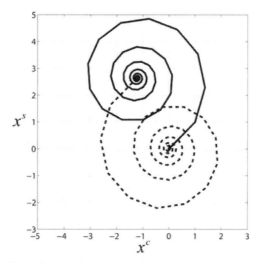

Figure 8.4   State of a complex mode during constant excitation and relaxation

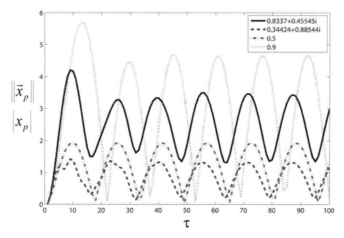

Figure 8.5   The norm of the state vector of modes under harmonic excitation

Let us consider the change in the norm of the state vector of various modes. We will take a system of two real and two complex modes: $\{\lambda\} = \{0.5,\ 0.9,\ 0.95e^{0.5i},\ 0.95e^{1.2i}\}$. Figure 8.5 shows dependency $\|\vec{x}_p\|$ and $|x_p|$ on time for input $u(t) = \sin(0.2t)$.

Excitation of the system by a harmonic input signal leads all modes to forced oscillations with the same frequency, but out of phase.

Figure 8.6 shows dependency $\|\vec{x}_p\|$ and $|x_p|$ on time for input $u(t) \equiv 1$.

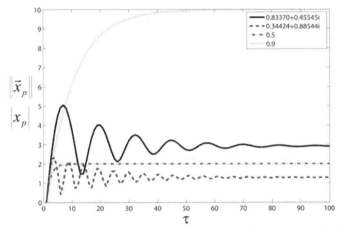

Figure 8.6 The norm of the state vector of modes under constant excitation

Excitation of the system by a constant input signal leads the modes to a certain limiting excited state.

Figure 8.7 shows dependency $\|\vec{x}_p\|$ and $|x_p|$ on time for rectangular pulse with length 20.

When the control is turned off, the norm of the mode state vector decreases exponentially.

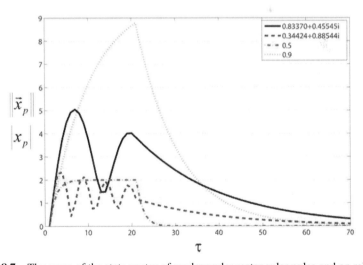

Figure 8.7 The norm of the state vector of modes under rectangular pulse and on relaxation

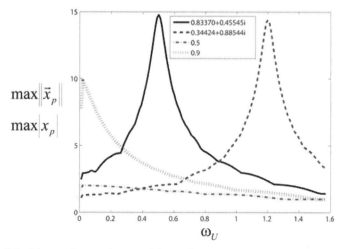

$$\max\|\vec{x}_p\|$$

$$\max|x_p|$$

Figure 8.8 The maximum of norm of the modes state vectors under harmonical input

Let us consider to what extent the modes can be excited as much as possible under harmonic excitation. We will apply harmonics to the input of the same system by changing the frequency. Figure 8.8 shows dependency maximum of $\|\vec{x}_p\|$ and $|x_p|$ on input frequency $\omega_U \in (0, {}^\pi/_2)$.

Complex modes are better excited on resonant frequencies. Real modes are better excited with low frequencies.

Simulation has shown that the described set of input signals of harmonic and rectangular pulses allows maximum excitation of all modes in separate experiments.

8.4.3 Effect of Using the Signal Part of Matrices

Two alternatives to the identification algorithm are considered:

1. To use or not to use the signal matrix $Y_{relax}^S(n)$ instead of matrix $\tilde{Y}_{relax}$ when forming the system of the regression coefficients $\vec{a}$.
2. To use or not to use the signal matrices $\tilde{Y}_{forced}^S$ and $\tilde{U}^S$ instead of matrices $\tilde{Y}_{forced}$ and $\tilde{U}$ when forming the system of the regression coefficients $\vec{b}$.

We will conduct identification of a system, but apply the following configurations of the algorithm:

A. Do not use signal matrix at all.

B. Use when finding $\vec{a}$.

C. Use when finding $\vec{a}$ and $\vec{b}$.

We use system with modes $\{\lambda\} = \{0.5,\ 0.9,\ 0.95e^{0.5i}\}$, threshold for equations $K_{NS} = 100$ and 100 random realizations of noise with relative error $\varepsilon_U = 10^{-4}$, $\varepsilon_Y = 10^{-4}$ and $\varepsilon_U = 10^{-6}$, $\varepsilon_Y = 10^{-4}$. The quality of the model was checked by the error in estimated regression coefficients and by the model's response to a rectangular pulse of duration 50. Results are represented in Tables 8.1 and 8.2.

Table 8.1 Maximum and average errors of models for $\varepsilon_U = 10^{-4}$ and $\varepsilon_Y = 10^{-4}$

Algorithm configuration	Error in $\vec{a}$	Error in $\vec{b}$	Error by output
A	0.02263 (0.01980)	0.19602 (0.16965)	0.01854 (0.01625)
B	0.00281 (0.00091)	0.02635 (0.00853)	0.00234 (0.00074)
C	0.00281 (0.00091)	0.03752 (0.02049)	0.00332 (0.00225)

Table 8.2 Maximum and average errors of models for $\varepsilon_U = 10^{-6}$ and $\varepsilon_Y = 10^{-4}$

Algorithm configuration	Error in $\vec{a}$	Error in $\vec{b}$	Error by output
A	0.02246 (0.01985)	0.19539 (0.17013)	0.01835 (0.01631)
B	0.00283 (0.00092)	0.02718 (0.00820)	0.00219 (0.00074)
C	0.00283 (0.00092)	0.02239 (0.00677)	0.00216 (0.00068)

The extraction of a matrix $Y^S_{relax}(n)$ of a given rank from the matrix $\tilde{Y}_{relax}$ can significantly improve the quality of the resulting model. The selection of a matrices $\tilde{Y}^S_{forced}$ and $\tilde{U}^S$ from matrices $\tilde{Y}_{forced}$ and $\tilde{U}$ has little effect on the quality of the resulting model. When ε_U is relatively small, we can achieve some improvement. In other case we risk worsening the result.

8.4.4 Effect of Discarding Data Close to Noise

Here we consider the effect of discarding rows from a system of equations that are formed from noisy data with a low signal-to-noise ratio. This happens when we form a system from $Y^S_{relax}(n)$ to find $\vec{a}$, and when we form a system from $\tilde{Y}_{forced}$ and $\tilde{U}$ to find $\vec{b}$.

We use system with modes $\{\lambda\} = \{0.5,\ 0.9,\ 0.95e^{0.5i}\}$, variable threshold $K_{NS} = \overrightarrow{5,500}$ and random noise with relative error $\varepsilon_U = 10^{-4}$, $\varepsilon_Y = 10^{-4}$. The quality of the model was checked by the model's response to a rectangular pulse of duration 50.

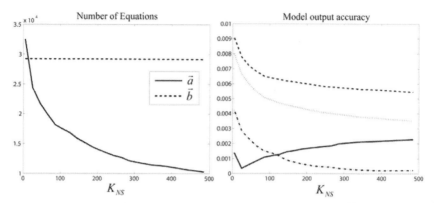

Figure 8.9   Average number of rows in overdetermined systems for $\vec{a}$ and $\vec{b}$ (left), the quality of the resulting model for different implementations of random noise (right)

Figure 8.9 shows the number of not discarded equations (left) and model quality (right) for varied coefficient K_{NS}.

The right figure shows the dependencies for four different variants of the implementation of random measurement noise. In one of these cases, dropping a large number of equations did not improve the result. An increase in the threshold SNR coefficient can significantly reduce the number of lines in the systems of equations and, in most cases, slightly increase the accuracy of the model.

8.4.5 The Best Systems for Identification

Let us carry out identification modeling for different systems of the same order to determine those of them that are identified most accurately.

We use threshold $K_{NS} = 100$ and noise with relative error $\varepsilon_U = 10^{-5}$, $\varepsilon_Y = 10^{-5}$.

Results are represented in Table 8.3.

Complex modes are identified better than real ones. The presence of close real or complex eigenvalues degrades the identification accuracy. The best system for identification is purely oscillatory, weakly damped, and does not contain close natural frequencies.

8.4.6 Model Order Determination

We carry out an experiment to determine the dimension of the system in terms of singular numbers. We use favorable configurations as initial systems: all

Table 8.3   The worst and average quality of models

Modes of a system	Error in $\vec{a}$	Error in $\vec{b}$	Error by output
$0.95e^{0.5i}$, $0.95e^{1.2i}$	2×10^{-6} (1×10^{-6})	5.6×10^{-5} (2.1×10^{-5})	5×10^{-6} (3×10^{-6})
$0.95e^{0.1i}$, $0.95e^{1.2i}$	4×10^{-6} (1×10^{-6})	4.3×10^{-5} (1.5×10^{-5})	1.9×10^{-5} (7×10^{-6})
$0.95e^{0.5i}$, $0.95e^{0.6i}$	3×10^{-5} (8×10^{-6})	5.7×10^{-5} (1.8×10^{-5})	2.6×10^{-4} (1.1×10^{-4})
$0.8e^{0.5i}$, $0.8e^{0.6i}$	2.7×10^{-4} (1×10^{-4})	5.2×10^{-4} (1.7×10^{-4})	2.2×10^{-3} (6×10^{-4})
0.2, 0.9, $0.95e^{0.5i}$	0.00046 (0.00010)	0.00054 (0.00012)	0.00007 (0.00002)
0.5, 0.9, $0.95e^{0.5i}$	0.00031 (0.00009)	0.00269 (0.00086)	0.00024 (0.00007)
0.2, 0.5, $0.95e^{0.1i}$	0.00217 (0.00070)	0.00165 (0.00055)	0.00069 (0.00024)
0.2, 0.3, $0.95e^{0.5i}$	0.02219 (0.00825)	0.02133 (0.00794)	0.00676 (0.00253)
0.2, 0.5, 0.85, 0.9	0.00682 (0.00227)	0.00629 (0.00208)	0.00363 (0.00117)
0.2, 0.5, 0.70, 0.9	0.01780 (0.00495)	0.01673 (0.00474)	0.00425 (0.00132)
0.2, 0.5, 0.89, 0.9	0.07777 (0.05265)	0.07756 (0.05248)	0.13216 (0.09800)

modes are complex, and are defined as

$$\lambda_p = 0.95e^{i\varphi_p}, \quad \varphi_p = 0.3 + 1.2 \cdot \left(\frac{p-1}{P-1} \right), \quad p = \overrightarrow{1, P},$$

that means the uniform distribution of p natural frequencies in the interval $[0.3, \ 1.5]$. The dimension n of the autoregressive part is equal to $2P$. We find the values of the first 30 singular values σ_i of the matrix $\tilde{Y}_{relax}$ for systems with different P.

Figure 8.10 shows σ_i, $i = \overrightarrow{1, 30}$ in the deterministic case $\varepsilon_Y = 0$. A sharp jump in values clearly indicates the true dimension of the system.

Now we consider the case with the presence of uncertainty. Figure 8.11 shows the same entities when $\varepsilon_Y = 10^{-3}$.

The gap in values has become much smaller, but can also be recognized.

Let us now consider how the magnitude of the first "noise" singular number σ_{2P+1} relates to the threshold value defined in (8.16). Since we set the measurement noise level as a relative value ε_Y, we must use normalization. We consider the ratio of singular values

$$\frac{\sigma_{2P+1}}{\sigma_1} = \frac{\left\| Y_{relax}^N \right\|_F}{\left\| \tilde{Y}_{relax} \right\|_F} \leq \frac{2\sqrt{Rl} \cdot \varepsilon_Y \cdot \|Y\|_{\max}}{\left\| \tilde{Y}_{relax} \right\|_F} \leq \frac{2\sqrt{Rl} \cdot \varepsilon_Y \cdot \|Y\|_{\max}}{\left\| \tilde{Y}_{relax} \right\|_{\max}}.$$

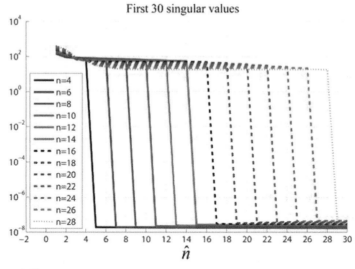

Figure 8.10   First singular values of the matrix Y_{relax} for systems of different order

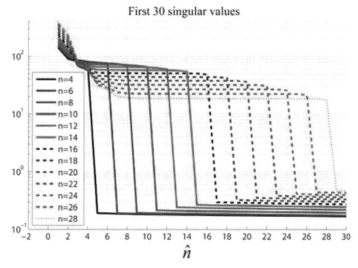

Figure 8.11   First singular values of the matrix $\tilde{Y}_{relax}$ for systems of different order. Case of relative error bound $\varepsilon_Y = 10^{-3}$

Let us put that $\|Y\|_{\max} \approx \left\|\tilde{Y}_{relax}\right\|_{\max}$, then $\frac{\sigma_{2P+1}}{\sigma_1} \leq 2\sqrt{Rl} \cdot \varepsilon_Y$. Therefore, let's build a graph of the singular values normalized to the first of them. Figure 8.12 shows normalized singular values for $\varepsilon_Y = 10^{-3}$ and a threshold.

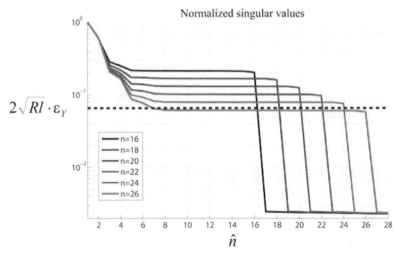

Figure 8.12 First normalized singular values of the matrix $\tilde{Y}_{relax}$ for systems of different order. Case of relative error bound $\varepsilon_Y = 10^{-3}$

As we can see, the guaranteed estimate $2\sqrt{Rl} \cdot \varepsilon_Y$ is overestimated due to the ratio of matrix norms $\| \cdot \|_F$ and $\| \cdot \|_{max}$. To get rid of this we check the possibility of determining the dimension directly in max norm. Figure 8.13 shows values $\tilde{\sigma}_{\hat{n}+1} = \dfrac{\left\| Y_{relax}^N(\hat{n}) \right\|_{max}}{\left\| \tilde{Y}_{relax} \right\|_{max}}$ for $\varepsilon_Y = 10^{-3}$ and a threshold $2\varepsilon_Y$.

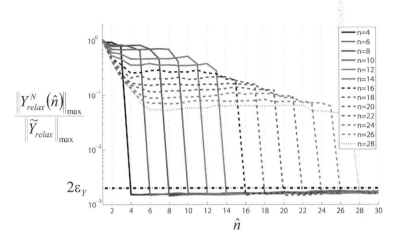

Figure 8.13 Relative max norm of the matrix Y_{relax}^N

The threshold $2\varepsilon_Y$ is much more accurate, but values $\tilde{\sigma}_{\hat{n}+1}$ are not monotonically decreasing. As we can see, this method can also be used to determine the dimension.

Thanks to a good selection of input signals, it is possible to accurately determine the order of the system up to high dimensions.

8.4.7 Maximum Dimension of Identifiable Model

Despite the fairly good possibilities of determining the exact dimension of the system, this is not a guarantee that the problem of identifying a model of this dimension will be correctly posed. Let us check the output error of full-size models identified under conditions of uncertainty in the data.

We use a set of favorable systems, similar to the one used in the previous paragraph. Identification is carried with next settings: use the signal matrix $Y_{relax}^{S}(n)$ is on, $K_{NS} = 100$, 50 realizations of noise with a relative bound $\varepsilon_Y = 10^{-10}$. Results are collected in a Table 8.4.

As we can see, the model of dimension 22 cannot be considered suitable in any way. This means that even in the case of determining the exact, but high dimension of the system, it is necessary to restrict ourselves only to models of reduced dimension. The limiting dimension, in addition to the characteristics of the system itself, depends on the magnitude of the noise.

Figure 8.14 shows values dependence of the accuracy of the model on its dimension and the magnitude of the noise.

Table 8.4   Quality of models found

P	Relative error by output	
	Maximum	Average
2	4.6773e-11	2.1630e-11
3	1.6977e-10	8.7567e-11
4	1.1200e-09	6.0051e-10
5	1.1383e-08	5.3145e-09
6	1.1548e-07	5.2235e-08
7	9.8088e-07	4.7196e-07
8	9.0444e-06	4.0171e-06
9	9.0678e-05	4.4297e-05
10	0.003637	0.002871
11	0.243185	0.228510

Model's output error

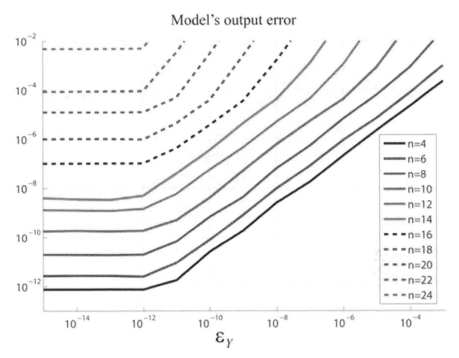

Figure 8.14 Accuracy of the model, depending on the dimension and the noise

8.5 Conclusion

In the classical statistical method, it is assumed that the implemented errors have the specified statistical properties, which ensure the consistency of the estimation, i.e. convergence to unbiased estimates of system parameters. In case of non-stochastic identification, arbitrary random errors in the data, including unfavorable ones, are allowed. In these cases, in order to ensure the quality of the models, during active experiments, it is necessary to pay serious attention to the information content of the system input. In this chapter, this issue is focused on. It is substantiated how the experiment should be planned so that the SNR for each of the modes of the system was the greatest. Numerical modeling has been carried out, which makes it possible to evaluate the effectiveness of the proposed approach to identification and interactively construct a mathematical model of the system, taking into account its structural features and experimental capabilities, on which the quality of the resulting models significantly depends.

References

[1] Yuriy P. Kondratenko, Arkadii A. Chikri, Vyacheslav F. Gubarev, Janusz Kacprzyk (Eds.), 'Advanced Control Techniques in Complex Engineering Systems: Theory and Applications', Dedicated to Professor Vsevolod M. Kuntsevich, Studies in Systems, Decision and Control, vol. 203, Springer Nature Switzerland AG, 337 pp., 2019.

[2] T.Söderström, & P.Stocia, 'System identification' U.K.: Prentice-Hall International, Hemel Hempstead, 612 pp., 1989.

[3] L.Ljung, 'System Identification: Theory for the user', Prentice-Hall, 1999.

[4] T. Söderström, 'Errors-in-variables methods in system identification', Springer, 2018.

[5] Van Huffel, S., & Lemmerling, P. (Eds.), 'Total least square and errors-in-variables modeling: Analysis, algorithms and applications', Springer, 2002.

[6] S. Beghelli, R.P. Guidorzi, U. Soverini, 'The frisch scheme in dynamic system identification', Automatica, vol. 26, pp. 171–176, 1990.

[7] M. Hong, T. Söderström, W.X. Zheng, 'Accuracy analysis of bias-eliminating least squares estimates for errors-in-variables systems', Automatica, vol. 43, pp. 1590–1596, 2007.

[8] T. Söderström, 'Accuracy analysis of the Frisch scheme for identifying errors-in-variables systems', IEEE Transactions on Automatic Control, vol. 52, no 6, pp. 985–997, 2007.

[9] T. Söderström, 'Errors-in-variables methods in system identification', Automatica, vol. 43, pp. 939–958, 2007.

[10] T. Söderström, U. Soverini, 'Errors-in-variables identification using maximum likelihood estimation in the frequency domain', Automatica, vol. 79, pp. 131–143, 2017.

[11] E. Zhang, R. Pintelon, 'Identification of multivariable dynamic errors-in-variables system with arbitrary inputs', Automatica, vol. 82, pp. 69–78, 2017.

[12] E. Zhang, R. Pintelon, 'Nonparametric identification of linear dynamic errors-in-variables systems', Automatica, vol. 94, pp. 416–425, 2018.

[13] Zheng, W.X. & Feng, C.B., 'Unbiased parameter estimation of linear systems in presence of input and output noise', International Journal of Adaptive Control and Signal Processing, vol. 3, pp. 231–251, 1989.

[14] Georgiou, T.T., & Smith, M.C., 'Optimal robustness in the gap metric', IEEE Transactions on Automatic Control, 35, pp. 673–686, 1990.

[15] Hyundeok Kang, Guoxiang Gu, Wei Xing Zheng, 'A graph subspace approach to system identification based on errors-in-variables system models', Automatica, Volume 109, 2019, 108535.

[16] Kuntsevich, V.M., Gubarev, V.F., Kondratenko, Y.P., Lebedev, D.V., Lysenko, V.P. (Eds), 'Control Systems: Theory and Applications'. River Publishers Series in Automation, Control and Robotics, 329 pp., 2018.

[17] Kondratenko, Y.P., Kuntsevich, V.M., Chikrii, A.A., Gubarev, V.F (Eds), 'Advanced Control Systems: Theory and Applications', River Publishers Series in Automation, Control and Robotics, 441 p., 2021.

[18] Verhaegen M. and Dewilde P, 'Subspace model identification. Part 1: The output-error state space model identification class of algorithms', International Journal of Control, Vol. 56, No 5, pp. 1187–1210, 1992.

[19] V.F. Gubarev, 'Modeling and Identification of Complex Systems', Naukova Dumka, 246 p. 2019 (in Ukrainian).

[20] G.H. Golub, Ch.F. Loan, 'Matrix Computations', The John Hopkins University Press, Baltimore, 780 p., 2013.

[21] A.N. Tikhonov and V.Y Arsenin, 'Solution of ill-posed problems', V.H. Winston & Sons, Washington, D.C.: John Wiley & Sons, New York, 258 p., 1977.

9

Fuzzy Systems Design: Optimal Selection of Linguistic Terms Number

Oleksiy Kozlov[1], Yuriy Kondratenko[1], Oleksandr Skakodub[1], and Zbigniew Gomolka[2]

[1]Petro Mohyla Black Sea National University, 10 68th Desantnykiv st., Mykolayiv, 54003, Ukraine
[2]University of Rzeszow, Department of Computer Engineering, Rzeszow, Poland
E-mail: kozlov_ov@ukr.net; y_kondrat2002@yahoo.com; aleksandrskakodub1996@gmail.com; zgomolka@ur.edu.pl

Abstract

This chapter considers developing and researching advanced information technology for fuzzy systems (FSs) design and structural optimization with optimal linguistic terms (LTs) numbers. The application of the proposed information technology allows increasing the FS accuracy and performance, reducing the computational costs spent on the generation of rule base (RB) and optimization of parameters, and simplifying its software and hardware implementation. The effectiveness study of the developed information technology is carried out on a specific example, particularly when designing and optimizing the fuzzy control system of a quadrotor unmanned aerial vehicle (QUAV). The research results obtained confirm the high effectiveness of the proposed information technology and the feasibility of its use in creating various types of fuzzy decision-making and control systems.

Keywords: A fuzzy system, design, and structural optimization, information technology, linguistic terms, fuzzy controller, quadrotor unmanned aerial vehicle.

9.1 Introduction

In today's world, artificial intelligence (AI) is the engine of overall progress. AI technologies can help address global challenges such as combating climate change and optimizing results in transport, medicine, agriculture, etc. [1, 2]. In particular, modern AI technologies are used for trading financial instruments on stock exchanges, carrying out crewless vehicles, establishing medical diagnoses, analyzing large amounts of data, recognizing and generating images, creating industrial and household robots, as well as high-precision autonomous weapons [3, 4]. AI processes vast amounts of data, making conclusions with the help of specific algorithms, which become the basis for independent decision-making.

Fuzzy logic is a relatively widespread and powerful AI tool with great potential [5–7]. It allows to operate effectively with expert knowledge, simulate human decision-making processes, and form linguistic models of complicated dynamic plants [8–10]. The most expedient is the application of methods and means of fuzzy logic to construct intelligent decision-making and control systems of different types. In particular, expert systems for decision-making based on fuzzy logic techniques are successfully used in conditions of uncertainty in many areas: financial management, technical and medical diagnostics, stock market forecasting, transport logistics, and others [11–13]. Also, fuzzy devices and inference systems are developed and frequently implemented as observers and adaptive devices, controllers and identifiers, units of tactical and strategic control for automation of complex nonstationary and nonlinear objects, such as power plants, robotic production lines, chemical reactors, spacecraft, mobile robots, satellites, marine ships, and floating structures, crewless underwater vehicles, drones and others [14–16].

A specific feature of units and systems based on fuzzy logic is a considerable number of structural elements and parameters that significantly influence their efficiency and should be tuned by the experts or automatically using specific progressive techniques during their design process [17–19]. In particular, the main adjustable parameters and elements are: (a) the shapes and parameters of the linguistic terms membership functions, (b) the number of LTs for input and output signals, (c) the types of operations of aggregation, activation, accumulation, and the defuzzification procedures, (d) the number of rules of the fuzzy rule base and its antecedents and consequents, (e) the adjustable input gains, etc. [20–22]. It gives the ability to create effective, flexible, and complicated strategies of decision-making and control but, at the same time, significantly complicates the development process of the given units and systems. Thus, in many cases, using expert assessments, fuzzy logic systems do not enormously increase the main quality indicators compared to similar systems based on conventional principles. Moreover, they may require more computational costs and more powerful hardware [23–25]. However, recently, the development and implementation of perspective progressive algorithms, approaches, and information technologies for fuzzy systems design based on specific optimization procedures are one of the leading directions of research in the FSs modern theory [26–28]. Modern fuzzy systems developed using these advanced optimization-based techniques can be successfully trained like neural networks, using training samples or objective functions for high accuracy attainment, and still maintain good interpretability and transparency [29–31].

This chapter focuses on developing and researching an advanced information technology for designing and optimizing fuzzy systems based on the optimal selection of linguistic terms number. The structure of the paper is organized as follows. Section 9.2 presents the brief literature survey, the statement of the research problem, and the primary purpose of this study. Section 9.3 sets out in detail the proposed information technology of the FS structural optimization based on the optimal selection of linguistic terms number. Section 9.4 presents the results of studying the effectiveness of the developed information technology on a specific example of a fuzzy control system for the quadrotor unmanned aerial vehicle with a detailed discussion of the results of computer simulation. Finally, Section 9.5 concludes the work and suggests the directions for future studies.

9.2 Related Works and Problem Statement

Many papers are currently published devoted to developing and approbating fuzzy control and decision-making systems of different types and fields [32–34]. Also, many studies focus on the challenges of synthesis and structural-parametric optimization of fuzzy systems that are solved through specifically developed methods, algorithms, and information technologies [35–37]. In particular, approaches and information technologies of design with optimization of the weight gains of rule bases and linguistic terms membership functions are discussed in papers [38–40]. In turn, methods of optimization of fuzzy systems' structures based on optimal selection of defuzzification procedures and RB interpolation and reduction are considered in [41–43]. The cutting-edge studies show that nature-inspired intelligent techniques are promising for synthesizing fuzzy systems and structural-parametric optimization. They have substantial advantages over classical optimization methods [44–46]. The main of them are: algorithms of ant colony optimization [47–49], particle swarm optimization [50–52], biogeography based optimization [53–55], grey wolf optimization [56–58], genetic methods [59–61] and evolutionary strategies [62], immune methods [63–65] and others [66–69]. The main benefits of these intelligent techniques are as follows: 1) the opportunity of effective optimization of fuzzy systems of large dimension; 2) the possibility of detailed studying of large, multimodal, and nonsmooth search space, excluding looping at local minima; 3) absence of any restrictions on the FS objective functions [70–72].

When analyzing the task of developing and optimizing a fuzzy system using the given above nature-inspired intelligent methods and information technologies, it is necessary to pay special attention to the structural optimization issue [39, 73, 74]. The given challenge is the most important since the optimal variant of the system's structure found in the search process allows not only to improve the accuracy, interpretability and transparency. It also reduces the computational costs spent on the RB creation and parametric optimization and the complexity of the FS software and hardware implementation. The best structure of the fuzzy inference system is such a variant of structure that provides: (a) attainment of the high accuracy, interpretability and transparency; (b) reasonable computational costs spent on the RB creation and optimization of parameters; (c) the admissible complexity of the FS software and hardware implementation [39, 73, 75]. Therefore, studies directed towards developing and enhancing approaches and information technologies for obtaining the fuzzy systems' optimal structures are relevant and urgent for the advanced theory of fuzzy control and decision-making systems.

The scope of main tasks of structural optimization includes the following tasks: determination of the LTs optimal number and membership functions types; determination of the optimal procedures of fuzzy inference engine (FIE), which include aggregation, activation and accumulation; determination of the optimal defuzzification procedure; reduction of the RB antecedents and number of rules [73, 76, 77]. Herewith, reduction of the rule base antecedents and number of rules can be performed separately after the generation of the structure and all the FS parameters to increase its interpretability and transparency and simplify further software and hardware implementation [41, 42, 46]. Also, these optimization procedures do not affect FS parametric optimization and optimization of other structural elements. Moreover, the determination of the optimal procedures of a fuzzy inference engine and defuzzification procedure can also be performed separately after synthesizing the FS structure and all the parameters. It increases its performance and accuracy and does not affect FS parametric optimization and optimization of other structure elements [28, 39]. In turn, the optimization task of the LT membership functions types can be solved at the determined number of terms and at a composed rule base for improving FS performance and accuracy and simplifying its subsequent parametric optimization procedures [73]. Thus, conducting this optimization procedure significantly affects the further parametric optimization of the FS, namely, the number of optimized parameters of LTs, which depends on the shapes of their MF. As for the problem of determining the optimal number of linguistic terms, the efficiency of its solution significantly affects the performance and accuracy and the complexity of the subsequent procedures of parametric optimization and software and hardware implementation of the FS. The setting of various numbers of linguistic terms allows implementing different strategies of control and decision-making at forming the rule bases of FSs. So, the number of terms for input and output signals affects the initial antecedents sets, the number of fuzzy rules and RB possible consequents, and the number of terms parameters that should be further optimized during the FS development process. In addition, it is necessary to compose a new rule base to calculate the FS efficiency for each new variant obtained in the process of the LTs number optimization, which requires high computational costs. Therefore, selecting the optimal number of linguistic terms is the most complicated and vital task of structural optimization of fuzzy systems, the solution of which is the focus of the present work. For detailed studying of the problem of selecting the LTs optimal number, next, consider the generalized MIMO fuzzy system with the specialized structural optimization unit.

The functional structure of the generalized MIMO fuzzy control/decision-making system with structural optimization unit that implements the optimal selection of LTs number is presented in Figure 9.1, where the following notations are accepted: $\mathbf{X}$ is the vector of FS inputs; $\mathbf{Y}$ is the vector of FS outputs; $\mathbf{S}$ is the vector that determines numbers of LTs for fuzzy system inputs and outputs; $\mathbf{R}$ is the vector that determines RB rules that consists of the corresponding antecedents and consequents; $\mathbf{LT}$ is the vector that determines certain LTs for FS inputs and outputs, based on which the antecedents and consequents of the rules are formed; J is the objective function that evaluates the FS performance and accuracy; $\mathbf{Z}$ is the vector of the outputs

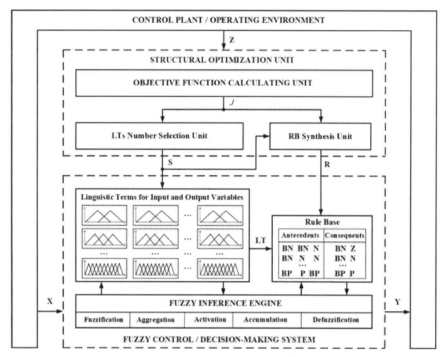

Figure 9.1 Structure of the generalized MIMO fuzzy control/decision-making system with structural optimization unit

of the control plant/operating environment that characterize its performance and quality indicators.

The presented generalized fuzzy system (Figure 9.1) implements the following nonlinear dependence f_{FS} [29, 45]

$$\mathbf{Y} = f_{FS}(\mathbf{X}),$$
$$\mathbf{Y} = (y_1, y_2, ..., y_j, ..., y_m), \mathbf{X} = (x_1, x_2, ..., x_i, ..., x_n), \quad (9.1)$$

where $\mathbf{X}$ is the vector of FS n input variables $x_1, x_2, \ldots, x_i, \ldots, x_n$; $\mathbf{Y}$ is the vector of FS m output variables $y_1, y_2, \ldots, y_j, \ldots, y_m$.

The fuzzy inference engine transforms input signals $\mathbf{X}$, performing sequential operations (fuzzification, aggregation, activation, accumulation, and defuzzification) to obtain the output variables $\mathbf{Y}$ [29, 78, 79]. Specific sets of linguistic terms (vector $\mathbf{LT}$) are used for fuzzification and defuzzification of FS inputs and outputs. They depend on the selected LTs numbers $\mathbf{S}$.

The calculation of the fuzzy values of the system outputs is performed based on the set of rules stored in the rule base and correspond to the vector **R**. Each fuzzy rule consists of the specific antecedent and consequent and, for example, can be defined by the expression [29]

$$
\begin{aligned}
&\text{IF } "x_1 = A_1" \text{ AND } "x_2 = A_2" \text{ AND } \dots \text{ AND} \\
&"x_i = A_3" \dots \text{ AND} \dots \text{ AND } "x_n = A_4" \\
&\text{THEN } "y_1 = B_1" \text{ AND } "y_2 = B_2" \text{ AND } \dots \text{ AND} \\
&"y_j = B_3" \dots \text{ AND} \dots \text{ AND } "y_m = B_4",
\end{aligned}
\tag{9.2}
$$

where A_1, A_2, A_3, A_4, B_1, B_2, B_3 and B_4 are corresponding LTs for the FS input and output variables.

To solve the task of selecting the optimal number of linguistic terms, the given fuzzy system (Figure 9.1) uses the specialized structural optimization unit that consists of the objective function calculation unit, the unit for LTs optimal number selection, and the unit that implements the automatic RB synthesis. In turn, the unit for LTs number optimal selection should provide finding such a vector **S** that will allow the FS high performance, accuracy and quality indicators, and admissible complexity of the FS software and hardware implementation. In this case, the vector **S** is presented in the following form

$$
\mathbf{S} = \{S_{xi}, S_{yj}\}, i = (1, 2, \dots, n), j = (1, 2, \dots, m),
\tag{9.3}
$$

where S_{xi} and S_{yj} are selected numbers of LTs for ith input and jth output FS variables.

For example for fuzzy system with 3 input variables ($n = 3$) and 2 output variables ($m = 2$) the following numbers of LTs are selected for the given variables: $S_{x1} = 5$, $S_{x2} = 3$, $S_{x3} = 3$, $S_{y1} = 5$, $S_{y2} = 7$. Then, the vector **S** has the following form

$$
\mathbf{S} = \{5, 3, 3, 5, 7\}.
\tag{9.4}
$$

In turn, for each found variant of the vector **S** by the unit of LTs number optimal selection, the new variant of the rule base (vector **R**) should be synthesized. Only then, the objective function J can be calculated for evaluation FS with obtained configuration. Thus, the task of the RB synthesis is the subtask of the LTs number optimal selection, which should be solved each time at every iteration during the implementation of the LTs number optimization process. The given subtask should be solved in automatic mode by the RB synthesis unit. In addition, in the optimization process, both units of the LTs number optimal selection and the RB synthesis use objective function J to evaluate the FS performance, accuracy, quality indicators, and complexity of software and hardware implementation. The objective function calculation unit should be applied to calculate the current value of the objective function J based on the vector $\mathbf{Z}$ of the control plant's outputs.

Thus for effective solving the presented complicated task of the optimal selection of the number of the linguistic terms for fuzzy control and decision-making systems, the given structural optimization unit should implement the specifically developed advanced information technology based on the bioinspired intelligent optimization methods.

The primary purpose of this work is the development and research of advanced information technology for fuzzy systems design and structural optimization with optimal selection of the number of linguistic terms based on the bioinspired intelligent methods.

9.3 Information Technology for Fuzzy Systems Design and Structural Optimization with Optimal Selection of Linguistic Terms Number

In this work, we developed two modifications of the information technology for fuzzy systems design and structural optimization with optimal selection of linguistic terms number: 1) based on the sequential search approach; 2) based on the random search approach. Moreover, each modification implements a nature-inspired technique of ant colony optimization (ACO) [29] and a method of sequential search [21] for automatic development of the rule base that corresponds to each new generated vector $\mathbf{S}$ that defines linguistic terms number. The first modification based on the sequential search uses the approach of sequential selection of the best values of the number of the terms for the FS, starting from the first FS input x_1 and ending with its last output y_m. This modification consists of the following successive stages.

Stage 1. Choosing of the FS inputs and outputs. At the given stage, for the developed FS n inputs and m outputs are chosen. In turn, the total number of the system's variables $n + m$ is selected, taking into account the peculiarities of the specific problem that FS should solve.

Stage 2. The setting of the minimum and maximum values of FS inputs and outputs. At this stage, the operating ranges between the minimum and maximum values are set for all n input and m output variables. For instance, if all signals come to the system's input in relative units to their maximum values, the operating ranges should be set from –1 to 1 [80, 81]. In addition, if some inputs or outputs can be only positive by nature (electric heating power, water or gas consumption, and others), then their changing ranges should be from 0 to 1.

Stage 3. Choosing the shapes of LT membership functions for FS inputs and outputs. The shapes of linguistic terms MFs for all n input and m output variables are chosen. Often, the same shapes are selected for all LTMFs of the system's variables at the initial stage of the designing process. For instance, Gaussian, trapezoidal or triangular shapes may be chosen [82, 83]. Also, the initial values of their parameters for each FS input and output should be set in an automatic mode so that the LTs would be evenly distributed over the operating ranges (between the minimum and maximum values) for any specified LTs number.

Stage 4. Determination of the vector **S** initial structure. The vector **S** that defines LTs numbers for FS input and output variables is formed at this stage, taking into account the chosen at the Stage 1 n inputs and m outputs. It has the following form

$$\mathbf{S} = \{S_{x1}, S_{x2}, ..., S_{xi}, ..., S_{xn}, S_{y1}, S_{y2}, ..., S_{yj}, ..., S_{ym}\}. \qquad (9.5)$$

The given vector includes variables $S_{xi,yj}$ that define the following numbers of the LTs for the fuzzy system's variables. These variables are located in order, starting from the first input x_1 and ending with the final output y_m.

Stage 5. The setting of vector **S** constraints. The constraints $\mathbf{S}_{min}$ and $\mathbf{S}_{max}$ on the LTs number are selected for each FS input and output. For example, the minimum value of the number of the terms may be selected equal to 2 ($S_{min} = 2$), and the maximum value can be chosen equal to 7 ($S_{max} = 7$) for each FS input [28, 39]. In turn, for the FS outputs, the minimum and maximum values can be equal to 3 ($S_{min} = 3$) and 9 ($S_{max} = 9$), respectively [82]. Moreover, at this stage, some additional constraints may be selected. For instance, for fuzzy control systems, if their variables can vary from positive to negative values within the symmetric operating ranges, the additional constraints may be set, that the number of the linguistic terms should be only odd [39]. Therefore, in this case, in addition to basic constraints ($\mathbf{S}_{min}$ and $\mathbf{S}_{max}$), the LTs number maybe only be as follows: 3, 5, and 7 – for FS inputs ($S_{xi} \in \{3, 5, 7\}$); 3, 5, 7, and 9 – for output variables ($S_{yj} \in \{3, 5, 7, 9\}$).

Stage 6. Selection of the FS complex objective function J_C. The main parameters and boundary (optimal) value of the complex objective function

J_C for evaluating the effectiveness of the developed FS with a certain number of LTs are selected at this stage. As the number of rules, the antecedents sets, and possible consequents of the system's RB directly depend on the determined number of linguistic terms, then two criteria should be used in the vector **S** structural optimization process. The criterion J_1, which evaluates the developed control or decision-making FS effectiveness, and criterion J_2, which considers the complexity of the designed rule base and its software and hardware realization [73]. Therefore, the problem of the optimal LTs number determining is the multi-criteria optimization problem [73, 84, 85], solve which it is necessary to find a compromise minimum of two objective functions J_1 and J_2. At solving this problem, the prior approach for multi-criteria optimization with aggregation of several objective functions is most advisable [73, 85]. According to this approach, two criteria J_1 and J_2, with the initial evaluation of their importance, are combined into one global criterion the complex objective function J_C. Thus, the current value of the complex objective function J_C should be calculated during the search process based on the following equation

$$J_C = J_1 + k_{J2} J_2, \tag{9.6}$$

where k_{J2} is the scale coefficient at J_2 that defines the significance of considering this objective function during the optimization process of the number of linguistic terms of the fuzzy system.

For example, suppose the fuzzy system is designed based on a training sample. In that case, the mean squared error (MSE) may be used as the criterion J_1, which measures the average of the squares of the errors between the estimated and actual FS outputs. In the case of the fuzzy control system, the objective function J_1 can be selected as an integral quadratic error of control or the integral quadratic deviation between the outputs of the reference model and the actual control plant [29, 30, 86]. In turn, for the fuzzy decision-making system, the criterion J_1 may be calculated as the percentage of wrong decisions obtained for the test dataset.

As for the values of the objective function J_2, which evaluates the complexity of the designed rule base as well as software and hardware realization of the fuzzy system, they should be calculated based on the expression [45]

$$J_2 = r_{\max} \left(\sum_{j=1}^{m} S_{yj} \right) = \left(\prod_{i=1}^{n} S_{xi} \right) \left(\sum_{j=1}^{m} S_{yj} \right), \tag{9.7}$$

where r_{max} is the total number of the RB rules, defined by the number of all possible combinations of the LTs of the system's inputs.

Moreover, the boundary (optimal) values of the functions J_{1opt} and J_{Copt} and the corresponding coefficient k_{J2} are preliminarily selected at this stage. It is necessary to consider the main requirements and peculiarities of the problem of fuzzy system design. For example, the optimal value of the objective function J_{Copt} and the scaling factor k_{J2} can be selected based on the following conditions

$$\frac{J_{Copt}}{J_{1opt}} < 2, \frac{J_{Copt}}{J_{1opt}} > 2, \frac{J_{Copt}}{J_{1opt}} = 2. \tag{9.8}$$

Stage 7. Determination of the vector **S** initial value. The initial values of LTs numbers (vector $\mathbf{S}_0$) of the FS inputs and outputs are defined at this stage, which are the starting points for the optimization process. In turn, the vector $\mathbf{S}_0$ should be set as its minimum value $\mathbf{S}_{min}$ that can be determined as constraints at Stage 5

$$\mathbf{S}_0 = \mathbf{S}_{min} = \{S_{x1\,min}, ..., S_{xi\,min}, ..., S_{xn\,min},$$
$$S_{y1\,min}, ..., S_{yj\,min}, ..., S_{ym\,min}\}. \tag{9.9}$$

At the following stages of the proposed information technology, the sequential search of the numbers of linguistic terms of the FS is performed, starting from the first FS input x_1 and ending with its last output y_m.

Stage 8. Selection of the first FS variable. At this stage, the transition to the first input variable x_1 of the fuzzy system is conducted to start the iterative sequential search for obtaining the optimal vector $\mathbf{S}_{opt}$ of the number of linguistic terms. Furthermore, the iterative search procedure starts from the initial value $\mathbf{S}_0$, previously determined at Stage 7.

Stage 9. Checklist check. Each variant of vector **S**, for which the fuzzy system RB is already developed with calculating the complex objective function J_C current value in the given information technology implementation, is recorded to the previously created Checklist. Also, the corresponding rule base and value of the system's objective function J_C enter into the Checklist. At this stage, the Checklist check is implemented to avoid the repeated rule base generation and additional calculations of its objective function J_C for the system with the same vector **S** of linguistic terms number. It gives the opportunity not to perform unnecessary iterations with rule base development and calculation of the complex objective function J_C, which number is $n +$

$m - 1$. If the current vector of the LTs number **S** is already recorded to the Checklist, then transition to Stage 13 is performed. In the opposite case, go to Stage 10.

Stage 10. Development of the rule base for the system with the current vector **S** and calculation of its complex objective function. The rule base generation is conducted at this stage through the ACO-based method or technique of the sequential search, designed and approbated in papers [29] and [21], respectively. Also, the complex objective function J_C value is calculated for this designed rule base. As the previous comprehensive research [45] shows, it is expedient to choose the sequential search technique or the nature-inspired ACO-based method for performing an effective automatic RB design, depending on its size and complexity, defined with the help of criterion J_2 (9.7). Herewith, if the criterion $J_2 \leq 600$, the sequential search approach should be used. If $J_2 > 600$, then the nature-inspired ACO-based technique will be more efficacious [45]. Consequently, at this stage, the criterion J_2 (9.7) for the current vector **S** is first calculated, and only then, based on its defined value, a suitable method for RB design is chosen (ACO-based method or sequential search). Further, the rule base development is performed with the calculation of the criterion J_1 using the chosen technique. After that, the value of the complex objective function J_C is finally calculated by the equation (9.6) using values of the criteria J_1 and J_2. In turn, the adjustable parameters of the ACO-based and sequential search algorithms (number of rounds of sequential search l, number of agents in the population Z_{max}, number of elite agents e, maximum number of iterations N^*_{max}, etc.) are previously set based on conducted experimental studies and obtained recommendations in the previous research [29, 45]. The antecedents of the rule base for the current vector **S** are synthesized in an automatic mode as all possible combinations of LTs of the system's inputs. As for their consequents, they are determined during the design process through the above sequential search method or ACO-based approach.

Stage 11. Check for the attainment of the complex objective function optimal value. At this stage, the achievement of the optimal value of the complex objective function J_{Copt} is checked for the obtained vector **S** of the LTs number. If the given check is positive, then transition to Stage 16 is carried out. Otherwise, go to Stage 12.

Stage 12. Record the current vector **S**, the corresponding rule base, and the system's value of the complex objective function to the Checklist. The current vector **S**, the corresponding developed rule base, and the system's current value of the complex objective function J_C are recorded to the Checklist at this stage.

Stage 13. Check the optimization process completion for the FS current variable. Optimization calculations are considered complete for each FS variable if the RBs were developed for every existing variant of the number of the linguistic terms $S_{xi,yj}$ within the set constraints $[S_{xi,yjmin}, S_{xi,yjmax}]$ for the given variable. If this check is positive, then transition to Stage 14 is carried out. In the opposite case, one is added to the current LTs number $S_{xi,yj}$ for the given variable ($S_{xi,yj} + 1$), and then return to Stage 9 is performed.

Stage 14. Choosing the best value of the number of linguistic terms for the current system's variable. Determination of the number of terms $S_{xi,yjbest}$ is performed at this stage, for which the complex objective function J_C is the smallest (J_{Cbest}) among all generated during the search process for this specific variable. Then, this variant of the LTs number $S_{xi,yjbest}$ is set for the given variable.

Stage 15. Check the optimization process completion for all the system's variables. The calculations of structural optimization are considered complete at this stage if the linguistic terms numbers were optimized and the best variants S_{best} were selected and set for all $n + m$ variables (starting with the first input x_1 and ending with the last output y_m). If the given check is positive, then the transition to Stage 16 is performed. Otherwise, go to the next FS variable ($i, j + 1$) and return to Stage 9.

Stage 16. Completion of the optimization process of the LTs number. When the LTs number optimization process is complete, the additional parametric or structural optimization of the FS membership functions can be carried out and the system's implementation using proper hardware and software [87]. In turn, the software and hardware realization will be significantly simplified due to the obtained optimal structure of the fuzzy system and the optimal number of rules of its rule base.

Figure 9.2 shows a flowchart of the first modification of the proposed information technology for fuzzy systems development and structural optimization with optimal selection of LTs number.

At the implementation of the given modification of the proposed information technology, the maximum number of iterations N_{max} is determined based on the expression

$$N_{\max} = \sum_{i=1}^{n} (\Delta S_{xi} + 1) + \sum_{j=1}^{m} (\Delta S_{yj} + 1) - (n + m - 1), \qquad (9.10)$$

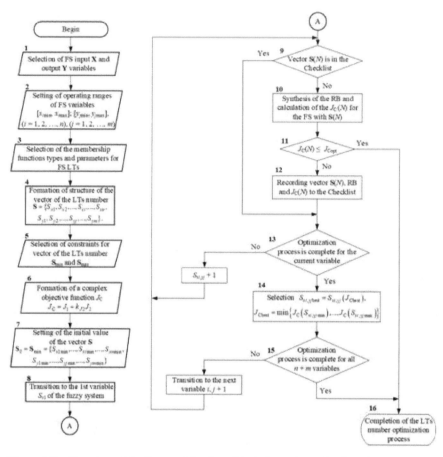

Figure 9.2   Flowchart of the first modification of the information technology for FS structural optimization based on optimal selection of linguistic terms number

where

$$\Delta S_{xi} = S_{xi\max} - S_{xi\min}, i = 1, 2, ..., n; \qquad (9.11)$$

$$\Delta S_{yj} = S_{yj\max} - S_{yj\min}, j = 1, 2, ..., m. \qquad (9.12)$$

Moreover, each N-th iteration of the presented information technology includes conducting N^*_{max} iterations of the ACO-based method or sequential search technique at Stage 10.

The second modification based on the random search approach uses the concept of creating H random variants of vectors of the LTs number and choosing the best from them. The first six stages of this modification are the same as for the first modification of the information technology. Next, consider the rest of the stages of this modification, starting from Stage 7.

Stage 7. Generation of H different variants of vector $\mathbf{S}$ in a random way. At this stage, H different variants of the vector of LTs number are randomly generated. In turn, each variable $S_{xi,yj}$ of each h-th vector $\mathbf{S}_h$ is generated randomly within the constraints $S_{xi,yjmin}$ and $S_{xi,yjmax}$, that are set at Sage 5. Moreover, the number of random variants H is previously selected before implementing this modification of the information technology. Herewith, if there are some identical vectors among the generated variants, then they are deleted, and new ones are generated in their place until all H vectors $\mathbf{S}$ are different. The rule bases will be synthesized, and the complex objective function J_C values will be calculated for each h-th vector of the given generated vectors at the further stages of the given modification of the information technology.

Stage 8. Selection of the first variant of the generated vectors $\mathbf{S}$. The first variant ($h = 1$) of the generated vectors of the LTs number $\mathbf{S}_1$ is selected at this stage to start the procedures of the optimization process.

Stage 9. Development of the rule base for the system with the current vector $\mathbf{S}$ and calculation of its complex objective function. The given stage of this modification is the same as Stage 10 for the first modification of the information technology.

Stage 10. Check for the attainment of the complex objective function optimal value. The given stage of this modification is the same as Stage 11 for the first modification of the information technology. If this check is positive, then the transition to Stage 13 is carried out. In the opposite case, go to Stage 11.

Stage 11. Check for the optimization process completion of the LTs number for the FS. The calculations of structural optimization are considered to be complete at this stage if the design of the rule base and determination of the complex objective function J_C is conducted for all H vectors of the LTs numbers generated at Stage 7 ($h = H$). If the given check is positive, then the transition to Stage 12 is performed. Otherwise, go to the next ($h + 1$) generated vector $\mathbf{S}_{h+1}$ and return to Stage 9.

Stage 12. Selection of the best variant of the generated vectors $\mathbf{S}$. At this stage, the best variant of the H generated vectors $\mathbf{S}$ is selected based on their values of the complex objective function J_C calculated at Stage 9.

Stage 13. Completion of the LTs number optimization process. The given stage of this modification is the same as Stage 16 for the first modification of the information technology.

Figure 9.3 shows a flowchart of the second modification of the proposed information technology for fuzzy systems development and structural optimization with optimal selection of LTs number.

At the implementation of the given modification of the proposed information technology, it is advisable to set the number of generated vectors H the same as the maximum number of iterations N_{max} for the first modification for the objectivity of comparing the computational costs of these two modifications.

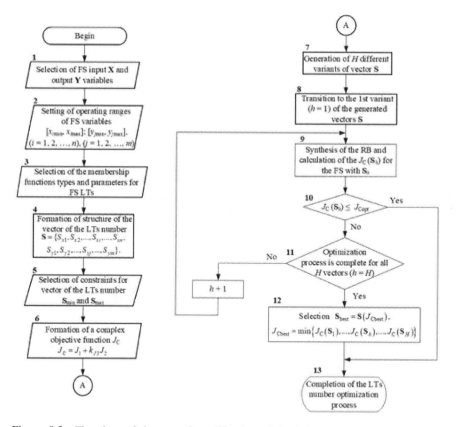

Figure 9.3 Flowchart of the second modification of the information technology for FS structural optimization based on optimal selection of linguistic terms number

from the previous stages

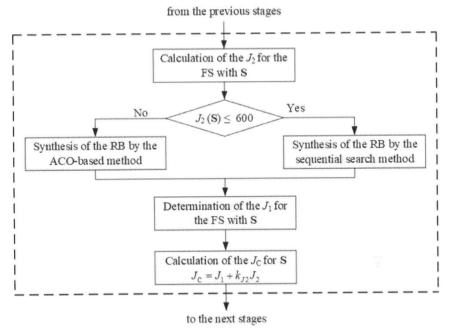

Figure 9.4 Flowchart of the stage of design of the RB and calculation of the complex objective function for the FS with the current vector **S**

Moreover, the flowchart of the stage of design of the RB and determination of the complex objective function for the fuzzy system with the current vector **S** (Stage 10 of the first modification, Stage 9 of the second modification) for both considered modifications has the form presented in Figure 9.4.

The main features of the sequential search and ACO-based methods for automatic rule base synthesis of different types of fuzzy systems are presented in works [21, 29, 45]. In turn, the sequential search method allows to carry out synthesis and optimization of highly efficient rule bases with an optimal set of consequents $\mathbf{R}_{opt}$, at which the value of the objective function J_1 of control or decision-making processes based on the developed FS will be optimal [21]. Antecedents of rules are formed as all possible combinations of linguistic terms of input variables of FS. The formation of the optimal vector of the RB consequents $\mathbf{R}_{opt}$ is carried out using an iterative search based on sequential enumeration of the consequents of each RB rule and calculation of the corresponding values of the objective function J_1.

As for the ACO-based method for automatic rule base synthesis, it is based on the principles of ant colony optimization and allows to represent the structure of the fuzzy system RB as a specific graph, on the edges and nodes of which the agents-ants will move [29].

The route of every Z-th agent of the colony runs through specific edges and nodes of the given graph at every separate iteration with number $N*$ [29, 88, 89]. After each Z-th agent passes through the whole graph at every $N*$-th iteration, the corresponding RB of fuzzy system with a certain consequents vector $\mathbf{R}^Z(N*)$ $(Z = 1, ..., Z_{max}; N* = 1, ..., N^*_{max})$ is formed. The route length that a specific ant has traveled in the current iteration depends on the value of the FS objective function (first criterion) $J_1^Z(N*)$.

During the movement of all ants along the graph, the probability $P_{r\gamma j}^Z(N*)$ of the Zth agent passing along the $r\gamma j$th edge from the node with number $r(j{-}1)$ to the rj-th node $(r \in \{1, ..., s\}; \gamma \in \{1, ..., S_{yj}\}; j \in \{1, ..., m\})$ in the $N*$-th iteration $(N* = 1, ..., N_{max})$ is calculated based on the equation [88]

$$P_{r\gamma j}^Z(N^*) = \frac{[\tau_{r\gamma j}(N^*)]^\alpha [\eta_{r\gamma j}]^\beta}{\sum_{r\gamma j=r1j}^{rvj} [\tau_{r\gamma j}(N^*)]^\alpha [\eta_{r\gamma j}]^\beta}, \qquad (9.13)$$

where r is the current rule number of the RB; γ is the current consequent number of the jth output variable; $\tau_{r\gamma j}(N^*)$ is the pheromone intensity on the $r\gamma j$th edge at the iteration N^*; $\eta_{r\gamma j}$ is an inverse of the relative length $D_{r\gamma j}$ of the $r\gamma j$th edge; α is an adjustable parameter which determines the importance of the pheromone trace considering on edge; β is the parameter that regulates importance of the relative length $D_{r\gamma j}$ of the $r\gamma j$th edge.

Other equations and main steps of the ACO-based method and sequential search method for automatic rule base design of fuzzy control and decision-making systems are considered in detail in [29, 89].

The effectiveness study of the developed information technology is performed in this work on a specific example, particularly when designing the fuzzy control system for the quadrotor unmanned aerial vehicle.

9.4 Design and Structural Optimization of the Fuzzy Control System for the Quadrotor Unmanned Aerial Vehicle

Quadcopters belong to the micro-class crewless aerial vehicles, which have recently become very popular and are widely used to implement monitoring and inspection tasks in various areas of the civil and military sector [90–92]. QUAVs have many advantages over other types of crewless aerial vehicles, namely: high maneuverability, high efficiency, the ability to take off and land vertically and hovering in space, ease of maintenance, and others. However, quadcopters are complicated control plants, for automation of which it is advisable to use intelligent automatic control systems, particularly fuzzy control systems [93–95]. The stabilization and automatic control of the QUAV flight altitude when moving over terrain with a complex relief is one of the most important tasks in automating its spatial motion. In this work, the fuzzy control system of the QUAV flight altitude is designed to test the effectiveness of the proposed information technology for FSs structural optimization with optimal selection of linguistic terms number.

The dynamics of spatial motion of the quadrotor unmanned aerial vehicle is described by the following system of basic equations [90]:

$$
\begin{cases}
\ddot{x} = \Big((F_1 + F_2 + F_3 + F_4)\,(\cos\phi\sin\theta\cos\psi + \sin\phi\sin\psi) \\
\quad - k_{x1}\,(\dot{x})^2 - k_{x2}F_x \Big)/m; \\[4pt]
\ddot{y} = \Big((F_1 + F_2 + F_3 + F_4)\,(\sin\phi\sin\theta\cos\psi + \cos\phi\sin\psi) \\
\quad - k_{y1}\,(\dot{y})^2 - k_{y2}F_y \Big)/m; \\[4pt]
\ddot{z} = \Big((F_1 + F_2 + F_3 + F_4)\cos\theta\cos\psi - F_g - k_{z1}\,(\dot{z})^2 - k_{z2}F_z \Big)/m; \\[4pt]
\ddot{\psi} = l\left(-F_1 + F_2 + F_3 - F_4 - k_{\psi 1}\left(\dot{\psi}\right)^2 - k_{\psi 2}F_\psi \right)/I_\psi; \\[4pt]
\ddot{\theta} = l\left(-F_1 - F_2 + F_3 + F_4 - k_{\theta 1}\left(\dot{\theta}\right)^2 - k_{\theta 2}F_\theta \right)/I_\theta; \\[4pt]
\ddot{\phi} = C_\phi\left(F_1 - F_2 + F_3 - F_4 - k_{\phi 1}\left(\dot{\phi}\right)^2 - k_{\phi 2}F_\phi \right)/I_\phi,
\end{cases}
$$

$$(9.14)$$

where x, y and z are QUAV's longitudinal-horizontal and transverse-horizontal coordinates, as well as flight altitude; ψ, θ and φ are the roll, pitch and yaw angles; m is the QUAV mass; F_1, F_2, F_3 and F_4 are the corresponding values of the rotors' lifting force; F_x, F_y, F_z, F_ψ, F_θ, and F_φ are the corresponding values of disturbing effects of wind; F_g is the gravity force; I_ψ, I_θ and I_φ are the QUAV's moments of inertia about the longitudinal, transverse and vertical axes; l is the distance from the rotor to the center of mass; C_φ is the yaw moment coefficient; k_{x1}, k_{x2}, k_{y1}, k_{y2}, k_{z1}, k_{z2}, $k_{\psi 1}$, $k_{\psi 2}$, $k_{\theta 1}$, $k_{\theta 2}$, $k_{\varphi 1}$ and $k_{\varphi 2}$ are the QUAV's model coefficients.

Since the time constants of the quadcopter drives are significantly less than the time constants of the quadcopter itself, then as an assumption, the dynamics of the rotor drives can be neglected. Thus, the control of the quadcopter is carried out using four control forces: u_1, u_2, u_3, and u_4 [90]. In turn, these forces are calculated by the equations

$$\begin{aligned}
u_1 &= F_1 + F_2 + F_3 + F_4; \\
u_2 &= -F_1 - F_2 + F_3 + F_4; \\
u_3 &= -F_1 + F_2 + F_3 - F_4; \\
u_4 &= F_1 - F_2 + F_3 - F_4.
\end{aligned} \tag{9.15}$$

The development and structural optimization with optimal selection of LTs number is conducted for the QUAV with the following main parameters: QUAV's total mass is 0.32 kg, distance from the rotor to center of mass is 0.209 m, total maximum lift force of 4 propellers is 6 N.

Figure 9.5 presents the functional structure of the FS for control of the QUAV flight altitude, where the following abbreviations are accepted: CSTL is the QUAV control system of tactical-level; CSHC is the control system of horizontal coordinates; SD is the setting device; AS is the altitude sensor, which directly determines the distance of the quadcopter from the terrain surface; R1, R2, R3, and R4 are the QUAV's 1st, 2nd, 3rd and 4th rotors; z_S and z_R are the flight altitude set and actual values relative to the terrain surface; z_T and z_{RT} are the absolute altitude values of the terrain surface and the QUAV, respectively; u_{SD} and u_{AS} are the SD and AS output signals; ε_z is the altitude control error of the QUAV flight.

The given control system (Figure 9.5) uses the Mamdani-type fuzzy controller (FC) for stabilizing and automatic control of the QUAV flight altitude, which inputs are: altitude control error ε_z, its derivative $\frac{d\varepsilon_z}{dt}$ and integral $\int \varepsilon_z dt$. In turn, the FC output is u_1. For a more objective study of the efficiency of the designed fuzzy control system, in this work, the process of QUAV's flight at a fixed altitude over mountainous terrain with a complex relief is considered. In this case, the quadcopter moves at a set constant speed $\dot{x}_s$ along the longitudinal-horizontal axis (x coordinate) over the mountainous terrain. Herewith, the value of the y coordinate remains unchanged, and the task of the fuzzy control system of the altitude is to keep the QUAV at a fixed distance from the surface z_S, bypassing around all the unevenness of the mountainous terrain. In this work, for simulating, we took a 1-kilometer long section with a complex relief that is randomly generated [96] and is presented by the dependence $z_T(x)$ and shown in Figure 9.6.

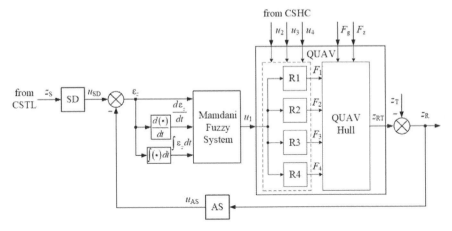

Figure 9.5 The fuzzy control system of the QUAV flight altitude

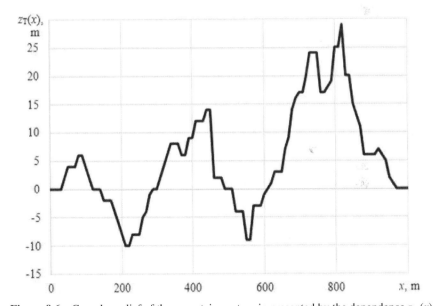

Figure 9.6 Complex relief of the mountainous terrain presented by the dependence $z_T(x)$

Then, we conducted the development and structural optimization with optimal selection of LTs number of the FC alternately using first and second modifications of the proposed information technology. First, the first six stages of information technology are performed, which are the same for both modifications.

At the first stage, we chose three input variables and one output variable for the designed fuzzy controller: ε_z, $\frac{d\varepsilon_z}{dt}$, $\int \varepsilon_z dt$, u_1. Therefore, the total number of the controller's variables is equal to 4. At Stage 2, the FC inputs and output operating ranges are determined from -1 to 1. Then, at the third stage, triangular shapes of the MFs are chosen for all LTs of the controller's inputs and output. The values of LTMF parameters are automatically set to be evenly distributed over the operating ranges for any specified number for each input and controller at the further optimization stages. At Stage 4, the vector **S**, which defines the numbers of LTs for the controller input and output variables, is formed in the following way

$$S = \{S_{x1}, S_{x2}, S_{x3}, S_{y1}\}. \tag{9.16}$$

At Stage 5, the vector **S** constraints on the LTs number are determined in the following way

$$S_{xi} \in \{3, 5, 7\}, i = 1, 2, 3; \\ S_{y1} \in \{5, 7, 9\}. \tag{9.17}$$

Furthermore, for all input and output variables of the fuzzy altitude controller using 3, 5, 7, and 9 terms, the following sets of LT with the triangular shape are implemented:

$$\{N, Z, P\};$$

$$\{BN, SN, Z, SP, BP\};$$

$$\{BN, N, SN, Z, SP, P, BP\};$$

$$\{VBN, BN, N, SN, Z, SP, P, BP, VBP\},$$

where VBN is very big negative; BN is big negative; N is negative; SN is small negative; Z is zero; SP is small positive; P is positive; BP is big positive; VBP is very big positive.

At the sixth stage, the complex objective function J_C for evaluating the efficiency of the synthesized QUAV's fuzzy control system is determined, which is calculated by the expression (9.6). In turn, the criterion J_1 is the mean integral quadratic deviation between the desired motion trajectory $z_D(x)$ and the actual motion trajectory of the QUAV $z_R(x, \dot{x}_s, S)$

$$J_1(x, \dot{x}_s, S) = \frac{1}{x_{\max}} \int_0^{x_{\max}} \left[(z_D(x) - z_R(x, \dot{x}_s, S))^2 \right] dx, \tag{9.18}$$

where x_{max} is the maximum flight length along the x axis; the desired motion trajectory $z_D(x)$ is as follows

$$z_D(x) = z_S + z_T(x). \tag{9.19}$$

Moreover, the criterion J_2 is presented by the equation (9.7). As the optimal values of the criteria J_C and J_1 the given values are chosen: $J_{Copt} = 0.3$; $J_{1opt} = 0.15$. The scaling factor k_{J2} for this function is equal to 0.00033.

Then, the seventh stage of the first modification of the proposed information technology is conducted, at which the initial value of the vector S for the fuzzy controller is selected. In this case, the initial value of the vector S is equal to its minimum possible value S_{min} according to constraints set at the fifth stage

$$S_0 = S_{\min} = \{3, 3, 3, 5\}. \tag{9.20}$$

Then, the iterative sequential optimization procedure starts with the first input variable ε_z and ends with the output variable u_1, according to all the remaining stages of the first modification (from 8th to 16th). Herewith, the adjustable parameters of the ACO-based technique and sequential search method for the rule base design at Stage 10 are previously selected in the following way. The number of rounds of sequential search l equals 3 for each possible vector S of the number of the linguistic terms at performing the sequential search algorithm. As for the ACO-based the population $e = 10$, the maximum number of iterations $N*_{max} = 100$, the values of the other adjustable parameters are as follows: $\alpha = 2$; $\beta = 1$; $Q = 0.1$; $\rho = 0.5$.

When conducting the given modification of the proposed information technology, the maximum number of iterations N_{max} equals 9 (determined based on the equation (9.10) and selected constraints).

Table 9.1 presents the obtained results (by the first modification) during the vector S structural optimization for the QUAV altitude fuzzy controller.

Figure 9.7 shows the changing curves of the complex objective function J_C and separately the criteria J_1 and J_2 at implementing the first modification

Table 9.1 Optimization results using the first modification

Iteration N	Vector S	Criterion J_1	Criterion J_2	Complex Objective Function J_C
1	{3, 3, 3, 5}	0.671	135	0.715
2	{5, 3, 3, 5}	0.249	225	0.323
3	{7, 3, 3, 5}	0.208	315	0.311
4	{5, 5, 3, 5}	0.217	375	0.340
5	{5, 7, 3, 5}	0.202	525	0.375
6	{5, 3, 5, 5}	0.231	375	0.354
7	{5, 3, 7, 5}	0.214	525	0.387
8	{5, 3, 3, 7}	0.119	315	0.223
9	{5, 3, 3, 9}	0.103	405	0.236

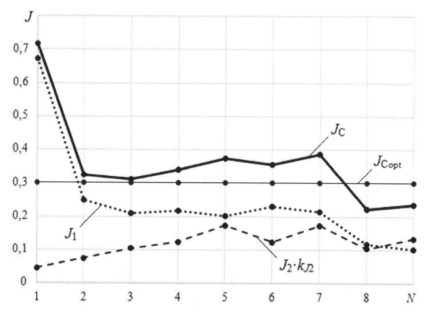

Figure 9.7   Changing curves of the complex objective function J_C and the criteria J_1, J_2 values during the structural optimization process (first modification)

of the information technology during the structural optimization process of the QUAV's altitude FC.

As can be seen from Table 9.1 and Figure 9.7, the optimal value of the complex objective function has been achieved ($J_C \leq J_{Copt}$) at the eighth iteration of the given information technology (first modification). Herewith, the optimization process could be completed according to Stage 11. However, to conduct a complete study, calculations were carried out for the ninth iteration as well. In this case, the obtained at the eighth iteration optimal vector of the number of the linguistic terms has the following form

$$\mathbf{S}_{\text{opt}} = \{5, 3, 3, 7\}\,.$$

In turn, the vector $\mathbf{S}$, obtained at the ninth iteration, has less value of the criterion J_1 ($J_1 = 0.103$) than $\mathbf{S}_{\text{opt}}$. However, its criterion J_2 is significantly bigger ($J_2 = 405$), which ultimately gives a bigger value of the complex objective function J_C ($J_C = 0.236$). In addition, for the vector $\mathbf{S}$ obtained at the first iteration, the value of criterion J_2 is the smallest ($J_2 = 135$), but its criterion J_1 has the worst value ($J_1 = 0.671$), which as a result gives the worst value of the complex objective function J_C ($J_C = 0.715$).

Table 9.2 RB fragment for the vector $\mathbf{S}_{opt}$ obtained using the first modification

Rule Number	Input and Output Variables			
	ε_z	$\frac{d\varepsilon_z}{dt}$	$\int \varepsilon_z \, dt$	u_1
1	BN	N	N	BN
5	BN	Z	Z	BN
12	SN	N	P	N
18	SN	P	P	SN
23	Z	Z	Z	Z
27	Z	P	P	P
30	SP	N	P	SP
36	SP	P	P	BP
41	BP	Z	Z	BP
45	BP	P	P	BP

Also, a particular feature of these calculations (carried out using the first modification) is that the controller RB was developed using the sequential search technique without using the ACO-based method during all iterations. Since, for all obtained values of the vector $\mathbf{S}$, criterion J_2 was less than 600.

Table 9.2 presents the RB fragment developed during the optimization process for the found optimal vector $\mathbf{S}_{opt}$. The given RB has 45 rules.

Also, the entire vector $\mathbf{R}$ of the RB consequents has the following form: $\mathbf{R}$ = (BN, BN, BN, BN, BN, BN, BN, BN, BN, BN, BN, N, N, N, SN, SN, SN, SN, N, N, SN, SN, Z, Z, SP, SP, P, SP, SP, SP, SP, P, P, SP, P, BP, BP, BP, BP, BP, BP, BP, BP, BP, BP).

Further, the remaining stages of the second modification of the proposed information technology are carried out (from 7th to 13th).

At Stage 7 of the second modification, H different variants of vector $\mathbf{S}$ of LTs number are randomly generated, considering constraints selected at Stage 5. In this case, the number of generated random variants H is equal to 9 ($H = 9$) to be the same as the maximum number of iterations N_{max} for the first modification for the objectivity of comparing the computational costs of these two modifications. The rule bases are synthesized, and the complex objective function J_C values are calculated for each h-th vector $\mathbf{S}_h$ ($h = 1$, 2, ..., H) of the given generated vectors at the further stages of the given modification. In turn, the RBs are developed through sequential search and ACO-based methods, depending on the current value of the criterion J_2. The authors set the same adjustable parameters of the sequential search method and ACO-based technique as for the first modification.

Table 9.3   Optimization results using the second modification

Iteration N	Vector S	Criterion J_1	Criterion J_2	Complex Objective Function J_C
1	$\{7, 5, 7, 7\}$	0.087	1715	0.653
2	$\{3, 7, 7, 5\}$	0.341	735	0.583
3	$\{7, 7, 3, 9\}$	0.08	1323	0.516
4	$\{5, 7, 3, 9\}$	0.095	945	0.407
5	$\{7, 5, 3, 9\}$	0.084	945	0.396
6	$\{3, 3, 7, 9\}$	0.237	567	0.424
7	$\{3, 5, 5, 7\}$	0.203	525	0.376
8	$\{5, 3, 7, 9\}$	0.098	945	0.409
9	$\{7, 7, 5, 5\}$	0.196	1225	0.601

Table 9.3 presents obtained results during the optimization of the vector S for the QUAV altitude fuzzy controller using the second modification.

Figure 9.8 shows the changing curves of the complex objective function J_C and separately the criteria J_1 and J_2 at implementing the second modification of the information technology during the structural optimization process of the QUAV's altitude FC.

As shown from Table 9.3 and Figure 9.8, the optimal value of the complex objective function J_{Copt} has not been achieved for any generated variant during the implementation of the given (second) modification of the information technology. The best variant (7th variant) of the nine generated vectors is selected that has the lowest value of the complex objective function ($J_C = 0.376$). It has the following form

$$S_{best} = \{3, 5, 5, 7\}.$$

Also, for most of the generated variants of vector S, the synthesis of the rule base was carried out using the ACO-based approach since the value of criterion J_2 was more than 600. In turn, the sequential search method was used only for the sixth and seventh variants.

In general, with the help of this modification, more complex configurations of fuzzy systems with a larger number of LTs for inputs and output were found in comparison with the variants obtained using the first modification of the information technology.

The found more complex variants made it possible to achieve lower values of criterion J_1 (especially 1, 3, 4, 5, 8). However, they have sufficiently large values of criterion J_2, which gave unsatisfactory values of the complex objective function J_C for all obtained vectors S. We can assume that with an increase in the number H of randomly generated variants of vectors of the number of the linguistic terms, it would be possible to find the optimal

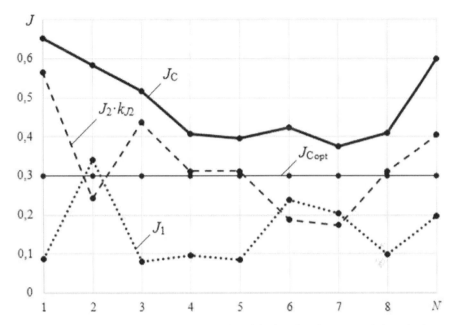

Figure 9.8 Changing curves of the complex objective function J_C and the criteria J_1, J_2 values during the structural optimization process (second modification)

vector $\mathbf{S}_{opt}$ (with the optimal value of the complex objective function J_{Copt}) using this second modification. Nevertheless, this would require more computational costs in comparison with the first modification of the information technology.

Thus, the studies carried out show that the first modification, in general, has a higher efficiency than the second since it allows to find the optimal configurations of fuzzy systems (by the LTs number of input and output variables) using lower computational costs. Also, we can conclude that when creating fuzzy systems, for which simplicity of configuration, the small size of the rule base, and simplicity of software and hardware implementation are essential, it is more expedient to use the first modification of this information technology. Since it allows synthesizing the simplest and enough effective fuzzy systems, and, at the same time, uses low computational costs. In turn, the second modification can be quite effectively used to design fuzzy systems, for which high accuracy and efficiency of solving the assigned problem are paramount, and the configuration complexity, the RB size, and the complexity of the software and hardware implementation are of less importance.

Table 9.4 RB fragment for the vector S_{best} obtained by the second modification

Rule Number	Input and Output Variables			
	ε_z	$\frac{d\varepsilon_z}{dt}$	$\int \varepsilon_z dt$	u_1
1	N	BN	BN	BN
12	N	Z	SN	BN
20	N	SP	BP	BN
28	Z	BN	Z	SN
38	Z	Z	Z	Z
46	Z	BP	BN	SP
54	P	BN	SP	BP
61	P	Z	BN	BP
68	P	SP	Z	BP
75	P	BP	BP	BP

The RB fragment developed in the optimization process for the obtained best variant of the vector of LT number S_{best} is presented in Table 9.4. The given RB has 75 rules.

Furthermore, the entire vector **R** of the RB consequents has the following form:

R = (BN, BN, BN, BN, BN, BN, BN, BN, BN, BN, BN, BN, BN, BN, BN, BN, BN, BN, BN, BN, BN, BN, BN, BN, BN, N, N, SN, SN, Z, SN, SN, SN, SN, Z, SN, SN, Z, SP, SP, Z, SP, SP, SP, P, SP, P, P, BP, BP, BP, BP, BP, BP, BP, BP, BP, BP, BP, BP, BP, BP, BP, BP, BP, BP, BP, BP, BP, BP, BP, BP, BP, BP).

Comparing two rule bases developed for vectors S_{opt} and S_{best} obtained using the first and second modifications, it is evident that for the vector S_{opt} the rule base has fewer rules and better value of the complex objective function J_C and better value of criterion J_1. It suggests that the correctly chosen number of linguistic terms for each specific variable at the synthesis of the optimal rule base has a more significant impact on the efficiency of a fuzzy system than the total number of rules.

To compare the effectiveness of the developed fuzzy controllers using the first and the second modifications of the proposed information technology, Figure 9.9 shows the flight trajectories over the mountainous terrain $z_T(x)$ (absolute altitude values) of the QUAV with a control system based on the given FCs. In turn, the detailed QUAV's flight trajectories are presented in Figure 9.10 ($z = -2\ldots 12$ m; $x = 0\ldots 100$ m) and Figure 9.11. ($z = 5\ldots 35$ m; $x = 780\ldots 900$ m).

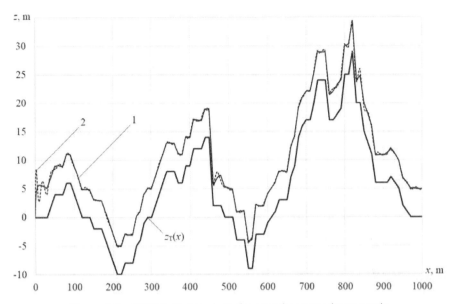

Figure 9.9 QUAV's flight trajectories over the mountainous terrain

Also, Figure 9.12 presents the transients of the flight control system of the quadcopter (changing of the real values of flight altitude relative to the terrain surface) with developed fuzzy controllers at flying over the mountainous terrain with complex relief $z_T(x)$ (Figure 9.6). In turn, the detailed transients of the flight control system are presented in Figure 9.13 ($t = 0...40$ s) and Figure 9.14 ($z_R = 3.5...6.5$ m; $t = 100...220$ s). The simulation results presented in Figures 9.9–9.14 are obtained for the QUAV with a flight altitude control system based on the FCs: 1 – developed by the first modification; 2 – developed by the second modification. The flight altitude set value z_S (relative to the terrain surface) is 5 m, the QUAV speed $\dot{x}_S$ along the longitudinal-horizontal axis x over the mountainous terrain equals 4.3 m/s.

As Figure 9.9–9.14 shows, the QUAV's flight control system based on the FC developed using the first modification has higher accuracy and quality indicators of control compared to the same system based on the fuzzy controller designed by the second modification. Moreover, the FC developed using the first modification has a more straightforward structure, fewer rules in the RB, and less computational costs were spent during its development, which in general confirms the high effectiveness of this modification of the proposed information technology.

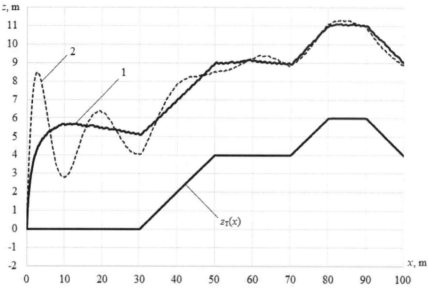

Figure 9.10   Detailed QUAV's flight trajectories ($z = -2 \ldots 12$ m; $x = 0 \ldots 100$ m)

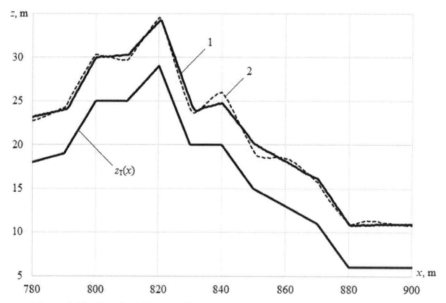

Figure 9.11   Detailed QUAV's flight trajectories ($z = 5 \ldots 35$ m; $x = 780 \ldots 900$ m)

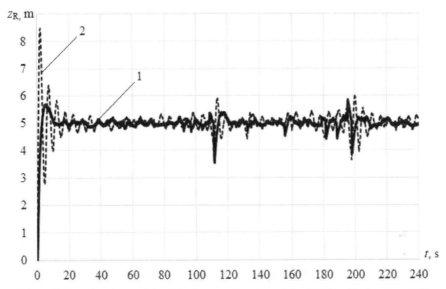

Figure 9.12 Transients of the altitude control system of the QUAV with developed FCs

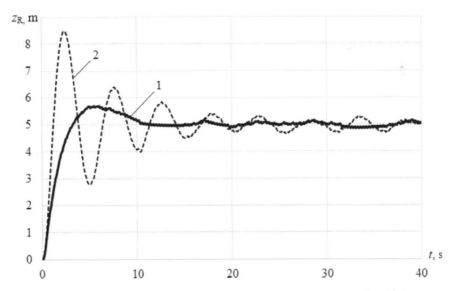

Figure 9.13 Detailed transients of the altitude control system ($t = 0\ldots40$ s)

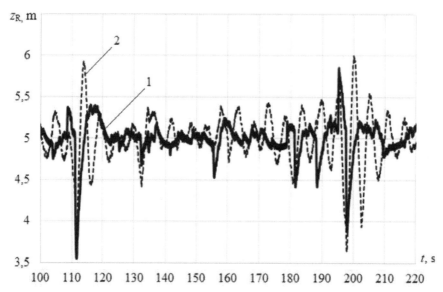

Figure 9.14 Detailed transients of the altitude control system (z_R = 3.5...6.5 m; t = 100...220 s)

In turn, if there is a need for additional improvement of the developed fuzzy controllers after their synthesis and optimization of vector **S** of the LTs number, it is possible to perform further structural and parametric optimization. In particular, it is advisable to perform rule base reduction and optimize the membership functions types and parameters [30, 39, 97].

9.5 Conclusion

The authors presented the development and study of the advanced information technology for design and structural optimization of fuzzy systems with optimal selection of linguistic terms number in this chapter. The application of the developed information technology increases the FS accuracy and performance, reduces the computational costs spent on the rule base generation and optimization of parameters, and simplifies its software and hardware realization. In particular, in this chapter, two modifications of the given information technology for fuzzy systems development and structural optimization with optimal selection of LTs number are developed and studied: 1) based on the sequential search approach; 2) based on the random search approach. In turn, each modification uses a bioinspired ACO-based method and sequential search technique for automatic rule base design for each new obtained variant of the vector **S** that defines numbers of linguistic terms. The first modification uses the concept of sequential selection of the best values of the number of LTs for the FS, starting from the first input x_1 and ending with the last output y_m. The second modification uses the concept of creating H random variants of vectors of the LTs number and choosing the best from them.

The effectiveness study of both modifications of the developed information technology is performed on a specific example, namely when designing the fuzzy control system for the quadrotor unmanned aerial vehicle. In particular, the development and structural optimization with optimal selection of LTs number of the fuzzy controller for QUAV's altitude control system is carried out alternately using first and second modifications of the proposed information technology. The conducted studies confirm, in general, the high effectiveness of the presented information technology since its application made it possible to develop a highly efficient QUAV's fuzzy control system with a fairly simple configuration and RB. The high efficiency is defined by the obtained flight trajectories of the QUAV over the mountainous terrain with complicated relief and control transients and low computational costs spent at developing the studied fuzzy control system.

Moreover, when comparing the first and the second modifications with each other, it is evident that the first modification has a higher efficiency than the second since it allows to find the optimal configurations of fuzzy systems (by the number of LTs for input and output variables) using lower computational costs. Also, we can conclude that when designing fuzzy systems, for which simplicity of configuration, the small size of the RB, and simplicity of software and hardware implementation are more important, it is more advisable to use the first modification of this information technology. Since it allows developing enough effective fuzzy systems with the simplest structure and uses low computational costs at the same time. In turn, the second modification can be quite effectively used to design FSs, for which high accuracy and efficiency of solving the assigned problem (control or decision-making) are paramount. Also, the configuration complexity, the RB size, and the complexity of these systems' software and hardware implementation are of less importance.

Further research should be conducted towards developing and studying a new specific modification of the proposed information technology for design and structural optimization with optimal selection of LTs number for fuzzy decision-making and control systems of the Takagi–Sugeno type.

References

[1] K. Knight, C. Zhang, G. Holmes, M.-L. Zhang (Eds.), 'Artificial Intelligence,' Second CCF International Conference, ICAI 2019, Xuzhou, China, August 22–23, 2019, Proceedings, Springer Singapore, 2019. DOI 10.1007/978-981-32-9298-7.

[2] S. N. Vassilyev, A. Yu. Kelina, Y. I. Kudinov, F. F. Pashchenko, 'Intelligent Control Systems,' in Procedia Computer Science, Vol. 103, pp. 623–628, 2017.

[3] C.C. Aggarwal, 'Artificial Intelligence,' Springer International Publishing, 2021. DOI 10.1007/978-3-030-72357-6.

[4] G. Krivulya, I. Skarga-Bandurova, Z. Tatarchenko, O. Seredina, M. Shcherbakova, E. Shcherbakov, 'An intelligent functional diagnostics of wireless sensor network,' in Proceedings of 2019 International Conference on Future Internet of Things and Cloud Workshops, FiCloudW 2019, pp. 135–139, 2019.

[5] L. A. Zadeh, 'Fuzzy Sets,' in Information & Control, 8, pp. 338–353, 1965.

[6] L. A. Zadeh, 'The role of fuzzy logic in modeling, identification and control,' in Modeling Identification and Control, 15 (3), pp. 191–203, 1994.

[7] L. A. Zadeh, A. M. Abbasov, R. R. Yager, S. N. Shahbazova, M. Z. Reformat, Eds., 'Recent Developments and New Directions in Soft Computing,' STUDFUZ 317, Cham: Springer, 2014.

[8] J. Kacprzyk, 'Multistage Fuzzy Control: A Prescriptive Approach,' John Wiley & Sons, Inc., New York, NY, USA, 1997.

[9] B. Kosko, 'Fuzzy Systems as Universal Approximators,' in IEEE Trans. on Computers, Vol. 43, Iss. 11, pp. 1329–1333, 1994.

[10] Y. P. Kondratenko, O. V. Korobko, O. V. Kozlov, 'Synthesis and Optimization of Fuzzy Controller for Thermoacoustic Plant,' in Recent Developments and New Direction in Soft-Computing Foundations and Applications, Studies in Fuzziness and Soft Computing 342, Lotfi A. Zadeh et al. (Eds.), Berlin, Heidelberg: Springer-Verlag, pp. 453–467, 2016.

[11] J. M. Mendel, 'Uncertain Rule-Based Fuzzy Systems, Introduction and New Directions,' Second Edition, Springer International Publishing, 2017.

[12] M. Solesvik, Y. Kondratenko, G. Kondratenko, I. Sidenko, V. Kharchenko, A. Boyarchuk, 'Fuzzy decision support systems in marine

practice,' in Fuzzy Systems (FUZZ-IEEE), 2017 IEEE International Conference on Fuzzy Systems, pp. 1–6, 2017.

[13] Y. P. Kondratenko, O. V. Kozlov, 'Mathematical Model of Ecopyrogenesis Reactor with Fuzzy Parametrical Identification,' in Recent Developments and New Direction in Soft-Computing Foundations and Applications, Studies in Fuzziness and Soft Computing 342, Lotfi A. Zadeh et al. (Eds.). Berlin, Heidelberg: Springer-Verlag, pp. 439–451, 2016.

[14] T. Bora, P. Chatterjee, S. Ghosh, 'Fuzzy Logic Based Control Of Variable Wind Energy System,' in 2020 5th IEEE International Conference on Recent Advances and Innovations in Engineering (ICRAIE), Jaipur, India, 2020, pp. 1–5. DOI: 10.1109/ICRAIE51050.2020.9358376.

[15] Y. P. Kondratenko, O. V. Kozlov, 'Mathematic Modeling of Reactor's Temperature Mode of Multiloop Pyrolysis Plant,' in a book: Lecture Notes in Business Information Processing: Modeling and Simulation in Engineering, Economics and Management, K. J. Engemann, A. M. Gil-Lafuente, J. M. Merigo, Eds., Vol. 115. Berlin, Heidelberg: Springer-Verlag, pp. 178–187, 2012.

[16] Y.P. Kondratenko, O.V. Korobko, O.V. Kozlov, 'Frequency Tuning Algorithm for Loudspeaker Driven Thermoacoustic Refrigerator Optimization,' in K. J. Engemann, A. M. Gil-Lafuente, J. M. Merigo (Eds.), Lecture Notes in Business Information Processing: Modeling and Simulation in Engineering, Economics and Management, Vol. 115, Berlin, Heidelberg: Springer-Verlag, 2012, pp. 270–279. doi.org/10.1007/978-3-642-30433-0_27.

[17] E. H. Mamdani, 'Application of fuzzy algorithms for control of simple dynamic plant,' in Proceedings of IEEE, Vol. 121, pp. 1585–1588, 1974.

[18] E. H. Mamdani, S. Assilian, 'An experiment in linguistic synthesis with a fuzzy logic controller,' in International journal of man-machine studies, No. 7(1), pp. 1–13, 1975.

[19] A. Piegat, 'Fuzzy Modeling and Control,' New York, Heidelberg: Physica-Verlag, 2001.

[20] R. Hampel, M. Wagenknecht, N. Chaker, 'Fuzzy Control: Theory and Practice, New York: Physika–Verlag,' Heidelberg, 2000.

[21] Y. P. Kondratenko, O. V. Kozlov, O. V. Korobko, 'Two Modifications of the Automatic Rule Base Synthesis for Fuzzy Control and Decision Making Systems,' in J. Medina et al. (Eds), Information Processing and Management of Uncertainty in Knowledge-Based Systems: Theory and Foundations, 17th International Conference, IPMU 2018, Cadiz, Spain,

Proceedings, Part II, CCIS 854, Springer International Publishing AG, pp. 570–582, 2018.

[22] Y. P. Kondratenko, O. V. Kozlov, G. V. Kondratenko, I. P. Atamanyuk, 'Mathematical Model and Parametrical Identification of Ecopyrogenesis Plant Based on Soft Computing Techniques,' in Complex Systems: Solutions and Challenges in Economics, Management and Engineering, Christian Berger-Vachon, Anna María Gil Lafuente, Janusz Kacprzyk, Yuriy Kondratenko, José M. Merigó, Carlo Francesco Morabito (Eds.), Book Series: Studies in Systems, Decision and Control, Vol. 125, Berlin, Heidelberg: Springer International Publishing, pp. 201–233, 2018.

[23] T. Takagi, M. Sugeno, 'Fuzzy identification of systems and its applications to modeling and control,' in IEEE Transactions on Systems, Man, and Cybernetics, SMC–15, N 1, pp. 116–132, 1985.

[24] Y.P. Kondratenko, O.V. Kozlov, 'Combined Fuzzy Controllers with Embedded Model for Automation of Complex Industrial Plants,' Shahnaz N. Shahbazova, Janusz Kacprzyk, Valentina Emilia Balas, Vladik Kreinovich (eds.) Recent Developments and the New Direction in Soft-Computing Foundations and Applications. Studies in Fuzziness and Soft Computing, Vol. 393, Springer, Cham, pp. 215–228, 2020.

[25] Z. Xiao, J. Guo, H. Zeng, P. Zhou, S. Wang, 'Application of Fuzzy Neural Network Controller in Hydropower Generator Unit,' in J. Kybernetes, Vol. 38, No. 10, pp. 1709–1717, 2009.

[26] O. Kozlov, G. Kondratenko, Z. Gomolka, Y. Kondratenko, 'Synthesis and Optimization of Green Fuzzy Controllers for the Reactors of the Specialized Pyrolysis Plants,' in Kharchenko V., Kondratenko Y., Kacprzyk J. (eds) Green IT Engineering: Social, Business and Industrial Applications, Studies in Systems, Decision and Control, Vol. 171, Springer, Cham, pp. 373–396, 2019.

[27] O. Kosheleva, V. Kreinovich, 'Why Bellman-Zadeh approach to fuzzy optimization,' in Appl. Math. Sci. Vol. 12, pp. 517–522, 2018.

[28] Y. P. Kondratenko, T. A. Altameem, E. Y. M. Al Zubi, 'The optimization of digital controllers for fuzzy systems design,' in Advances in Modelling and Analysis, AMSE Periodicals, Series A 47, pp. 19–29, 2010.

[29] Y. P. Kondratenko, A. V. Kozlov, 'Generation of Rule Bases of Fuzzy Systems Based on Modified Ant Colony Algorithms,' in Journal of Automation and Information Sciences, Vol. 51, Issue 3, New York: Begel House Inc., pp. 4–25, 2019.

[30] Y. P. Kondratenko, A. V. Kozlov, 'Parametric optimization of fuzzy control systems based on hybrid particle swarm algorithms with elite strategy,' in Journal of Automation and Information Sciences, Vol. 51, Issue 12, New York: Begel House Inc., pp. 25–45, 2019.

[31] V.M. Kuntsevich et al. (Eds), 'Control Systems: Theory and Applications,' Book Series in Automation, Control and Robotics, River Publishers, Gistrup, Delft, 2018.

[32] M. Jamshidi, V. Kreinovich, J. Kacprzyk, Eds., 'Advance Trends in Soft Computing,' Cham: Springer-Verlag, 2013.

[33] Y. Kondratenko, G. Kondratenko, I. Sidenko, 'Two-stage method of fuzzy rule base correction for variable structure of input vector,' in 2017 IEEE First Ukraine Conference on Electrical and Computer Engineering (UKRCON), Kyiv, Ukraine, pp. 1043–1049, 2017.

[34] E. Ferreyra, H. Hagras, M. Kern, G. Owusu, 'Depicting Decision-Making: A Type-2 Fuzzy Logic Based Explainable Artificial Intelligence System for Goal-Driven Simulation in the Workforce Allocation Domain', in 2019 IEEE International Conference on Fuzzy Systems (FUZZ-IEEE), New Orleans, LA, USA, pp. 1–6, 2019. DOI: 10.1109/FUZZ-IEEE.2019.8858933.

[35] W. A. Lodwick, J. Kacprzhyk, Eds., 'Fuzzy Optimization,' STUDFUZ 254, Berlin, Heidelberg: Springer-Verlag, 2010.

[36] Y. Kondratenko, V. Korobko, O. Korobko, G. Kondratenko, O., Kozlov, 'Green-IT Approach to Design and Optimization of Thermoacoustic Waste Heat Utilization Plant Based on Soft Computing,' Kharchenko V., Kondratenko Y., Kacprzyk J. (Eds), Green IT Engineering: Components, Networks and Systems Implementation. Studies in Systems, Decision and Control, Vol. 105, Springer, Cham, pp. 287–311, 2017. DOI: 10.1007/978-3-319-55595-9_14.

[37] K. Tanaka, H. O. Wang, 'Fuzzy Control Systems Design and Analysis: A Linear Matrix Inequality Approach,' John Wiley & Sons, New York, USA, 2001.

[38] D. Simon, 'H∞ estimation for fuzzy membership function optimization,' in International Journal of Approximate Reasoning, 40, pp. 224–242, 2005.

[39] Y. P. Kondratenko, D. Simon, 'Structural and parametric optimization of fuzzy control and decision making systems,' in Zadeh L., Yager R., Shahbazova S., Reformat M., Kreinovich V. (eds), Recent Developments and the New Direction in Soft-Computing Foundations and

Applications, Studies in Fuzziness and Soft Computing, Vol. 361, Springer, Cham, pp. 273–289, 2018.

[40] Y. P. Kondratenko, E. Y. M. Al Zubi, 'The Optimisation Approach for Increasing Efficiency of Digital Fuzzy Controllers,' in Annals of DAAAM for 2009 & Proceeding of the 20th Int. DAAAM Symp. "Intelligent Manufacturing and Automation," Published by DAAAM International, Vienna, Austria, pp. 1589–1591, 2009.

[41] B. Jayaram, 'Rule reduction for efficient inferencing in similarity based reasoning,' in International Journal of Approximate Reasoning 48, no. 1, pp. 156–173, 2008.

[42] Y. Yam, P. Baranyi, C.- T. Yang, 'Reduction of fuzzy rule base via singular value decomposition,' in IEEE Transactions on Fuzzy Systems, 7, no. 2, pp. 120–132, 1999.

[43] Y. P. Kondratenko, L. P. Klymenko, E. Y. M. Al Zu'bi, 'Structural Optimization of Fuzzy Systems' Rules Base and Aggregation Models,' in Kybernetes, Vol. 42, Iss. 5, pp. 831–843, 2013.

[44] D. Simon, 'Evolutionary Optimization Algorithms: Biologically Inspired and Population-Based Approaches to Computer Intelligence,' John Wiley & Sons, 2013.

[45] O.V. Kozlov, Y.P. Kondratenko, 'Bio-Inspired Algorithms for Optimization of Fuzzy Control Systems: Comparative Analysis,' in Yuriy P. Kondratenko, Vsevolod M. Kuntsevich, Arkadiy A. Chikrii, Vyacheslav F. Gubarev (Eds.) "Advanced Control Systems: Theory and Applications." Series in Automation, Control and Robotics, River Publishers, Gistrup, Denmark, pp. 83–128, 2021.

[46] W. Pedrycz, K. Li, M. Reformat, 'Evolutionary reduction of fuzzy rule-based models,' in Fifty Years of Fuzzy Logic and its Applications, STUDFUZ 326, Cham: Springer, pp. 459–481, 2015.

[47] M. Dorigo, M. Birattari, 'Ant Colony Optimization,' in Encyclopedia of Machine Learning, Sammut C., Webb G.I. (eds.), Springer, Boston, MA, 2011.

[48] B. Benhala, A. Ahaitouf, M. Fakhfakh, A. Mechaqrane, 'New Adaptation of the ACO Algorithm for the Analog Circuits Design Optimization' in International Journal of Computer Science (IJCSI), Vol. 9, no. 3, pp. 360–367, 2012.

[49] R. Gan, Q. Guo, H. Chang, Y. Yi, 'Improved Ant Colony Optimization Algorithm for the Traveling Salesman Problems,' in Journal of Systems Engineering and Electronics, pp. 329–333, 2010.

[50] A. Nabi, N.A. Singh, 'Particle swarm optimization of fuzzy logic controller for voltage sag improvement,' in Proceedings of 2016 3rd International Conference on Advanced Computing and Communication Systems (ICACCS), Vol. 01, pp. 1–5, 2016.

[51] S. Vaneshani, H. Jazayeri-Rad, 'Optimized Fuzzy Control by Particle Swarm Optimization Technique for Control of CSTR,' in International Journal of Electrical and Computer Engineering, Vol. 5, No:11, pp. 1243–1248, 2011.

[52] A. Engelbrecht, 'A study of particle swarm optimization particle trajectories,' in Information Sciences, No 176(8), pp. 937–971, 2006.

[53] D. Simon, 'Biogeography-Based Optimization,' in IEEE Transactions on Evolutionary Computation, 12(6), pp. 702–713, 2008.

[54] G. Thomas, P. Lozovyy, D. Simon, 'Fuzzy Robot Controller Tuning with Biogeography-Based Optimization,' in Modern Approaches in Applied Intelligence: 24th International Conference on Industrial Engineering and Other Applications of Applied Intelligent Systems, IEA/AIE 2011, Syracuse, NY, USA, June 28 – July 1, Proceedings, Part II, pp. 319–327, 2011.

[55] M. Huang, S. Shi, X. Liang, X. Jiao, Y. Fu, 'An Improved Biogeography-Based Optimization Algorithm for Flow Shop Scheduling Problem,' in 2020 IEEE 8th International Conference on Computer Science and Network Technology (ICCSNT), Dalian, China, pp. 59–63, 2020. DOI: 10.1109/ICCSNT50940.2020.9305008.

[56] S. Mirjalili, S. M. Mirjalili, A. Lewis, 'Grey wolf optimizer.' Adv. Eng. Software, Vol., 69, pp. 46–61, 2014.

[57] B.P. Sahoo, S. Panda, 'Improved grey wolf optimization technique for fuzzy aided PID controller design for power system frequency control,' J. Sustainable Energy, Grids and Networks, Vol. 16, pp. 278–299, 2018.

[58] T. Jayabarathi, T. Raghunathan, B.R. Adarsh, P.N. Suganthan, 'Economic dispatch using hybrid grey wolf optimizer,' Energy, Vol. 111, pp. 630–641, 2016.

[59] S. Khan, et al., 'Design and Implementation of an Optimal Fuzzy Logic Controller Using Genetic Algorithm,' in Journal of Computer Science, Vol. 4, No. 10, pp. 799–806, 2008.

[60] S. K. Oh, W. Pedrycz, 'The Design of Hybrid Fuzzy Controllers Based on Genetic Algorithms and Estimation Techniques,' in J. Kybernetes, Vol. 31, No. 6, pp. 909–917, 2002.

[61] R. Alcalá, J. Alcalá-Fdez, M. J. Gacto, F. Herrera, 'Rule base reduction and genetic tuning of fuzzy systems based on the linguistic 3-tuples representation', in Soft Computing 11, no. 5, pp. 401–419, 2007.

[62] F. L. Minku, T. Ludermir, 'Evolutionary strategies and genetic algorithms for dynamic parameter optimization of evolving fuzzy neural networks,' in Evolutionary Computation, The 2005 IEEE Congress, Vol. 3, pp. 1951–1958, 2005.

[63] R. T. Alves, M. R. Delgado, H. S. Lopes, A. A. Freitas, 'An Artificial Immune System for Fuzzy-Rule Induction in Data Mining,' in Yao X. et al. (eds) Parallel Problem Solving from Nature - PPSN VIII. PPSN 2004. Lecture Notes in Computer Science, Vol 3242. Springer, Berlin, Heidelberg, pp. 1011–1020, 2004.

[64] J. Zhu, F. Lauri, A. Koukam, V. Hilaire, 'Fuzzy Logic Control Optimized by Artificial Immune System for Building Thermal Condition,' in Siarry P., Idoumghar L., Lepagnot J. (eds) Swarm Intelligence Based Optimization, ICSIBO 2014, Lecture Notes in Computer Science, Vol. 8472. Springer, Cham, pp. 42–49, 2014.

[65] A. Prakash, S. G. Deshmukh, 'A multi-criteria customer allocation problem in supply chain environment: An artificial immune system with fuzzy logic controller based approach,' in Expert Systems with Applications, Volume 38, Issue 4, pp. 3199–3208, 2011.

[66] R. Menon, S. Menon, D. Srinivasan, L. Jain, 'Fuzzy logic decision-making in multi-agent systems for smart grids,' in Computational Intelligence Applications in Smart Grid (CIASG), 2013 IEEE Symposium, pp. 44–50, 2013.

[67] O. Castillo, P. Ochoa, J. Soria, 'Differential Evolution with Fuzzy Logic for Dynamic Adaptation of Parameters in Mathematical Function Optimization.' in Angelov P., Sotirov S. (eds) Imprecision and Uncertainty in Information Representation and Processing. Studies in Fuzziness and Soft Computing, Vol. 332. Springer, Cham, pp. 361–374, 2016.

[68] N. Quijano, K. M. Passino, 'Honey bee social foraging algorithms for resource allocation: theory and application,' in Columbus: Publishing house of the Ohio State University, 2007.

[69] D. H. Kim, C. H. Cho, 'Bacterial foraging based neural network fuzzy learning,' Proceedings of the 2nd Indian International Conference on Artificial Intelligence (IICAI – 2005), Pune: IICAI, pp. 2030–2036, 2005.

[70] A. Melendez, O. Castillo, 'Evolutionary optimization of the fuzzy integrator in a navigation system for a mobile robot,' in Recent Advances on Hybrid Intelligent Systems, pp. 21–31, 2013.

[71] H. Ishibuchi, T. Yamamoto, 'Fuzzy rule selection by multi-objective genetic local search algorithms and rule evaluation measures in data mining,' in Fuzzy Sets and Systems 141, no. 1, pp. 59–88, 2004.

[72] G. Narvydas, R. Simutis, V. Raudonis, 'Autonomous Mobile Robot Control Using Fuzzy Logic and Genetic Algorithm,' in IEEE International Workshop on Intelligent Data Acquisition and Advanced Computing Systems: Technology and Applications, Dortmund, Germany, pp. 460–464, 2007.

[73] O. Kozlov, 'Optimal Selection of Membership Functions Types for Fuzzy Control and Decision Making Systems,' in Proceedings of the 2nd International Workshop on Intelligent Information Technologies & Systems of Information Security with CEUR-WS, Khmelnytskyi, Ukraine, IntelITSIS 2021, CEUR-WS, Vol-2853, pp. 238–247, 2021.

[74] O. Cordon, F. Gomide, F. Herrera, F. Hoffmann, L. Magdalena, 'Ten Years of Genetic Fuzzy Systems: Current Framework and New trends,' in Fuzzy Sets and Systems, Vol. 141, Iss. 1, pp. 5–31, 2004.

[75] L. T. Koczy, K. Hirota, 'Size reduction by interpolation in fuzzy rule bases,' in IEEE Transactions on Systems, Man, and Cybernetics, Part B: Cybernetics, 27, no. 1, pp. 14–25, 1997.

[76] D. Driankov, H. Hellendoorn, M. Reinfrank, 'An introduction to fuzzy control,' Springer Science & Business Media, 2013.

[77] J. P. Fernández, M. A. Vargas, J. M. V. García, J. A. C. Carrillo, J. J. C. Aguilar, 'Coevolutionary Optimization of a Fuzzy Logic Controller for Antilock Braking Systems Under Changing Road Conditions,' IEEE Transactions on Vehicular Technology, 70, 2, pp. 1255–1268, 2021. DOI: 10.1109/TVT.2021.3055142.

[78] Q. Suna, R. Li, P. Zhang, 'Stable and Optimal Adaptive Fuzzy Control of Complex Systems Using Fuzzy Dynamic Model,' in J. Fuzzy Sets and Systems, Vol. 133, pp. 1–17, 2003.

[79] Y.P. Kondratenko, O.V. Kozlov, L.P. Klymenko, G.V. Kondratenko, 'Synthesis and Research of Neuro-Fuzzy Model of Ecopyrogenesis Multi-circuit Circulatory System,' Advance Trends in Soft Computing, M. Jamshidi, V. Kreinovich, J. Kazprzyk (Eds.), Series: Studies in Fuzziness and Soft Computing, Vol. 312, pp. 1–14, 2014. DOI: 10.1007/978-3-319-03674-8_1.

[80] C. Von Altrock, 'Applying fuzzy logic to business and finance,' in Optimus, 2, pp. 38–39, 2002.

[81] M. Pasieka, N. Grzesik, K. Kuźma, 'Simulation modeling of fuzzy logic controller for aircraft engines,' in International Journal of Computing, 16(1), pp. 27–33, 2017.

[82] A. P. Rotshtein, H. B. Rakytyanska, 'Fuzzy evidence in identification, forecasting and diagnosis,' Vol. 275, Heidelberg: Springer, 2012.

[83] D. Simon, 'Sum Normal Optimization of Fuzzy Membership Functions,' in International Journal of Uncertainty: Fuzziness and Knowledge-Based Systems, 10, pp. 363–384, 2002.

[84] A.M. Nazarenko, M.V. Karpusha, 'Modeling and Identification in the Problems of Multicriteria Optimization with Linear and Quadratic Performance Criteria under Statistical Uncertainty,' Journal of Automation and Information Sciences, Volume 46, Issue 3, pp. 17–29, 2014. DOI: 10.1615/JAutomatInfScien.v46.i3.30

[85] Y. Kondratenko, P. Khalaf, H. Richter, D. Simon, 'Fuzzy Real-Time Multiobjective Optimization of a Prosthesis Test Robot Control System,' in Yuriy P. Kondratenko, Arkadii A. Chikrii, Vyacheslav F. Gubarev, Janusz Kacprzyk (Eds) Advanced Control Techniques in Complex Engineering Systems: Theory and Applications, Dedicated to Professor Vsevolod M. Kuntsevich, Studies in Systems, Decision and Control, Vol. 203. Cham: Springer Nature Switzerland AG, pp. 165–185, 2019.

[86] J. Zhao, L. Han, L. Wang, Z. Yu, 'The fuzzy PID control optimized by genetic algorithm for trajectory tracking of robot arm,' in 2016 12th World Congress on Intelligent Control and Automation (WCICA), Guilin, China, pp. 556–559, 2016.

[87] Y.P. Kondratenko, O.V. Korobko, O.V. Kozlov, 'PLC-Based Systems for Data Acquisition and Supervisory Control of Environment-Friendly Energy-Saving Technologies,' chapter in a book: "Green IT Engineering: Concepts, Models, Complex Systems Architectures, Studies in Systems, Decision and Control," Vyacheslav Kharchenko, Yuriy Kondratenko, Janusz Kacprzyk (Eds.), Book Series: Studies in Systems, Decision and Control, Vol. 74, Berlin. Heidelberg: Springer International Publishing, pp. 247–267, 2017.

[88] Y. Khaluf, S. Gullipalli, 'An Efficient Ant Colony System for Edge Detection in Image Processing,' in Proceedings of the European Conference on Artificial Life, pp. 398–405, 2015.

[89] R.-M. Chen, Y.-M. Shen, C.-T. Wang, 'Ant Colony Optimization Inspired Swarm Optimization for Grid Task Scheduling,' in 2016 International Symposium on Computer, Consumer and Control (IS3C), pp. 461–464, 2016.

[90] V.L. Timchenko, D.O. Lebedev, 'Optimization of Processes of Robust Control of Quadcopter for Monitoring of Sea Waters,' Journal of Automation and Information Sciences, NY., Begell House, Vol. 51, Issue 2, pp. 1–10, 2019.

[91] N. Ben, S. Bouallègue, J. Haggège, 'Fuzzy gains-scheduling of an integral sliding mode controller for a quadrotor unmanned aerial vehicle,' Int. J. Adv. Comput. Sci. Appl., Vol. 9, no. 3, pp. 132–141, 2018.

[92] Y. Kondratenko, 'Robotics, Automation and information systems: Future perspectives and correlation with culture, Sport and life science,' Lecture Notes in Economics and Mathematical Systems, Vol. 675, Springer Verlag, pp. 43–55, 2014. DOI: 10.1007/978-3-319-03907-7_6.

[93] A. Eltayeb, M. F. Rahmat, M. A. M. Basri, M. A. M. Eltoum, S. El-Ferik, 'An Improved Design of an Adaptive Sliding Mode Controller for Chattering Attenuation and Trajectory Tracking of the Quadcopter UAV,' IEEE Access, Vol. 8, pp. 205968–205979, 2020.

[94] R. Duro et al. (Eds). 'Advances in Intelligent Robotics and Collaborative Automation.' River Publishers, Gistrup, 2015. DOI: https://doi.org/10.1 3052/rp-9788793237049.

[95] Y.P. Kondratenko, et al (Eds). 'Advanced Control Systems: Theory and Applications.' River Publishers, Gistrup, 2021.

[96] I.P. Atamanyuk, V.Y. Kondratenko, O.V. Kozlov, Y.P. Kondratenko, 'The Algorithm of Optimal Polynomial Extrapolation of Random Processes,' Lecture Notes in Business Information Processing: Modeling and Simulation in Engineering, Economics and Management., K. J. Engemann, A. M. Gil-Lafuente, J. M. Merigo (Eds.), Berlin, Heidelberg: Springer-Verlag, Vol. 115, pp. 78–87, 2012.

[97] D. Simon, 'Design and rule base reduction of a fuzzy filter for the estimation of motor currents,' in International Journal of Approximate Reasoning, 25, pp. 145–167, 2000.

10

Analysis of the Dynamics and Controllability of an Autonomous Mobile Robot with a Manipulator

Ashhepkova Natalja[1], Zbrutsky Alexander[2], and Koshevoy Nicolay[3]

[1]Department of Mechanotronics, Oles Honchar Dnipro National University, Ave. Gagarin, 72, Dnipro, Ukraine, 49010
[2]Department of Aircraft Control Systems, National Technical University of Ukraine "Igor Sikorsky Kyiv Polytechnic Institute", Str. Botkin, 1, Kyiv, Ukraine, 03056
[3]Department of Intelligent Visual Systems and Engineering of Quality, National Aerospace University "Kharkov Aviation Institute", st. Chkalov, 17, Kharkov, Ukraine, 61070
E-mail: ashchepkova.ftf.dnu@gmail.com; zbrutsky@cisavd.kpi.ua; kafedraapi@ukr.net

Abstract

The results of the study of the dynamics and control of an autonomous mobile robot with a manipulator are presented. An all-wheel drive model with a four-wheeled chassis layout and equipped with a four-link manipulator is considered. The design of the manipulator consists of a docking ring rotating around a vertical axis and three rod links of the arm, connected by rotary kinematic pairs of the fifth class. The mathematical model of controlled movement of autonomous mobile robot with a manipulator is proposed. With relative motion of the manipulator, the tensor of inertia of an autonomous mobile robot in the coordinate system associated with the chassis is non-diagonality and non-stationarity. Simplifications of the mathematical model for several driving modes are substantiated. An algorithm for integrating the equations of dynamics is developed, taking into account certain features.

Based on the results of mathematical modeling, an algorithm for adaptive control of autonomous mobile robot with a manipulator is proposed.

Keywords: autonomous mobile robot, non-diagonal tensor of inertia, dynamics, adaptive control.

10.1 Introduction

The development of new technologies makes it possible to use manipulators (M) for work in extreme conditions for humans. For example, the use of M and autonomous mobile robots (AMR) for eliminating the consequences of environmental and man-made disasters, the operation of remotely controlled M for working with radioactive substances, the use of M for assembling structures in space or on the seabed, etc. For work in extreme conditions, as a rule, remote-controlled robots and M. are used. In many cases, a remote-controlled arm-manipulator is installed on a controlled mobile platform. This expands the service characteristics of M, but imposes more stringent requirements on the design of control and monitoring systems. An example of such structures is: a manipulator arm is placed on a spacecraft, on a ship, on a wheeled chassis, etc. In all of the above cases, the movement of the base affects the kinematics and dynamics of the manipulator arm, increasing the positioning error and deviation of the gripper pole from the specified trajectory. Dynamic processes for M on a mobile platform can be divided into three modes of movement:

- controlled movement of the platform along a given route with "check-in" M;
- performing technological operations M with the platform stopped;
- M performs technological operations with a controlled movement of the platform.

The controlled movement of a platform with a "locked" manipulator has been well studied for various chassis designs. In [1], the dynamics of an all-wheel drive 4-wheeled bogie when turning is investigated. The synthesis of the law of control of the plane motion of a transport robot is a typical task: several methods of stabilization of motion along a given trajectory are known. In [2], the authors propose a modification of the system of dynamic equations with the change of coordinates and the transformation of outputs to chain forms. In [3], a transformation of the system of equations to the Cauchy normal form was applied with subsequent differentiation with respect to a new independent variable [3–5]. In [2, 6], the equations of the bogie dynamics are relative to road coordinates, which well fix deviations from a given trajectory.

When performing technological operations M, which is installed on a stopped platform, it is advisable to consider the equation of dynamics with respect to a fixed coordinate system associated with the docking point M and the platform. In this case, the solution of the problems of kinematic and dynamic analysis of M depends on the ratio of the inertial characteristics of the platform and M with a load. The load can be a manipulated object, probe, drill, bucket, etc. In [7–10], two approaches to formulating the equations of the dynamics of M are considered: the Newton–Euler method and the Lagrange–Euler method. To simulate the processes of dynamics and control of M, it is advisable to use the mathematical packages Mathcad [11–13] or Matlab [14–16]. For the effective execution of M technological operations in extreme conditions, it is necessary at the design stage to automatically carry out kinematic and dynamic analysis [11], determine the boundaries of the working space [12], analyze the optimality of configurations at the points of the working space, and ensure the positioning accuracy of the gripper [13]. In [14], pseudo symbolic dynamic modeling (PSDM) is presented for creating simplified dynamic models M, the design of which contains up to 7 degrees of mobility. The presented algorithm allows to generate code in real time, simulate dynamics, and improve the efficiency of the model by eliminating the least important elements. In addition, the authors of [14] have developed an implementation of the algorithm in the MATLAB environment, which is publicly available. The authors of [15] developed a virtual model of kinematics and dynamics M in MATLAB & SIMULINK. The presented model uses a PID controller, and the dynamics equations are obtained by the Lagrange–Euler method. The work [16] is devoted to the dynamic modeling of three-link spatial M using symbolic and numerical methods. An algorithm based on the Newton–Euler method is proposed to formulate the equations of dynamics in the form of a state space. The algorithm is implemented in the Maple system, the modeling is carried out in MATLAB & SIMULINK.

When developing the design of the AMR arm-manipulator, it is advisable to investigate the efficiency of technological operations with a movable platform. Let's analyze the methods for studying the dynamics and control of a manipulator arm on a mobile platform, for example, on a ship, spacecraft or ground transport. In [17–19], a multi-mass system is considered, which consists of a spacecraft and M. The manipulator arm mounted on the ship's deck was considered in [20–23]. It is generally believed that the M dynamics affects the platform, but does not cause it to move. However, the results presented in [17] prove that the controlled motions of M cause oscillations of the spacecraft and an increase of 43% in energy consumption for its angular stabilization. In [24, 25], the dynamic equations of a manipulator arm on a movable platform are derived. It is believed that the inertia of the platform is large enough and is not affected by the relative motion of the manipulator. In [26], the analysis of the inertia tensor AMR with M with different configurations of their mutual position showed that the values of the centrifugal moments of inertia of the system with relative motion M are comparable to the value of the axial moments of inertia of the system, even if the mass M with a load is less than 20% of the platform mass. So, the relative movement of the manipulator induces linear or angular movements of the platform. Therefore, in the general case, the tensor of inertia of such a system must be taken non-diagonality and non-stationarity, and the mathematical model of the dynamics of AMR with M must take into account the cross-coupling of control channels.

AMR with M, as a control object, is a multichannel, multiply connected, essentially nonlinear dynamic system. The dynamics and control of robots were studied by S. Burdakov, I. Miroshnik, E. Yurevich, Yu. Martynenko, B. Topchiev and others. However, taking into account the nonlinearity and complexity of the control system was carried out only for particular solutions. For AMR with an adaptive control system, operating under not predetermined environmental conditions, the driving system must be tested on a model. When designing AMR, it is economically expedient to carry out not only mathematical, but also simulation modeling.

Thus, the study of the dynamics of AMR with M by methods of mathematical and simulation modeling is an urgent scientific and applied problem.

10.2 Results of Study of Dynamics and Control of Autonomous Mobile Robots

10.2.1 Problem Statement

The expansion of the scope of application and the complication of tasks for modern AMR determines the presence of structural elements that are movable relative to the robot platform: a manipulator, a probe, a drill, a bucket, etc. During the operation of the AMR, these elements move relative to the platform, changing the geometry of the masses of the system. As an example, AMR of variable configuration, let's consider AMR with M. The peculiarities of the dynamics of AMR with M, as a system of bodies of variable configuration, are:

- change in the position of the center of mass of the system with relative movement of the manipulator,

- commensurability of non-diagonality and diagonal elements of the tensor of inertia, calculated with respect to the axes of the basic coordinate system $CX_CY_CZ_C$, associated with the center of mass of the AMR platform.

The AMR operation cycle can be divided into several stages:

- movement of the AMR platform from the start point S to the finish point F with a stationary manipulator in compliance with the requirements of optimality (speed, productivity or economy, etc.);
- performing technological operations with a manipulator while the platform is stopped;
- return of the AMR platform to the starting point S or movement to a predetermined end point K with a stationary manipulator.

At the first stage of work, AMR is a cart with an adaptive control system that moves in a predetermined environment. To ensure the stable AMR movement from the start point S to the finish point F with a stationary manipulator in compliance with the requirements of optimality, it is necessary to perform mathematical and simulation modeling.

When returning the AMR platform, it is necessary to take into account the change in the mass distribution of the structure relative to the center of mass of the platform due to the presence of objects to be manipulated, soil samples, water samples, etc. with a stationary manipulator.

Object of research: AMR with M, that is, a controlled wheeled cart, equipped with a movable M and information devices (camera, microphone, dosimeter, etc.). The control system allows to move in a limited space (working area) along a given trajectory using GPS navigation or orientation on a map or beacons. Within the limits of the working area, possible static and (or) moving obstacles that must be overcome with the least deviation from the specified route.

The aim of research is to increase the efficiency of using AMR with M, which acts in not predetermined environmental conditions.

To achieve the aim, the research objectives were formulated:

- determination of the features of AMR dynamics with a manipulator;
- development of an algorithm for integrating the current state of the equations of dynamics;
- assessment of the interconnection of control channels;
- development of an adaptive control algorithm.

Research methods: conclusions are formulated in the work, scientific assumptions and recommendations are based on the fundamental provisions of theoretical mechanics, the theory of differential equations and control theory.

10.2.2 Design of an Autonomous Mobile Robot with a Manipulator

A diagram of the AMR design with a manipulator is shown in Figure 10.1. The design consists of an all-wheel drive 4-wheeled platform AMR, and an anthropomorphic manipulator composed of a ring rotating around a vertical axis and rod links of the arm, connected by rotational kinematic pairs of the fifth class.

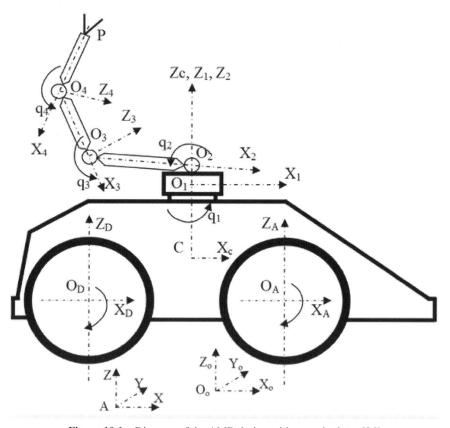

Figure 10.1 Diagram of the AMR design with a manipulator [26]

To analyze the results of mathematical modeling of the dynamics of AMR with M, let's assume the following:

- AMR platform—absolutely rigid body with uniform mass distribution with density ρ=const; development of an algorithm for integrating the current state of the equations of dynamics;
- the links of the manipulator are absolutely solid, rigid and have a uniform mass distribution with a density ρ=const.

Further, for definiteness, when creating a mathematical model of AMR of variable configuration, an all-wheel drive 4-wheeled platform AMR with a manipulator is considered as an example.

Let's introduce the following right coordinate systems (Figure 10.1):

$AXYZ$—inertial coordinate system.

$CX_CY_CZ_C$—movable base coordinate system. The origin is connected to the point C by the center of mass of the platform. The axes are parallel to the main central axes of inertia of the AMR platform. Here CZ_C is perpendicular to the platform movement plane, coincides with the local vertical and is directed upward to the manipulator ring, CX_C is located in the platform movement plane and is directed towards the movement, the CY_C axis is located in the platform movement plane and complements the coordinate system to the right.

$OX_0Y_0Z_0$—moving coordinate system. The origin is connected to point O–by the center of mass of the system of bodies. The axes are parallel to the axes of the $CX_CY_CZ_C$ coordinate system.

$O_1X_1Y_1Z_1$—moving coordinate system is connected. The origin is connected to the point O_1–by the center of mass of the ring. The axes coincide with the main central axes of inertia of the ring. In the initial position of the manipulator ring, the axes of coordinate systems $O_1X_1Y_1Z_1$ are parallel to the axes of the $CX_CY_CZ_C$ coordinate system.

$O_iX_iY_iZ_i$ (for $i = 2, 3, 4$)—connected moving coordinate systems. The origin is connected to the point O_i–by the center of the kinematic pair. The O_iX_i axes coincide with the longitudinal axes of the rod links of the manipulator arm, in the initial position of the manipulator links the axes of the $O_iX_iY_iZ_i$ coordinate systems are parallel to the axes of the $CX_CY_CZ_C$ coordinate system.

$MX_MY_MZ_M$—movable base coordinate system. The origin is connected to the point M—by the center of mass of the manipulator. In the initial position of the manipulator, the axes of the $MX_MY_MZ_M$ coordinate systems are parallel to the axes of the $CX_CY_CZ_C$ coordinate system.

10.2.3 Dynamics Analysis

The mathematical model is compiled taking into account the non-diagonality of the inertia tensor AMR with M. Assessment of the elements of the inertia tensor AMR with a manipulator with different configurations of their mutual position, if the mass of the manipulator with a load is from 10% to 30% of the mass of the AMR platform, carried out in [26].

Let's consider the controlled motion of the AMR with M taking into account the non-diagonality of the tensor of inertia, if the main central axes of inertia of the AMR change orientation relative to the vectors of the control torques. The motion of AMR with M is characterized by the speed of motion $\dot{v}_0$ along the trajectory of the pole O_0—the center of mass of the system and by the vector of angular velocity $\bar{\Omega}$.

The mathematical model of the motion of a dynamic system, taking into account the nonstationarity and non-diagonality of the tensor of inertia, describes:

- movement of the center of mass of the AMR system with M;
- angular movement AMR with M;
- relative movement of the manipulator;
- equations of forces acting on AMR with M;
- equations of executive bodies of the control system.

1) Equations of motion of the center of mass of the AMR system with M

The equations of motion of the center of mass of the AMR system with M when moving along the trajectory in the inertial coordinate system has the form:

$$m \cdot \left(\frac{d^2 \bar{r}_c}{dt^2} + \frac{\tilde{d}_0 \bar{\Omega}}{dt} \times \bar{p}_{\text{co}1} + \bar{\Omega} \times \left(\bar{\Omega} \times \bar{p}_{\text{co}1} \right) \right) + m_< \cdot \left[\frac{\tilde{d}_1^2 \bar{p}_{\text{o}1M}}{dt^2} + \right.$$

$$+ \left[2 \left(\bar{w}_1 + \bar{\Omega} \right) \times \frac{\tilde{d}_1 \bar{p}_{\text{o}1M}}{dt} + \frac{\tilde{d}_1 \bar{w}_1}{dt} \times \bar{p}_{\text{o}1M} + \bar{\Omega} \times \left(\bar{w}_1 \times \bar{p}_{\text{o}1M} \right) + \right.$$

$$+ \bar{\Omega} \times \left(\bar{w}_1 \times \bar{p}_{\text{o}1M} \right) + \left(\bar{\Omega} \times \bar{w}_1 \right) \times \bar{p}_{\text{o}1M} +$$
$$+ \bar{w}_1 \times \left(\bar{\Omega} \times \bar{p}_{\text{o}1M} \right) + \bar{w}_1 \times \left(\bar{w}_1 \times \bar{p}_{\text{o}1M} \right) \right] = \sum_v \bar{F}_v.$$

$$(10.1)$$

For the sections of the trajectory on which the AMR moves with an immovable manipulator, let's write formula (10.1) in the form:

$$m \cdot \left(\frac{d^2 \bar{r}_c}{dt^2} + \frac{\tilde{d}_0 \bar{\Omega}}{dt} \times \bar{p}_{\text{co}1} \right) + m_M \bar{\Omega} \times \left(\bar{\Omega} \times \bar{p}_{\text{CM}} \right) = \sum_v \bar{F}_v. \quad (10.2)$$

If the AMR moves with a slow angular motion of the manipulator, from formula (10.1) let's obtain:

$$m \cdot \left(\frac{d^2 \bar{r}_c}{dt^2} + \frac{\tilde{d}_0 \bar{\Omega}}{dt} \times \bar{p}_{col} \right) + m_M \cdot \left(\frac{\tilde{d}_1^2 \bar{p}_{o1M}}{dt^2} + 2 \left(\bar{w}_1 + \bar{\Omega} \right) \times \frac{\tilde{d}_1 \bar{p}_{o1M}}{dt} + \right.$$

$$\left. + \frac{\tilde{d}_1 \bar{w}_1}{dt} \times \bar{p}_{o1M} \right) = \sum_v \bar{F}_v. \tag{10.3}$$

Equation (10.1) is the equation of motion of the center of mass of the AMR system with a manipulator in the general case, and Equations (10.2) and (10.3) are separate cases if the manipulator does not move or moves with a limited speed.

2) The equation of AMR with M angular motion

The equations of angular motion of the AMR with M when moving along the trajectory in the inertial coordinate system has the form:

$$\bar{p}_{oo1} \cdot \left\{ \frac{\tilde{d}_0^2}{dt^2} \left(m_M \bar{p}_{o1M} + m_c \bar{p}_{o1c} \right) + \frac{\tilde{d}_0 \bar{\Omega}}{dt} \left(m_M \bar{p}_{o1M} + m_c \bar{p}_{o1c} \right) + \right.$$

$$+ 2\bar{\Omega} \times \frac{\tilde{d}_0}{dt} \left(m_M \bar{p}_{o1M} + m_c \bar{p}_{o1c} \right)$$

$$\left. + \bar{\Omega} \times \left[\bar{\Omega} \times \left(m_M \bar{p}_{o1M} + m_c \bar{p}_{o1c} \right) \right] \right\}$$

$$+ + \sum_j \bar{p}_{o1j} \times m_j \left[\frac{\tilde{d}_0^2 \bar{p}_{o1j}}{dt^2} + \frac{\tilde{d}_0 \bar{\Omega}}{dt} \times \bar{p}_{o1j} + 2\bar{\Omega} \times \frac{\tilde{d}_0 \bar{p}_{o1j}}{dt} + \bar{\Omega} \right.$$

$$\left. \times \left(\bar{\Omega} \times \bar{p}_{o1j} \right) \right]$$

$$+ + \sum_i \bar{p}_{o1i} \times m_i \left[\frac{\tilde{d}_0^2 \bar{p}_{o1i}}{dt^2} + \frac{\tilde{d}_0 \bar{\Omega}}{dt} \times \bar{p}_{o1i} + 2\bar{\Omega} \times \frac{\tilde{d}_0 \bar{p}_{o1i}}{dt} + \bar{\Omega} \right.$$

$$\left. \times \left(\bar{\Omega} \times \bar{p}_{o1i} \right) \right] = \bar{M}_o. \tag{10.4}$$

3) Equation of manipulator relative motion

Let's write the equation of the relative motion of the manipulator in the inertial coordinate system:

$$\sum_j \bar{p}_{o1j} \times m_j \left\{ \dot{\bar{v}}_o + \frac{\tilde{d}_0^2 \bar{p}_{oc}}{dt^2} + 2\bar{\Omega} \times \frac{\tilde{d}_0 \bar{p}_{oc}}{dt} + \frac{\tilde{d}_0 \bar{\Omega}}{dt} \times (\bar{p}_{oc} + \bar{p}_{o1j}) + \right.$$

$$+ \bar{\Omega} \times [\bar{\Omega} \times (\bar{p}_{oc} + \bar{p}_{o1j})] + \frac{\tilde{d}_1^2 \bar{p}_{o1j}}{dt^2} + \bar{\Omega} \times \frac{\tilde{d}_1 \bar{p}_{o1j}}{dt} + \frac{\tilde{d}_1 \bar{w}_1}{dt} \times \bar{p}_{o1j} +$$

$$\left. + 2\bar{w}_1 \times \frac{\tilde{d}_1 \bar{p}_{o1j}}{dt} + (\bar{\Omega} \times \bar{w}_1) \times \bar{p}_{o1j} + \bar{w}_1 \times [(\bar{w}_1 + \bar{\Omega}) \times \bar{p}_{o1j}] \right\}$$

$$= \bar{M}_{01}.$$

$$(10.5)$$

4) Equation of executive bodies of the control system

Angular stabilization of the AMR manipulator, as a system of bodies of variable configuration, occurs under the action of an active relay-type control system. The main vector of the control torque, relative to point C: $\bar{M}_c^U = \bar{M} \cdot \chi$,

where $\bar{M} = \left(M_x^U \quad M_y^U \quad M_z^U \right)^T$ – vector of moments of the control motors, χ – control function:

$$\chi = \begin{cases} 1, & j \le -1, \ \frac{dj}{dt} > 0; & j \le -m, \ \frac{dj}{dt} < 0; \\ 0, & -1 < j < m, \ \frac{dj}{dt} > 0; & -m < j < 1, \ \frac{dj}{dt} < 0; \\ -1, & j \ge m, \ \frac{dj}{dt} > 0; & j \ge 1, \ \frac{dj}{dt} < 0. \end{cases}$$

$$(10.6)$$

Vector of control actions $\bar{j} = T \cdot \bar{\Omega} + \bar{\xi}$, where $\bar{\Omega} = \begin{pmatrix} \Omega_x \\ \Omega_y \\ \Omega_z \end{pmatrix}$,

$\bar{\xi} = \begin{pmatrix} \xi_x = \varphi \\ \xi_y = \psi \\ \xi_z = v \end{pmatrix}$.

The equations for the on and off lines for each control channel are as follows:

$$\pm g_i = T_i \cdot \Omega_i + \xi_i;$$
$$\pm n_i \cdot g_i = T_i \cdot \Omega_i + \xi_i;$$
$$i = x, \ y, \ z,$$

where $0 \le n_i = const \le 1$ – the delay coefficient of the control system for each channel; $g_i = const$ – restriction on motion parameters for each control channel; T_i – time constants.

It was not possible to solve the system of equations of the mathematical model by standard methods of numerical integration, because, in addition to

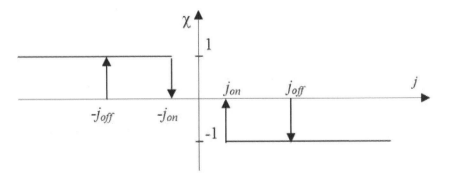

Figure 10.2 Relay static characteristic

the indicated features at the switching points j_{off}, j_{on}, the control functions χ_i have several values (Figure 10.2).

It is proposed to solve the system of equations by the Runge–Kutta–Fehlberg method [27]. In this case, on the sections of motion corresponding to $\chi_i = const$, numerical integration is carried out with a certain constant step h. If $|j-j_{off}| \leq E$ (or $|j-j_{on}| \leq E$, where $E = const$ – the vicinity of the switching point of the relay characteristic), then the integration step decreases and is equal to $h_1 = h/C$. The most common is the so-called double step method with $C = 2^n$, $n = const$, $n > 0$.

If the condition $|j-j_{off}| \leq E/C$ (or $|j-j_{on}| \leq E/C$) let's assume that the switching of the relay characteristic takes place and χ_i changes the value in accordance with the control law. The values of the constants h_1, h, E, C, n are chosen based on the initial conditions. The structural diagram of the algorithm for solving the system of equations by the Runge–Kutta–Fehlberg method is shown in Figure 10.3. The use of the Runge–Kutta–Fehlberg method makes it possible to create software for identifying the current dynamic state of AMR with M.

The results of mathematical modeling of the AMR dynamics without taking into account the non-diagonality and non-stationarity of the tensor of inertia are given in Table. 10.1 (t – time; $q_1(t)$ – the generalized coordinate of the relative motion of the manipulator; $\alpha_x(t)$, $\alpha_y(t)$, $\alpha_z(t)$ – the angular coordinates of the system of bodies relative to the coordinate system $CX_CY_CZ_C$; $\omega_x(t)$, $\omega_y(t)$, $\omega_z(t)$ – the angular velocities of the system of bodies relative to the coordinate system $CX_CY_CZ_C$; χ_x, χ_y, χ_z – control functions).

The results of mathematical modeling of the AMR dynamics taking into account the non-diagonality and non-stationarity of the tensor of inertia are

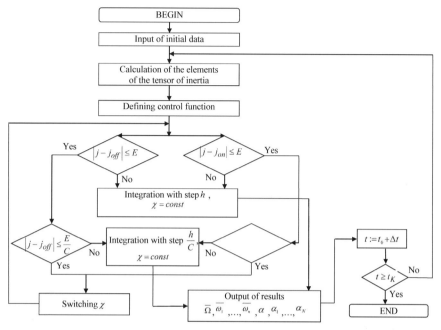

Figure 10.3 Algorithm for solving the problem of dynamics of AMR with M by the Runge–Kutta–Fehlberg method

given in Table 10.2(t – time; $q_1(t)$–generalized coordinate of the relative motion of the manipulator; $\alpha_x(t)$, $\alpha_y(t)$, $\alpha_z(t)$ – angular coordinates of the system of bodies relative to the coordinate system $CX_CY_CZ_C$; $\omega_x(t)$, $\omega_y(t)$, $\omega_z(t)$ – angular velocities of the system of bodies relative to the coordinate $CX_CY_CZ_C$; χ_x, χ_y, χ_z – control functions).

The non-diagonality of the tensor of inertia of a system of bodies relative to the base coordinate system determines:

- occurrence of inconsistency of the main central axes of inertia of the system of bodies with the axes associated with the center of mass of the AMR platform by the basic $CX_CY_CZ_C$ coordinate system,
- inconsistency of control actions with the directions of the main central axes of inertia of the system,
- interdependence of control channels.

Table 10.1 Results of mathematical modeling of AMR dynamics without taking into account the non-diagonality and non-stationarity of the tensor of inertia

t, c	q_1, rad	α_x, rad	$\omega_x.10^{-3}$, rad /s	χ_x	α_y, rad	$\omega_y.10^{-3}$, rad /s	χ_y	$\alpha_z.10^{-3}$ rad	$\omega_z.10^{-3}$ rad /s	χ_z
0	0	0	0	0	0	0	0	15,71	17,45	−1
0,03	0	0	−0,1	1	0	0,12	−1	15,71	−17,45	1
18,03	0,05	−0,24	−0,1	1	−0,18	−0,1	1	−15,71	−17,45	1
18,05	0,05	−0,24	1,0	−1	−0,18	0,05	0	−15,71	17,45	−1
36,05	0,11	1,47	1,0	−1	−0,05	0,09	0	15,71	17,45	−1
36,08	0,11	1,47	−0,1	1	−0,05	−0,1	1	15,71	−17,45	1
54,08	0,16	1,24	−0,1	1	−0,01	0,02	0	−15,71	−17,45	1
54,11	0,16	1,37	1,0	−1	−0,01	0,14	−1	−15,71	17,45	−1
72,11	0,22	3,08	1,0	−1	0,23	0,18	−1	15,71	17,45	−1
72,14	0,22	3,08	−0,1	1	0,23	0,06	0	15,71	−17,45	1
90,14	0,27	2,85	−0,1	1	0,38	0,11	−1	−15,71	−17,45	1

Table 10.2 Results of mathematical modeling of AMR dynamics taking into account the non-diagonality and non-stationarity of the tensor of inertia

t, c	q_1, rad	α_x, rad	$\omega_x.10^{-3}$, rad /s	χ_x	α_y, rad	$\omega_y.10^{-3}$, rad /s	χ_y	$\alpha_z.10^{-3}$ rad	$\omega_z.10^{-3}$ rad /s	χ_z
0	0	0	0	0	0	0	0	15,71	17,45	−1
0,03	0	0	−0,1	1	0	−0,1	1	15,71	−17,45	1
18,03	0,05	−0,24	−0,1	1	−0,02	−0,1	1	−15,71	−17,45	1
18,05	0,05	−0,24	0,1	−1	−0,02	0,1	−1	−15,71	17,45	−1
36,05	0,11	0,14	0,1	−1	0,01	0,1	−1	15,71	17,45	−1
36,08	0,11	0,14	−0,1	1	0,01	−0,1	1	15,71	−17,45	1
54,08	0,16	−0,14	−0,1	1	−0,01	−0,1	1	−15,71	−17,45	1
54,11	0,16	−0,14	1,0	−1	−0,01	1,0	−1	−15,71	17,45	−1
72,11	0,22	0,13	1,0	−1	0,01	1,0	−1	15,71	17,45	−1
72,14	0,22	0,13	−0,1	1	0,01	−0,1	1	15,71	−17,45	1
90,14	0,27	−0,12	−0,1	1	−0,01	−0,1	1	−15,71	−17,45	1

10.2.4 Analysis of AMR with M Controllability

The results of mathematical modeling of linear motion of AMR are given in [26] and prove that the growth of q_1 (t) from 0 to 0.27 rad causes an increase in angular coordinates and velocities in directions perpendicular to the direction of motion. The control system tries to work out the disturbances and prevent the growth of angular velocities in the directions perpendicular to the direction of motion. Despite the operation of the control system, due to the non-diagonality and non-stationarity of the tensor of inertia and cross-coupling of control channels, after 90 s of the AMR movement along the trajectory, the angular coordinates reach the values $\alpha_x = 2.85$ rad, $\alpha_y = 0.38$ rad. Thus, AMR loses its vertical position, in the conditions of autonomous operation it leads to the impossibility of performing technological operations.

If the magnitude and direction of action of the control actions is determined taking into account the cross-connection of control channels due to the non-diagonality and non-stationarity of the AMR inertia tensor relative to the coordinate system $CX_CY_CZ_C$, then the angular coordinates reach the values $\alpha_x = -0.12$ rad, $\alpha_y = -0.01$ rad. In this case, the growth of $q_1(t)$ from 0 to 0.27 rad also causes the occurrence of angular coordinates and velocities in the directions perpendicular to the direction of motion, but AMR does not lose its vertical position and continues to perform technological operations.

The effectiveness of manipulator control is largely determined by teaching methods, applied methods and means of adaptation. The control of the grab of the AMR manipulator is carried out by the operator remotely. The design provides for three data transmission channels (fiber optic wire, Wi-fi and protected radio communication). Correction of movements and control of the results of the manipulator's activity is carried out using a video surveillance system.

The operating conditions of AMR with a manipulator are often not only not known a priori, but can also vary unpredictably over a wide range. The reasons for the uncertainty and nonstationarity of these conditions are: 1) the lack of information about the properties of the external environment; 2) work in technical limitations, natural spread and drift of parameters of sensory and motor systems; 3) in the occurrence of obstacles and computational errors in communication and control channels.

Usually an adaptive control system has to adapt to the conditions of the external environment. That is why work with adaptive control and elements of artificial intelligence is significantly superior in terms of the possibilities of working with programmed control: they can adequately respond to changes in the external environment, adapt to the drift of environmental parameters, recognize and avoid obstacles, identify target objects, and determine their characteristics.

The dynamic characteristics of the AMR system with a manipulator change significantly during the operation cycle. Before generating a command to the actuators of the chassis and M, the control system has to identify the current state of AMR with M. After which the command is corrected and the distribution of influences of AMR with M on the actuators takes into account the cross-connection of the control channels due to the non-diagonality and non-stationarity of the inertia tensor.

In this regard, an algorithm for adaptive control of AMR with M is proposed, which provides for the study and assessment in real time of the dynamic characteristics of AMR (identification problems) and determination of control parameters from the obtained model (synthesis problem).

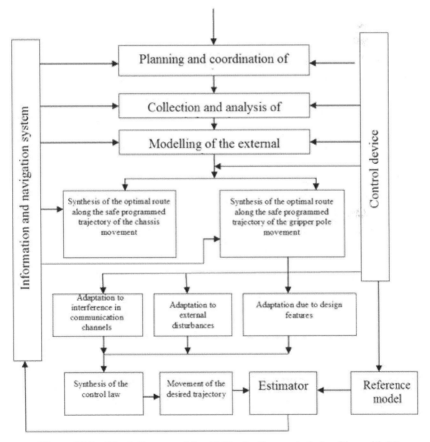

Figure 10.4   Block diagram of the AMR adaptive control algorithm with M

According to [28], the AMR model can be represented as an "input - output" equation:

$$y(t) = f_y \left[y(t-1), y(t-2), ..., u(t-1), u(t-2), ..Q(t), \Theta(t) \right],$$

where y – output variable u – control, Θ – disturbances, f_y – some function, Q – parameters of AMR with manipulator, $t=1,2,...n$ – discrete time.

The recurrent type adaptive control algorithm is written as

$$u(t) = f_u \left[u(t-1), y(t), U_e(t), Q(t) \right],$$

where Ue – desirable trajectory of AMR movements with manipulator, fu – control function depending on the introduced law.

The adaptation criterion is to minimize the deviation of the AMR state vector from the desired trajectory, which can be written in general form:

$$I = \Phi(U_e(t) - y(t)) \to \min.$$

The block diagram of the proposed algorithm for adaptive control of AMR with a manipulator is shown in Figure 10.4.

The proposed control algorithm will make it possible to equip work with means of automatic learning (self-learning) and adaptation, which turn their control system into an extremely flexible tool for organizing purposeful behavior under conditions of uncertainty.

10.3 Conclusion

The peculiarities of the dynamics of AMR with M are the non-diagonality and non-stationarity of the tensor of inertia of bodies relative to the basic coordinate system, which determine:

- occurrence of inconsistency of the main central axes of inertia of the system of bodies with the axes associated with the center of mass of the AMR platform by the basic $CX_cY_cZ_c$ coordinate system,
- inconsistency of control actions with the directions of the main central axes of inertia of the system,
- interdependence of control channels.

A mathematical model of controlled motion of AMR with M and an algorithm for solving the problem of dynamics by the Runge–Kutta–Fehlberg method are presented. Software has been developed for identifying the current dynamic state of AMR with M and assessing the mutual connection of control channels.

An algorithm for adaptive control of AMR with M has been developed, which provides for the simultaneous study and assessment in real time of the dynamic characteristics of AMR (identification problems) and determination of control parameters from the obtained model (synthesis problem). The proposed control algorithm makes it possible to equip work with means of automatic learning (self-learning) and adaptation, which turn their control system into an extremely flexible tool for organizing purposeful behavior under conditions of uncertainty.

The research carried out allows to get close to the creation of an adaptive multi-connected control system for AMR with M to increase its survivability and efficiency during autonomous operation in extreme conditions. Further research is aimed at developing predictive control [29], and control of the AMR group [30].

Acknowledgments

The results presented in the article were obtained by the authors in the course of research on the topic 0120U103294 "Universal anthropomorphic manipulator" and on the topic 0119U101151 "Applied research in mechanics and mechatronics."

References

[1] N. Ashchepkova, S. Ashchepkov, S. Kapera, 'Dynamics of transport robot model during the turns', Science and Education in New Dimension, pp. 26–29, VI(19), Issue 171, 2018. (in Ukrainian)

[2] B. Siciliano, O. Khatib, 'Handbook of Robotics', pp. 799–825, Springer, 2008.

[3] S. Tkachev, 'Stabilization of nonminimum-phase multi-input affine system', Science and Education. Scientific periodical of the Bauman MSTU, #8, 2012.

[4] O. Andrianova, 'Path following simulation of wheeled vehicle', Science and Education. Scientific periodical of the Bauman MSTU, #10, 2011.

[5] R. Gilimyanov, A. Pesterev, L. Rapoport, 'Motion control for a wheeled robot following a curvilinear path', Journal of Computer and Systems Science International, pp. 987–994, Vol. 47, #6, 2008.

[6] A. Kanatikov, T. Kasatkina, 'Features of transition to path coordinates in a problem of path stabilization', Science and Education. Scientific periodical of the Bauman MSTU, #7, 2012.

[7] S. Bai, L. Zhou, G. Wu, 'Manipulator Dynamics', in book: Handbook of Manufacturing Engineering and Technology, pp. 799–825, Springer, 2015 DOI: 10.1007/978-1-4471-4670-4_91

[8] M. Orsag, C. Korpela, P. Y. Oh, S. Bogdan, 'Aerial Manipulator Dynamics', in book: Aerial Manipulation, pp. 799–825, Springer, 2018, DOI: 10.1007/978-3-319-61022-1_5

[9] E. Jurevich, 'Basis of robot', 304 p., BHV - Petersburg, St. Petersburg, 2017. (in Russian)

[10] S. Kolyubin, 'Dynamics of the robots systems', 117 p. Publishing house of ITMO University, St. Petersburg, 2017. (in Russian)

[11] N. Ashchepkova, 'Mathcad in the kinematic and dynamic analysis of the manipulator', Eastern-European Journal of Enterprise Technologies, pp. 54–63, Vol. 5/7 (77), 2015. DOI: 10.15587/1729-4061.2015.51105 (in Russian)

[12] N. Ashchepkova, 'Divising a method to analyze the current state of the manipulator workspace', Eastern-European Journal of Enterprise Technologies, pp. 63–74, Vol. 1/7 (109), 2021. DOI: 10.15587/1729-4061.2021.225121

[13] N. Ashchepkova, 'Determination of optimal configurations of an antropomorphic manipulator with six degress of mobility', Herald of the National Technical University "KhPI". Series of: Informatics and Modeling. Vol. 28(1353), pp. 94–107, December 2019, DOI: 10.20998/2411-0558.2019.28.01, (in Russian)

[14] S. Lloyd, R. Irani, M. Ahmadi, 'A numeric derivation for fast regressive modeling of manipulator dynamics', Mechanism and Machine Theory,#156:104149, February 2021. DOI: 10.1016/j.mechmachtheory.2020.104149

[15] J. Khurpade, S.S. Dhami, s. S. Banwait, 'A Virtual Model of 2D Planar Manipulator Dynamics' International Conference on Smart Systems and Inventive Technology (ICSSIT), December 2018, DOI: 10.1109/ICSSIT.2018.8748674

[16] S.X. Tian, S.Z. Wang, 'Dynamic Modeling and Simulation of a Manipulator with Joint Inertia', International Symposium on Information and Automation, ISIA 2010: Information and Automation, pp. 10–16, 2010.

[17] N. Ashchepkova, 'Analysis of vibrations round centre-of-mass system the 'Small space vehicle with a manipulator', Herald of the Dnepr National University. Series of: Space-rocket, pp. 11–17, Vol. 4/1 (13), 2009. ISSN 9125 0912, (in Russian)

[18] S. Dubowsky, E. Papadopoulos, 'The Kinematic, Dynamic and Control of Free Flying Systems', IEEE Transactions on Robotic and Automation, Vol. 9(5), 1993.

[19] P. Efimova, A. Shimanchuk, 'Design of motion of space manipulation robot', Problems of mechanics and contral: Nonlinear dynamic systems, pp. 20–30, Vol. 46, 2014. (in Russian)

[20] L. Love, J. Jansen, F. Pin, "On the modeling of robots operating onships", in Robotics and Automation, Proc. In IEEE International Conference (ICRA'04), pp. 2436–2443, 2004.

[21] L. Love, J. Jansen, F. Pin, 'Compensation of wave-induced motion and force phenomena for ship-based high performance robotic and human amplifying systems', Technical Report ORNL/TM-2003/233. Oak Ridge National Laboratory, Oak Ridge, Tenn, 2003.

[22] P. J. From, V. Duindam, J. T. Gravdahl, 'Modeling and motion planning for mechanisms on a non-inertial base', Proc. In IEEE International Conference on Robotics and Automation - 2009, pp. 3320–3326, 2009.

[23] E. Sadraei, M. M. Moghaddam, 'On a Moving Base Robotic Manipulator Dynamics', International Journal of Robotics, Vol. 4, No. 3, October 2015, pp. 66–74, 2015.

[24] C. M. Wronka, M. W. Dunnigan, 'Derivation and analysis of a dynamic model of a robotic manipulator on a moving base', Robotics and Autonomous Systems, Vol. 59, pp. 758–769, 2011.

[25] W. Vereecken, et. al., 'Energy Efficiency in thin client solutions', ICST Int. Conf. on Networks for Grid Applic., Athens, 2009.

[26] N. Ashchepkova, A. Zbrutsky, 'Modeling the dynamics of an autonomous mobile robot with a manipulator', Herald of the National Technical University "KhPI". Series of: Informatics and Modeling. Vol. 2(4), pp. 34–45, December 2020, DOI: 10.20998/2411-0558.2020.01.0 3. (in Ukrainian)

[27] N. Ashchepkova, O Bulaniy,. 'A decision of equalizations of dynamics of space vehicle is with the undiagonal tensor of inertia', Question of optimization of calculations, Proceedings of the international science and technology conference, NAS of Ukraine, pp. 12–15, Kyiv, October 6–8, 1997. (in Russian)

[28] M. Brdni, Adaptive control of manipulator robots: Author's thesis, Moscow, 18 p., 1993.

[29] V.M. Kuntsevich, et. al (Eds). *Control Systems: Theory and Applications.* Series in Automation, Control and Robotics, River Publishers, Gistrup, Delft, 2018.

[30] Y.P. Kondratenko, V.M. Kuntsevich, A.A. Chikrii, V.F. Gubarev, (Eds). *Advanced Control Systems: Theory and Applications.* Series in Automation, Control and Robotics, River Publishers, Gistrup, 2021.

11

Safe Navigation of an Autonomous Robot in Dynamic and Unknown Environments

Yuriy P. Kondratenko[1], Arash Roshanineshat[2], and Dan Simon[3]

[1]Petro Mohyla Black Sea National University, Intelligent Information Systems Dept., Mykolayiv, 54003, Ukraine
[2]The University of Arizona, Electrical & Computer Engineering Dept., Tucson, AZ, 85721, USA
[3]Cleveland State University, Electrical Engineering and Computer Science Dept., Cleveland, OH, 44115, USA
E-mail: y_kondrat2002@yahoo.com; aroshanineshat@email.arizona.edu; d.j.simon@csuohio.edu

Abstract

For robots to become an efficient means for performing tasks, they need to navigate safely and effectively without supervision. This chapter is devoted to the problem of fuzzy control system design for mobile robots operating in dynamic environments with obstacles. The decision-making method, based on the transformation of radar-sensor information to fuzzy set "Obstacles" and robot's heading course to fuzzy set "Direction to Target Point," is considered. We also address the safety aspects of the proposed approach.

The different scenarios of the working environment (with static and dynamic obstacles), the adjusted speed of the autonomous robot, and peculiarities of "robot-obstacle" interaction are the issues for discussion. The simulation results confirm the efficiency and safety of the robot's autonomous navigation in dynamic and unknown environments.

Keywords: The autonomous robot, unknown environment, static and dynamic obstacles, fuzzy set, membership function, adjusting speed, safety

11.1 Introduction

Service robots acting in uncertain environments have become very popular in the last few years [1–5]. Many robots and robotic systems successfully work in hospitals, office buildings, department stores, museums, enterprises, etc. [6–8].

For robots to become capable team members and helpful assistants, especially in dynamic human-populated environments, they need to navigate efficiently and safely [9–12] in the target area.

In many cases, the efficiency of moving vehicles or robot missions depends on the properties of the sensor and control systems in providing the obstacles avoiding [13–20] (spacecraft, ships, crewless underwater vehicles, etc.). The target area, especially for mobile and underwater robots, often consists of static and dynamic obstacles. Thus, it is necessary to consider various environmental uncertainties. The obstacle avoidance with a priori unknown parameters is the peculiar feature of the control systems for above mentioned autonomous vehicles and robots.

Intelligent systems based on fuzzy sets and fuzzy logic [3, 7, 11, 12, 21-34] can be a good solution for this problem, especially when a system's mathematical model is either unavailable or too complex. Many publications [1, 3, 9, 11, 12, 35–45] are devoted to robot trajectory planning in the working environment with obstacles. However, this problem still requires developing new models and algorithms that optimize and improve the process of automatic path planning in real-time.

The authors of this chapter develop and investigate intelligent algorithms, providing the efficient planning and optimization of the robot's trajectory in uncertain and dynamic environments. The proposed algorithm is based on the representation of radar-distance observations and direction to the target point by specially constructed fuzzy sets with adjusted parameters of membership functions according to the current "robot-obstacle" situations.

This chapter pays special attention to an autonomous robot's efficient and safe navigation in a dynamic or a priori unknown environment. Efficient navigation means the ability of the robot to reach a target point from a starting position in a priori unknown environment, and the safety measure corresponds to the number of the robot collisions with dynamic obstacles during its navigation to the target point.

11.2 Related Works

Robot motion planning in dynamic environments has recently received substantial attention due to the advent of autonomous cars and growing interest in social, service, and assistive robots.

Deepu et al. [46] focus on providing a path for a robot to ease movement, detecting and avoiding obstacles in an environment using a single camera and a laser source. In [47] authors propose a potential hybrid field, which can be computed in real-time, to navigate a robot in the dynamic environment with 50 randomly moving obstacles. A fuzzy-inference system with an accelerate/brake module is developed in [48] for real-time navigation of autonomous underwater vehicles in both static and dynamic three-dimensional environments with automatically avoiding the dynamic obstacles using sonar model, virtual acceleration, and velocity in both horizontal and vertical planes. This chapter [49] presents a fuzzy truck control system for obstacle avoidance with a reasonably good trajectory, using 33 fuzzy inference rules for an autonomous vehicle's steering control and 13 rules for speed control. In [50], two fuzzy logic controllers for an autonomous vehicle's steering and velocity control consist of seven control modules. It provides (a) motion to the target point, (b) vehicle's final orientation, (c) avoiding collision with obstacles, (d) driving the vehicle through mazes, and (e) control the velocity of the movement toward the target point in the neighborhood of obstacles or when the vehicle turns sharp corners. The authors consider in [51] a reactive strategy for navigating a mobile robot in dynamic a priori unknown environments densely cluttered with moving and deforming obstacles.

Learning is essential for cognitive robots in adapting the human experience to intelligent robot control in the natural environment [52–57]. Learning can assist in future robot decisions for both efficiency and robustness. In [58], authors propose to form a robot's experience online and to transfer knowledge among appropriate contexts using the learning paradigm of inductive logic programming to frame hypotheses. Using first-order logic for such hypotheses representation is helpful for further reasoning and planning processes.

Combining artificial neural networks and evolutionary algorithms is an effective method for providing robustness in different navigation conditions. An example is an autonomous robot "Khepera" [59], with a sensory-motor model, based on the neural network and genetic algorithm, with the fitness evaluation in terms of the navigation performance in a maze course. In [60] authors present the framework for the navigation and target tracking system for the mobile robot, which uses a Microsoft Xbox Kinect sensor. The main task of the fuzzy controller is providing control of the robot in obstacle avoidance and target following. The designed multi-robot system [60] can work autonomously in an outdoor environment. In [61], the various robot motion planning schemes, including a genetic-fuzzy system, genetic-neural system, and a conventional potential field approach, have been compared in terms of traveling time taken by the robot, robustness, adaptability, goal-reaching capability, and repeatability. Neural networks, genetic algorithms and genetic programming are augmented with fuzzy logic-based schemes to enhance artificial intelligence of mobile robots [14, 16, 29, 30, 34, 35, 39, 42, 62–67]. In [68], the authors propose a neuro-fuzzy system architecture for behavior-based control of a mobile robot in unknown environments. The methodology of the behavior-based control approach includes two steps: (a) to analyze and to decompose a complex task based on stimulus-response behavior, (b) to formulate (quantitatively) each type of behavior with a simple feature by fuzzy sets and fuzzy rules, and (c) to coordinate conflicts and competition among multiple types of behavior by fuzzy reasoning.

An integrated representation of the environment with an approximation of obstacles' shapes by discs or polygons [69] is the base for collision-free navigation of a mobile robot in complex dynamic environments with moving obstacles. The navigation algorithm presented in [69] provides a short robot path planning through the crowd of moving or steady obstacles.

An efficient stereovision-based motion compensation method for moving robots, discussed in [70], uses the disparity map and three modules: segmentation, feature extraction, and estimation. In the segmentation module, the authors propose using extended type-2 fuzzy sets to extract the objects. Fuzzy logic is used for the robot in [71] to implement the behavior design and coordination during the realization of the "memory grid" and "minimum risk method." The robot can choose the safest region to avoid colliding with obstacles in different scenarios, particularly in the long wall, large concave, recursive UU-shaped, unstructured, cluttered, maze-like, and dynamic indoor environments. This chapter [72] presents the fuzzy-logic controller based on the Mamdani-type fuzzy inference engine for robot navigation and obstacles avoidance in a cluttered environment. The fuzzy controller with three inputs and single output provides safe navigation of the robot in a static environment considering the accuracy of absolute measurements of its position, obstacle distances, goal distance, velocity, orientation, and rate of change in its heading angle. This chapter [73] describes a fast and reliable obstacle avoidance method for ground mobile robots in both outdoor and indoor navigation. The process realizes two contradictory approaches (non-complex implementation and human-like smooth steering) and applies in different mobile robotic systems regardless of used sensors. In [74], a conceptual fuzzy logic approach is the basis for solving a multi-link robot's local navigation and obstacle avoidance problem. The proposed method is the core of developing an online local navigation system for generating instantaneous collision-free trajectories.

Thus, the analysis of last publications on autonomous and mobile robots and their motion control in the environment with obstacles shows [49, 60, 61] that researchers continue to develop new avoiding obstacles' algorithms and new devices' solutions using well-known and new design methods, approaches, and methodologies. It deals, first of all, with the different and specific missions of the mobile robots; the high level of information uncertainty concerning the nature of obstacles and the character of their appearance; changes of obstacles' movement velocities; a large set of the various geometrical shapes of obstacles; unknown level of the danger in the case of an accident, etc.

Additional design requirements also stimulate researchers to develop new approaches, methods, and algorithms for avoiding obstacles by a mobile robot. Among such most important requirements are:

(a) restricting the design space in the robot's construction for corresponding navigation and control devices' placement (that is a vital requirement for design processes in the underwater and space robotics);

(b) simplifying the information and signal processing algorithms for embedded robot's navigation and control systems with providing the trajectory planning in real-time [75];

(c) reconfiguring the information processing algorithms and corresponding systems [76] based on the modern field-programmable gate array - FPGA [63, 77–79];

(d) providing structural-parametrical optimization and high reliability of the computerized control systems [30, 32, 34, 52, 80–85] for robot navigation, etc.

The stringent requirements to the robot's navigation and control systems' parameters allow realizing efficient trajectory planning in real-time. In particular, it requires the minimization of such parameters as:

• *the response time* between the moment of obstacle emergence in the "vision" zone of the mobile robot and the moment of its recognition by the robot's navigation system;

- *the time of the fuzzy information processing* for the calculation and correction of the robot motion's course providing the obstacle avoidance;
- *the time for calculating and correcting the robot motion's velocity* to secure the save motion near the obstacle.

Therefore, the actual research task is to modify the existing and create new, more efficient navigation and control systems algorithms for their successful implementation in mobile robots and autonomous vehicles.

As discussed above, the artificial intelligence methods, including the application of fuzzy sets and fuzzy logic, are powerful tools for solving such problems as planning efficient trajectories for the safe movement of mobile robots in an uncertain environment with obstacles. Choosing a corresponding artificial intelligence method for implementation in a robot navigation-control system (providing efficient fuzzy information processing) requires taking into account: (a) all specific robot's environment conditions and peculiarities; (b) an extensive spectrum of the possible missions for a mobile robot; (c) the universality for different real working scenarios with static and dynamic obstacles and other uncertainties.

Thus, it is reasonable to develop new design methods, algorithms, and models that provide efficient fuzzy information processing. It will simultaneously improve the design processes for robots' navigation-control systems and enhance the control indicators of their functioning in uncertain environments with obstacles.

The main idea of this chapter is an investigation of the fuzzy approach for mobile robots' trajectory planning in an environment with uncertain obstacles, using proposed by authors fuzzy reasoning algorithm. This algorithm will provide: (a) the increasing levels of universality for different uncertain scenarios with obstacles, and (b) simplicity by decreasing the complexity of the fuzzy information processing and by reducing intricacy of the corresponding device's synthesis process for hardware realization, as well as, (c) increasing the reliability and (d) decreasing the time for decision making in different conflict situations.

Finally, let us formulate the aims of this chapter as:

- introducing the fuzzy reasoning algorithm [21–25] instead of fuzzy-rule-base inference engine for increasing efficiency of robot's navigation in the uncertain environment with unknown obstacles;
- the investigation of proposed by authors "radar-fuzzy-set" fuzzy processing approach, concerning the continuous adjusting of robot movement parameters and parameters of fuzzy reasoning algorithm, based on the current information of the radar-sensory system;

- developing the structured multi-agent software for modeling and simulation of the different scenarios of robot's navigation with confirmation of the efficiency of the developed fuzzy processing algorithms for robot's navigation in uncertainty.

The rest of the chapter covers multiple aspects related to the topic under discussion. Section 11.3 considers a general representation of the proposed method of fuzzy information processing for robot navigation in the environment with unknown obstacles, based on the intersection of two dynamic fuzzy sets, "Environment with obstacles" and "Fuzzy direction to the target point." In Section 11.4, the authors provide a more detailed description of the fuzzy-based path planning for investigation of the main properties of the proposed robot's navigation system. Section 11.5 deals with the developed algorithm for robot speed adjusting in a dynamic environment with obstacles. Section 11.6 represents the modeling and simulation results for the different mobile robot's navigation scenarios in an uncertain environment with the radar-sensory system for obstacles detection. This chapter ends with a conclusion in Section 11.7.

11.3 Problem Statement

Figure 11.1 shows a sample of a working environment in a two-dimensional space. The area is a priori unknown to the autonomous robot. SP is the initial (start) position of the robot, and TP is the target point.

The robot has a 360° radar sensor with a limited range L_{max}. The problem is that the robot should autonomously synthesize the non-linear trajectory for efficient movement from SP to TP, avoiding all the static and dynamic obstacles and providing (a) the minimum length of the trajectory, (b) the minimum time for robot motion on the trajectory, and (c) high reliability for movement in the dynamic working environment.

Online trajectory planning can be done through continuous fuzzy information processing of the radar sensor data and identifying the robot's direction at the beginning of each motion step on the planning trajectory.

The following section considers the proposed algorithm for such real-time trajectory planning in detail. It is supposed that the mobile robot's current position (coordinates) SP and the coordinates of the target point TP in the target area are known or can be measured at every step of the trajectory planning.

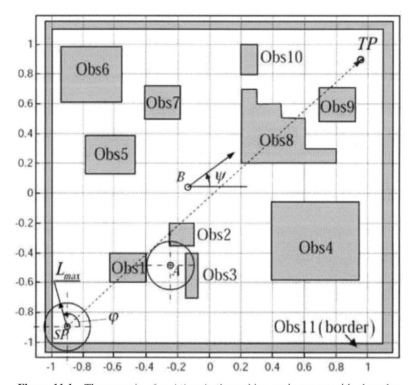

Figure 11.1   The scenario of mobile robot's working environment with obstacles

Table 11.1   Table of Notations

φ_i	The angle of the radar beam
$\Delta\varphi$	Step of the radar beam
L_i	Distance to the object in a φ_i direction
L_{max}	Maximum distance of radar beam
$\underset{\sim}{E}$	Radar output fuzzy set
$\mu_{\underset{\sim}{E}(\varphi_i)}$	Radar output membership function
α	The angle between robot position and target point
$\underset{\sim}{G}$	Our introduced rectangular fuzzy set
$\mu_{\underset{\sim}{G}}(\varphi_i)$	Chosen rectangular membership function
$\underset{\sim}{R}$	Resulted fuzzy set in intersection stage
$\mu_{\underset{\sim}{R}}(\varphi_i)$	Resulted membership function from intersecting rectangular membership function and radar output
ψ	Correct angle to choose by the robot

11.4 Fuzzy based Path Planning

The robot can detect the obstacles within its radar sensor [3, 12, 86], scanning the working space by a beam of a limited length $L_{\max}$. Such beam is turning in a positive direction φ_i with a discrete step $\Delta\varphi = \frac{2\pi}{n}$:

$$\varphi_{i+1} = \varphi_i + \Delta\varphi = (i+1)\,\frac{2\pi}{n}, \quad i = 0, 1, 2, ...n - 1,$$

where n is the number of discrete steps for a scanning beam; L_i is a distance between the robot and the nearest obstacle at every ith step φ_i of the scanning beam, $L_i/L_{\max}$, $L_i \in [0, 1]$.

Figure 11.2 illustrates the circular scan (radar-beam-diagram) for the configuration of the obstacles Obs2 and Obs3 at the mobile robot's trajectory point A (Figure 11.1). The distance $L_i, (i = 1, 2, ..., n)$ between the robot

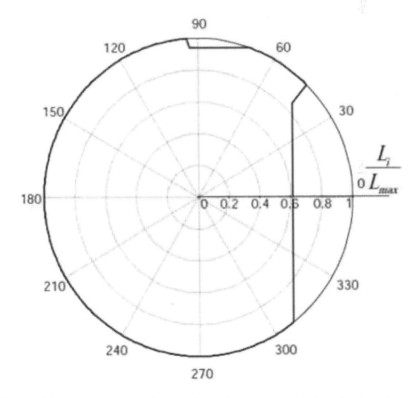

Figure 11.2   The radar-beam-diagram of the robot sensor system in point A on the motion trajectory, presented in Figure 11.1

(point 0 in the center of the circle diagram, Figure 11.2) and any obstacle on the corresponding beam is presented in the relative form $L_i/L_{\max} \in [0, 1]$.

The proposed algorithm for fuzzy information representation and processing consists of the following stages:

Stage 1. Scanning radar area. The interpretation of the scanned results of the robot's visible zone (domain) $\{L_0, L_1, ..., L_i, , ..., L_n\}$ for the working environment with detected obstacles as the fuzzy set $\underset{\sim}{E}$.

Such fuzzy set $\underset{\sim}{E}$ (for fuzzy interpretation of the radar information shown in Figure 11.2) is shown in Figure 11.3, considering the following dependence

$$\mu_{\underset{\sim}{E}}(\varphi_i) = \frac{L_i}{L_{\max}}, \tag{11.1}$$

where $\mu_{\underset{\sim}{E}}(\varphi_i)$ is a value of the membership function $\mu_{\underset{\sim}{E}}$ of the fuzzy set $\underset{\sim}{E}$, which corresponds to the value of beam's angle position φ_i, included to the support $S(\underset{\sim}{E})$ [87] of the fuzzy set $\underset{\sim}{E}$:

$$\forall \varphi_i : \varphi_i \in S(\underset{\sim}{E}) = \text{supp}(\underset{\sim}{E}) = \left\{\varphi_i : \mu_{\underset{\sim}{E}}(\varphi_i) > 0, \varphi_i \in [0, 2\pi]\right\}.$$

If any obstacle is absent on the ith scanning beam, the value of membership function $\mu_{\underset{\sim}{E}}(\varphi_i) = \frac{L_i}{L_{\max}} = 1$, because $L_i = L_{\max}$. If the scanning beam meets the obstacle, then $\mu_{\underset{\sim}{E}}(\varphi_i) = \frac{L_i}{L_{\max}} \in [0, 1]$, because $L_i < L_{\max}$. If the robot contacts a corresponding obstacle or robot situated at the minimal admissible distance $L_{\min}$ to the obstacle (and the further movement in the current direction is impossible), then $\mu_{\underset{\sim}{E}}(\varphi_i) = \frac{L_i}{L_{\max}} = 0$, as $L_i = 0$ or $L_i \leq L_{\min}$.

It is evident that the membership function of the fuzzy set $\underset{\sim}{E}$ has a rectangular form for the case if no obstacle is in the radar's visible zone

$$\forall \varphi_i \in [0, 2\pi] : \quad \mu_{\underset{\sim}{E}}(\varphi_i) = 1.$$

It is possible to calculate the recommended value of the robot course (motion direction) φ^*, that has the optimal distance from the visible obstacles using any defuzzification algorithm [88–90] for the interpreted fuzzy set (target area). The value of the recommended course $\varphi^* = 0.99\pi$, presented in Figure 11.3, was calculated using the gravity center method for the defuzzification of the fuzzy set $\underset{\sim}{E}$ (Figure 11.3). This recommended value φ^* secures only the safe course, but it is not related to the final target position of the robot after its accomplishing mission.

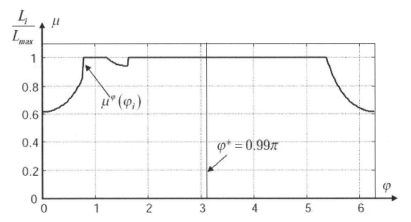

Figure 11.3 The fuzzy set $\underset{\sim}{E}$, which illustrates the proposed fuzzy interpretation of the radar-beam-diagram, presented in Figure 11.2

Stage 2. Introducing a suitable membership function. This stage deals with the interpretation of the desired (straight directed) course α to the target point TP from the current position of the mobile robot by corresponding fuzzy set $\underset{\sim}{G}$ "Fuzzy direction to the target point" with membership function $\mu_G(\varphi_i)$, $\varphi_i \in [0, 2\pi]$, where $\mu_G(\varphi_i) = 1$ in the condition $\alpha - c \leq \varphi_i \leq \alpha + c$.

Different types and shapes of membership functions can present the fuzzy set $\underset{\sim}{G}$ and its membership function $\mu_G(\varphi_i)$. We consider the rectangular membership function to represent the fuzzy set $\underset{\sim}{G}$ in the following form:

$$\mu_G(\varphi_i) = \begin{cases} 1 & if \ \alpha - c \leq \varphi_i \leq \alpha + c \\ 0 & otherwise, \end{cases} \tag{11.2}$$

where c is a parameter of membership function $\mu_G(\varphi_i)$, which influences the wideness of the rectangular membership function.

The membership function $\mu_G(\varphi_i)$ presents (Figure 11.4) the example of the fuzzy set $\underset{\sim}{G}$ for the desired course $\alpha = 0$ and $|c| < 1$.

Stage 3. Interacting membership function and radar data. We can find at this stage the resulting fuzzy set $\underset{\sim}{R}$, which corresponds to the fuzzy representation of the mobile robot's path with securing direction to the target point and avoiding obstacles. It is possible to calculate the parameters of this fuzzy set $\underset{\sim}{R}$ as intersection of two normal fuzzy sets $\underset{\sim}{G}$ and $\underset{\sim}{E}$:

$$\underset{\sim}{R} = \underset{\sim}{G} \bigcap \underset{\sim}{E}, \tag{11.3}$$

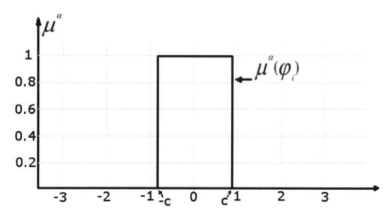

Figure 11.4 The fuzzy set $\underset{\sim}{G}$ with rectangular membership function (11.2)

where $\bigcap$ is a fuzzy intersection operator, for example, t-norm operator [87, 90, 91, 92].

The membership function $\mu_{\underset{\sim}{R}}(\varphi_i)$ of the fuzzy set $\underset{\sim}{R}$ can be calculated using the algorithm

$$\mu_{\underset{\sim}{R}}(\varphi_i) = \mu_{\underset{\sim}{G}}(\varphi_i) \bigcap \mu_{\underset{\sim}{E}}(\varphi_i). \tag{11.4}$$

Usually, a fuzzy set with a membership function $\mu_{\underset{\sim}{R}}(\varphi_i)$ has a one-interval or multi-intervals core [87] if an obstacle exists in the radar range of the robot.

Figure 11.5 illustrates the resulting fuzzy set $\underset{\sim}{R}$ with membership function $\mu_{\underset{\sim}{R}}(\varphi_i)$ for interpreting the robot's heading course ψ as an intersection of the two fuzzy sets $\underset{\sim}{G}$ and $\underset{\sim}{E}$ according to (11.4).

The shape of the membership function $\mu_{\underset{\sim}{R}}(\varphi_i)$ mostly depends on the shapes of the membership functions $\mu_{\underset{\sim}{G}}(\varphi_i)$ and $\mu_{\underset{\sim}{E}}(\varphi_i)$, but the choice of the intersection operator ($\bigcap$) also influences the final form.

It is possible to use different types of t-norm operators ($\bigcap$) as intersection operators from two families [87] of non-parameterized and parameterized t-norm operators. All t-norm operators satisfy the special conditions of t-norm: mapping space, zeroing, the identity of unity, commutativity, union, monotonicity.

For example, we can use the non-parameterized minimum intersection operator (*MIN*-operator) for the calculation of the values $\mu_{\underset{\sim}{R}}(\varphi_i)$ according to (11.4) in the following way:

$$\mu_{\underset{\sim}{R}}(\varphi_i) = \mu_{\underset{\sim}{G}}(\varphi_i) \bigcap \mu_{\underset{\sim}{E}}(\varphi_i) = MIN\left(\mu_{\underset{\sim}{G}}(\varphi_i), \mu_{\underset{\sim}{E}}(\varphi_i)\right). \tag{11.5}$$

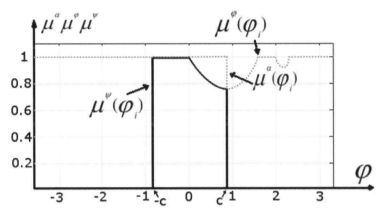

Figure 11.5 The resulting fuzzy set $\underset{\sim}{R}$ with one-interval core for the determination of mobile robot's heading course according to (11.3), (11.4)

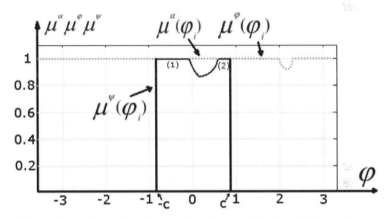

Figure 11.6 The multimodal resulting fuzzy set $\underset{\sim}{R}$ with multi-intervals core

At the same time, it is possible to use (alternatively) in (11.4) the different parameterized *mean* operators as intersection operators [87] (which are not satisfying with special *t*-norm conditions).

Figure. 11.6 shows the membership function $\mu_{\underset{\sim}{R}}(\varphi_i)$ with the multi-intervals core of the resulting fuzzy set $\underset{\sim}{R}$.

The next step is finding sections where there are no obstacles. Values $\mu_{\underset{\sim}{R}}(\varphi_i) = 1$ mean that there is no obstacle in the direction φ_i. In Figure 11.6, highlighted segments with (1) and (2) show no obstacles in these areas. The robot should choose a segment that has less probability of collision with an obstacle.

First, the robot finds all sections where the radar distance value is more than a threshold for finding a segment with less probability. Since there are measurement noises, radar may not have the exact value of 1.

Determining *Obstacles* in $\mu_{\underset{\sim}{R}}(\varphi_i) =$	a) *No Obstacles*, if $\mu_{\underset{\sim}{R}}(\varphi_i) \leq$ threshold b) *Obstacles*, if $\mu_{\underset{\sim}{R}}(\varphi_i) >$ threshold

Then the robot chooses the segment which is the longest. Figure 11.6 shows that segment (1) is longer than segment (2), so the robot will select a part (1).

Defuzzification of this segment will return the most appropriate direction for the robot. In some cases, the choice of the defuzzification method is limited by the necessity to take into account the multi-intervals core of the resulting fuzzy set $\underset{\sim}{R}$ and its membership function $\mu_{\underset{\sim}{R}}(\varphi_i)$ that practically does not meet in the majority of the Mamdani-type fuzzy inference systems. The multimodality presence explains that if the direct way to the target point is blocked for the mobile robot by an obstacle, there can be some alternate route concerning the shape of the specified obstacle.

Stage 4. Defuzzification. The last stage is the determination of the crisp value of the robot's direction ψ by defuzzification of the resulting fuzzy set $\underset{\sim}{R}$ with membership function $\mu_{\underset{\sim}{R}}(\varphi_i)$, presented as an example in Figure 11.5:

$$\psi = Arg\left[Defuzzification\left(\mu_{\underset{\sim}{R}}(\varphi_i)\right)\right] \tag{11.6}$$

Different defuzzification algorithms can be used [87–90], particularly the first maxima, the last maxima, the middle maxima, the center of gravity, the center of sum, the bi-sector, the height method.

The current value of the robot's heading course ψ should be calculated in real-time by the robot during its navigation to the target point TP avoiding any obstacles taking into account the changeable character of both fuzzy sets $\underset{\sim}{G}$ and $\underset{\sim}{E}$.

Thus as mathematical modeling shows, situations are frequent when two or more course angles have an equivalent grade of the membership to the fuzzy set $\underset{\sim}{R}$. The inadequacy of such commonly used defuzzification methods as a center of gravity or bisector for a case of multimodal membership functions is shown in [75].

Figure 11.7 shows the highlighted area for defuzzification of the resulting fuzzy set $\underset{\sim}{R}$ with the multi-intervals core.

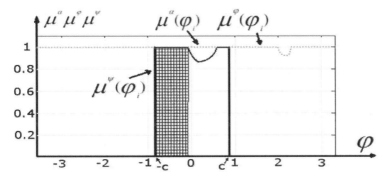

Figure 11.7   The highlighted area for defuzzification

11.5 Adjusting Speed

Some situations (in dynamic environments with obstacles and real-time control of mobile robots movement) need increasing or decreasing robot motion speed (speed control).

In general, the mobile robot has enough time for re-planning its path if the robot detects a static obstacle. Thus, we will consider the situations with changing the speed of the robot only if dynamic obstacles [3, 12, 86, 93, 94] are detected.

Figure 11.8 shows different divisions we have considered for the speed of the autonomous robot.

There are five-speed parts in total (Figure 11.8) according to the different descriptions in Table 11.2.

11.6 Experiments and Evaluations

We have experimented with our method on four maps and six times for each of them. In three of the experiments, safe navigation was activated, and in the other three experiments, safe navigation was disabled. Numbers (Table 11.3) are the absolute values of the average of three experiments for both safe and unsafe experiments.

Dynamic obstacles have pink and static obstacles have black borders (Figure 11.9).

The size of the maps is 500x400 pixels, and the radar radius is 30 pixels.

Since there is no dynamic obstacle in Map 1 (Figure 11.9), the robot could reach the target point without colliding with any obstacle in safe and unsafe modes.

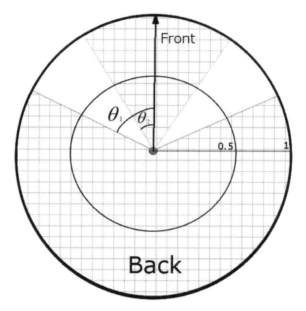

Figure 11.8   The robot radar division for different speeds

Table 11.2   Table of the robot speed

No.	Speed	Condition 1	Condition 2
1	Maximum Speed	$360 - \theta_1 > \min(\varphi_i) > \theta_1$	$L_{\min(\varphi_i)} < 0.5$
2	High Speed	$360 - \theta_1 > \min(\varphi_i) > \theta_1$	$0.5 < L_{\min(\varphi_i)} < 1$
3	Low Speed	$\theta_2 < \min(\varphi_i) < \theta_2$	$0.5 < L_{\min(\varphi_i)} < 1$
4	Stop	$\theta_2 < \min(\varphi_i) < \theta_2$	$L_{\min(\varphi_i)} < 0.5$
5	Normal	Otherwise	

Table 11.3   Result of experiments

	The Average Number of Collisions in Safe Navigation	The Average Number of Collisions in Unsafe Navigation
Map 1	0	0
Map 2	0	8
Map 3	0	16
Map 4	0	33

In Map 2, Map 3, and Map 4 (Figure 11.9), in all three corresponding safe mode experiments, the robot reached the target point without colliding with any obstacle, but in unsafe experiments, the robot was hit with dynamic obstacles (Table 11.3).

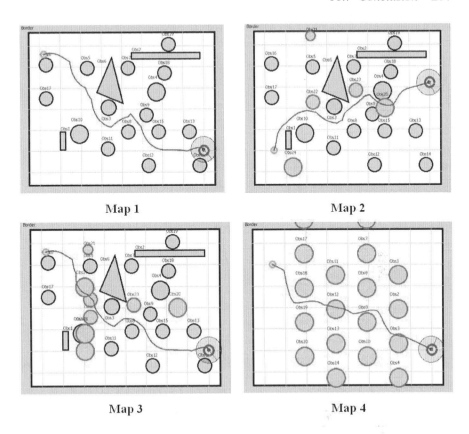

Figure 11.9 Maps of unknown environments used for experiments

11.7 Conclusion

We propose a fuzzy-based approach enabling robots to navigate efficiently and safely in uncertain environments.

Experimental results show the applicability of the proposed approach for autonomous robot navigation in dynamic environments where moving obstacles are avoided efficiently, and the robot manages to reach its destination.

In our experiments, the autonomous robot chooses a safe path for navigation. In situations where it needs to avoid collisions with dynamic obstacles, the robot can adjust its speed to escape from dangerous situations as safely as possible. We showed that changing rate resulted in having no collisions between the robot and obstacles.

For future work, we would like to extend this work to a multi-robot setting and investigate the robustness of sensor-based multi-robot control in an uncertain environment [95–99] using hybrid artificial intelligence methods.

Acknowledgments

The authors thank the Fulbright Program (USA) for supporting Prof. Y. P. Kondratenko with a Fulbright scholarship and making it possible for this team to conduct research together in the USA at the Cleveland State University, Electrical Engineering, and Computer Science Department.

References

[1] M. Derkach, D. Matiuk, I. Skarga-Bandurova, 'Obstacle Avoidance Algorithm for Small Autonomous Mobile Robot Equipped with Ultrasonic Sensors,' in IEEE 11th Int. Conf. on Dependable Systems, Services, and Technologies, DESSERT 2020, pp. 236–241, 2020.

[2] A.N. Tkachenko, N.M. Brovinskaya, Y.P. Kondratenko, 'Evolutionary adaptation of control processes in robots operating in non-stationary environments,' Mechanism and Machine Theory, Vol. 18, No. 4, pp. 275–278, 1983. DOI: 10.1016/0094-114X(83)90118-0

[3] Y. Kondratenko, G. Khademi, V. Azimi, D. Ebeigbe, M. Abdelhady, S. Fakoorian, T. Barto, A. Roshanineshat, I. Atamanyuk, D. Simon, 'Robotics and Prosthetics at Cleveland State University: Modern Information, Communication, and Modeling Technologies,' in Information and Communication Technologies in Education, Research, and Industrial Applications, ICTERI 2016. Communications in Computer and Information Science, Vol. 783. Springer, Cham, pp. 133–155, 2017. DOI: https://doi.org/10.1007/978-3-319-69965-3_8

[4] Y. Kondratenko, L. Klymenko, V. Kondratenko, G. Kondratenko, E. Shvets, 'Slip Displacement Sensors for Intelligent Robots: Solutions and Models', in Proc. of the 2013 IEEE 7th Int. Conf. on Intelligent Data Acquisition and Advanced Computing Systems (IDAACS), Berlin, vol. 2, Sept. 12–14, pp. 861–866, 2013. DOI: 10.1109/IDAACS.2013.6663050

[5] U. Patel et al., 'Beam: A Collaborative Autonomous Mobile Service Robot,' in Proceedings of the AAAI Fall Symposium on Artificial Intelligence for Human-Robot Interaction (AI-HRI), 2017.

[6] Z. Li and Z. Huang, 'Design of a type of cleaning robot with ultrasonic,' in Journal of Theoretical and Applied Information Technology, Vol. 47, No. 3, pp. 1218–1222, 2013.

[7] Y.P. Kondratenko, 'Robotics, Automation and Information Systems: Future Perspectives and Correlation with Culture, Sport and Life Science,' in Decision Making and Knowledge Decision Support Systems, Lecture Notes in Economics and Mathematical Systems, Vol. 675, A. M. Gil-Lafuente, C. Zopounidis, Eds. Springer International Publishing Switzerland, pp. 43–56. 2015. DOI: 10.1007/978-3-319-03907-7_6

[8] M.O. Taranov, et al., 'Models of Robot's Wheel-Mover Behavior on Ferromagnetic Surfaces,' in International Journal of Computing, Vol.17, Issue 1, Open Access, pp. 8–14, 2018. DOI: https://doi.org/10.47839/ijc.17.1.944

[9] M. D'Arcy, P. Fazli, D. Simon, 'Safe Navigation in Dynamic, Unknown, Continuous, and Cluttered Environments,' in Proceedings of the IEEE International Symposium on Safety, Security, and Rescue Robotics (SSRR), 2017.

[10] M. Taranov, et al., 'Simulation of Robot's Wheel-Mover on Ferromagnetic Surfaces,' in IEEE 9th Int. Conf. IDAACS, Bucharest, Romania, vol. 1, Sept. 21–23, pp. 283–288, 2017. DOI: 10.1109/IDAACS.2017.8095091

[11] D. Driankov, A. Saffiotti (Eds.), 'Fuzzy Logic Techniques for Autonomous Vehicle Navigation,' Physica, 2013.

[12] Y.P. Kondratenko, S.A. Sydorenko, 'Fuzzy-based trajectory planning in environment with obstacles', in Proceedings of East-West Fuzzy Colloquium 2006, 13th Zittau Fuzzy Collquium, Institut fur Prozesstechnik, Prozessautomatisierung und Messtechnik, Zittau, Sept. 13–15, pp. 40–48, 2006.

[13] Y.P. Kondratenko, V.L. Timchenko, 'Increase in Navigation Safety by Developing Distributed Man-Machine Control Systems,' in Third International Offshore and Polar Engineering Conference, Singapore, Vol. 2, pp. 512–519, 1993, Paper Number: ISOPE-I-93-173.

[14] Y.P. Kondratenko, J. Rudolph, O.V. Kozlov, Y.M. Zaporozhets, O.S. Gerasin, 'Neuro-fuzzy observers of clamping force for magnetically operated movers of mobile robots,' in Technical Electrodynamics, No. 5, pp. 53–61, 2017 (in Ukrainian). DOI: 10.15407/techned2017.05.053

[15] V.L. Timchenko, et al. 'Robust stabilization of Marine Mobile Objects on the Basis of Systems with Variable Structure of Feedbacks,' in Journal of Automation and Information Sciences, Vol. 43, No. 6. –

New York: Begel House Inc., pp. 16–29, 2011. DOI: 10.1615/JAutomat-InfScien.v43.i6.20

[16] R. Duro, Y. Kondratenko (Eds.), 'Advances in Intelligent Robotics and Collaborative Automation', in Book Series: River Publishers Series in Automation Control and Robotics, 2015. DOI: https://doi.org/10.13052/rp-9788793237049

[17] O.S. Gerasin, et al. 'Remote IoT-based control system of the mobile caterpillar robot,' in 16th International Conference on ICT in Education, Research and Industrial Applications. Integration, Harmonization, and Knowledge Transfer, Volume I: Main Conference, ICTERI 2020, Kharkiv, CEUR Workshop Proceedings, Vol. 2740, pp. 129–136, 2020. CEUR-WS.org/Vol-2740/20200129.pdf

[18] T. Huntsberger, A. Hrand, E. Baumgartner, S. Paul, 'Behavior-based control systems for planetary autonomous robot outposts,' in Aerospace Conference Proceedings, 2000 IEEE, vol. 7, IEEE, pp. 679–686, 2000.

[19] D. Simon, 'Neural Networks for Optimal Robot Trajectory Planning,' in E. Fiesler, R. Beale (Eds), Handbook of Neural Computation, Institute of Physics Publishing, pp. G2.5:1–8, 1997.

[20] T. Huntsberger, 'Biologically inspired autonomous rover control,' in Autonomous Robots 11, No. 3, pp. 341–346, 2001.

[21] L.A. Zadeh, 'Fuzzy sets. Information and control', No. 8, pp. 338–353, 1965.

[22] L.A. Zadeh, 'Outline of a new approach to the analysis of complex systems and decision processes,' in IEEE Transactions on Systems, Man and Cybernetics, Vol. 3, No. 1, pp. 28–44, 1973.

[23] L.A. Zadeh, 'Inference in fuzzy logic,' in IEEE Proceedings, Vol. 68, pp. 124–131, 1980.

[24] L.A. Zadeh, 'From computing with number to computing with words – from manipulations of measurements to manipulation with perceptions,' in IEEE Transactions on Circuits and Systems Part I, Vol. 4, no. 3, pp. 105–109, 1999.

[25] J. Kacprzyk, 'Multistage Fuzzy Control: A Model-Based Approach to Control and Decision-Making', in Wiley, Chichester, 1997.

[26] Y.P. Kondratenko, N.Y. Kondratenko, 'Reduced library of the soft computing analytic models for arithmetic operations with asymmetrical fuzzy numbers,' in Alan Casey (Ed), Soft Computing: Developments, Methods, and Applications. Series: Computer Science, technology and applications, NOVA Science Publishers, Hauppauge, New York, pp. 1–38, 2016.

[27] W. Lodwick, J. Kacprzhyk (Eds), 'Fuzzy Optimization,' in STUDFUZ 254, Berlin, Heidelberg: Springer-Verlag, 2010.

[28] R. R. Yager, D. P. Filev, 'Unified structure and parameter identification of fuzzy models,' in Systems, Man and Cybernetics, Vol. 23, issue 4, 1993.

[29] V.M. Kuntsevich et al. (Eds), 'Control Systems: Theory and Applications,' in Book Series in Automation, Control and Robotics, River Publishers, Gistrup, Delft, 2018. ISBN: 9788770220248

[30] Y.P. Kondratenko, A.V. Kozlov, 'Parametric optimization of fuzzy control systems based on hybrid particle swarm algorithms with elite strategy,' in Journal of Automation and Information Sciences 51(12), pp. 25–45, 2019. DOI: 10.1615/JAutomatInfScien.v51.i12.40

[31] V. Shebanin et al., 'Application of Fuzzy Predicates and Quantifiers by Matrix Presentation in Informational Resources Modeling,' in Perspective Technologies and Methods in MEMS Design: Proceedings of the International Conference MEMSTECH-2016. Lviv-Poljana, Ukraine, April 20–24, pp. 146–149, 2016. DOI: 10.1109/MEMSTECH.2016.7507536

[32] Y. Kondratenko, D. Simon, 'Structural and parametric optimization of fuzzy control and decision making systems,' in Recent Developments and the New Direction in Soft-Computing Foundations and Applications. Studies in Fuzziness and Soft Computing, Zadeh L., Yager R., Shahbazova S., Reformat M., Kreinovich V. (eds), Vol. 361, pp. 273–289, 2018. Springer, Cham. DOI: https://doi.org/10.1007/978-3-319-75408-6_22

[33] A. Gozhyj et al., 'The method of web-resources management under conditions of uncertainty based on fuzzy logic,' in Proceedings of the 2018 IEEE 13th International Scientific and Technical Conference on Computer Sciences and Information Technologies, CSIT 2018, 11–14 Sept. 2018, Lviv, Ukraine, pp. 347–352, 2018. DOI: 10.1109/STC-CSIT.2018.8526761

[34] Y.P. Kondratenko, A.V. Kozlov 'Generation of Rule Bases of Fuzzy Systems Based on Modified Ant Colony Algorithms', in Journal of Automation and Information Sciences, Vol. 51, Issue 3, New York: Begel House Inc., pp. 4–25, 2019. DOI: 10.1615/JAutomatInfScien.v51.i3.20

[35] K.C. Ng, M.M. Trivedi, 'A neuro-fuzzy controller for mobile robot navigation and multirobot convoying', in Systems, Man, and Cybernetics, Part B: Cybernetics, IEEE Transactions on 28, no. 6, pp. 829–840, 1998.

[36] A. Ollero, A. García-Cerezo, J. Martínez, A. Mandow, 'Fuzzy tracking methods for mobile robots,' in Applications of fuzzy logic: Towards high machine intelligence quotient systems 9, pp. 347–364, 1997.

[37] P. Pirjanian, M. Matarić. 'Multiple objectives vs. fuzzy behavior coordination,' in Fuzzy logic techniques for autonomous vehicle navigation, Physica-Verlag HD, pp. 235–253, 2001.

[38] A. Saffiotti et al., 'Using fuzzy logic for mobile robot control,' in Practical applications of fuzzy technologies. Springer US, pp. 185–205, 1999.

[39] Y. Kondratenko, E. Gordienko, 'Neural Networks for Adaptive Control System of Caterpillar Turn,' in Annals of DAAAM for 2011 & Proceeding of the 22nd Int. DAAAM Symp. Intelligent Manufacturing and Automation, (20–23 Oct. 2011, Vienna, Austria), Published by DAAAM International, Vienna, Austria, pp. 305–306, 2011.

[40] Y. Kondratenko, E. Shvets, O. Shyshkin, 'Modern Sensor Systems of Intelligent Robots Based on the Slip Displacement Signal Detection,' in Ann. of DAAAM for 2007 & Proc. of the 18th Int. DAAAM Symp. Intelligent Manufacturing and Automation. Vienna, pp. 381–382, 2007.

[41] A. Das, R. Fierro, V. Kumar, B. Southall, J. Spletzer, C. Taylor, 'Real-time vision based control of a nonholonomic mobile robot,' in IEEE Intl. Conf. Robotics and Automation (ICRA 2001), Seoul, Korea, pp. 1714–1719, 2001.

[42] J. Mbede, X. Huang, M. Wang, 'Robust neuro-fuzzy sensor-based motion control among dynamic obstacles for robot manipulators,' in Fuzzy Systems, IEEE Transactions on 11, no. 2, pp. 249–261, 2003.

[43] A. Saffiotti, 'Fuzzy logic in autonomous navigation,' in Fuzzy Logic Techniques for Autonomous Vehicle Navigation. Physica-Verlag HD, pp. 3–24, 2001.

[44] H. Surmann, L. Peters, 'Robot with Fuzzy Controlled Behaviour,' in Fuzzy Logic Techniques for Autonomous Vehicle Navigation, Physica-Verlag HD, pp. 343–365, 2001.

[45] G. Fu, A. Menciassi, P. Dario, 'Development of a low-cost active 3D triangulation laser scanner for indoor navigation of miniature mobile robots,' in Robotics and Autonomous Systems, Vol. 60, Issue 10, pp. 1317–1326, 2012.

[46] R. Deepu, B. Honnaraju, S. Murali, 'Path generation for robot navigation using a single camera,' in Procedia Computer Science, Vol. 46, pp. 1425–1432, 2015.

[47] S. Ratering, M. Gini, 'Robot navigation in a known environment with unknown moving obstacles,' in Autonomous Robots, 1, pp. 149–165, 1995.

[48] B. Sun, D. Zhu, L. Jiang, S.X. Yang, 'A novel fuzzy control algorithm for three-dimensional AUV path planning based on sonar model,' in Journal of Intelligent & Fuzzy Systems 26, pp. 2913–2926, 2014.

[49] D.H. Kim, K.B. Kim, E.Y. Cha, 'Fuzzy truck control scheme for obstacle avoidance,' in Neural Computing & Applications, 18, pp. 801–811, 2009.

[50] N.E. Hodge, L.Z. Shi, M.B. Trabia, 'A distributed fuzzy logic controller for an autonomous vehicle,' in Journal of Robotic Systems 21(10), pp. 499–516, 2004.

[51] A.S. Matveev, M.C. Hoy, A.V. Savkin, 'A globally converging algorithm for reactive robot navigation among moving and deforming obstacles,' in Automatica, Vol. 54, pp. 292–304, April 2015.

[52] Y.P. Kondratenko, N.Y. Kondratenko, 'Soft Computing Analytic Models for Increasing Efficiency of Fuzzy Information Processing in Decision Support Systems,' in Decision Making: Processes, Behavioral Influences, and Role in Business Management, R. Hudson (Ed.), Nova Science Publishers, New York, pp. 41–78, 2015.

[53] D. Simon, A. Shah, C. Scheidegger, 'Distributed learning with biogeography-based optimization: Markov modeling and robot control," Swarm and Evolutionary Computation, vol. 10, pp. 12–24, June 2013.

[54] M. Hamandi, M. D'Arcy, P. Fazli. 'Learning to Navigate Like Humans,' in Proceedings of the Workshop on Learning and Inference in Robotics, Robotics: Science and Systems Conference (RSS), 2018.

[55] M. Hamandi, P. Fazli, 'Learning Human Navigational Intentions,' in Proceedings of the Workshop on Towards a Framework for Joint Action: What about Theory of Mind?, Robotics: Science and Systems Conference (RSS), 2018.

[56] M. Hamandi, M. D'Arcy, P. Fazli, 'DeepMoTIon: Learning to Navigate Like Humans,' in Proceedings of the IEEE International Conference on Robot and Human Interactive Communication (RO-MAN), 2019.

[57] M. Hamandi, P. Fazli, 'Online Learning of Human Navigational Intentions,' in Proceedings of the International Conference on Social Robotics (ICSR), pp. 1–10, 2018.

[58] S. Karapinar, S. Sariel, 'Cognitive robots learning failure contexts through real-world experimentation,' in Autonomous Robots, 39, pp. 469–485, 2015.

[59] T. Hoshino, D. Mitsumoto, T. Nagano, 'Fractal fitness landscape and loss of robustness in evolutionary robot navigation,' in Autonomous Robots 5, pp. 199–213, 1998.

[60] P. Benavidez, M. Jamshidi, 'Mobile robot navigation and target tracking system,' in 6th International Conference on Systems of Systems Engineering, Albuquerque, New Mexico, USA, pp. 299–304, 2011.

[61] N.B. Hui, D.K. Pratihar, 'A comparative study on some navigation schemes of a real robot tackling moving obstacles,' in Robotics and Computer-Integrated Manufacturing, Vol. 25, Issues 4–5, pp. 810–828, 2009.

[62] M.R. Akbarzadeh, K. Kumbla, E. Tunstel, M. Jamshidi, 'Soft computing for autonomous robotic systems,' in Int. J. Computers and Electrical Engineering, vol. 26, 1, pp. 5–32, 2000.

[63] Y. Kondratenko, E. Gordienko, 'Implementation of the neural networks for adaptive control system on FPGA,' in Annals of DAAAM for 2012 & Proceeding of the 23rd Int. DAAAM Symp. "Intelligent Manufacturing and Automation," Volume 23, No.1, B. Katalinic (Ed.), Published by DAAAM International, Vienna, Austria, EU, pp. 389–392, 2012.

[64] Y.P. Kondratenko, O.V. Kozlov, O.S. Gerasin, Y.M. Zaporozhets, 'Synthesis and research of neuro-fuzzy observer of clamping force for mobile robot automatic control system,' in IEEE First International Conference on Data Stream Mining& Processing (DSMP), pp. 90–95, 2016. DOI: 10.1109/DSMP.2016.7583514

[65] O. Gerasin et al., 'Neural controller for mobile multipurpose caterpillar robot,' in Proceedings of the 2019 10th IEEE International Conference on Intelligent Data Acquisition and Advanced Computing Systems: Technology and Applications, IDAACS, pp. 222–227, 2019. DOI: 10.1109/IDAACS.2019.8924321

[66] D. Simon, 'The Application of Neural Networks to Optimal Robot Trajectory Planning,' Robotics and Autonomous Systems, vol. 11, no. 1, pp. 23–34, May 1993.

[67] C. Churavy, et al., 'Effective Implementation of a Mapping Swarm of Robots,' IEEE Potentials, Vol. 27, no. 4, pp. 28–33, July 2008.

[68] W. Li, C. Ma, F.M. Wahl, 'A neuro-fuzzy system architecture for behavior-based control of a mobile robot in unknown environments,' in Fuzzy Sets and Systems, Vol. 87, Issue 2, pp. 133–140, 1997.

[69] A.V. Savkin, C. Wang, 'Seeking a path through the crowd: Robot navigation in unknown dynamic environments with moving obstacles based on an integrated environment representation,' in Robotics and Autonomous Systems, Vol. 62, Issue 10, pp. 1568–1580, October 2014.

[70] T.K. Kang, et al., 'Ego-motion-compensated object recognition using type-2 fuzzy set for a moving robot', in Neurocomputing, Vol. 120, pp. 130–140, 2013.

[71] M. Wang, J.N.K. Liu, 'Fuzzy logic-based real-time robot navigation in an unknown environment with dead ends,' in Robotics and Autonomous Systems, Vol. 56, Issue 7, pp. 625–643, 2008.

[72] A. Pandey, D.R. Parhi, 'MATLAB Simulation for mobile robot navigation with hurdles in cluttered environment using minimum rule based fuzzy logic controller,' in Procedia Technology, Vol. 14, pp. 28–34, 2014.

[73] P. Bigai, J. Bartoszek, 'Low time complexity collision avoidance method for autonomous mobile robot,' in Intelligent Systems 2014, Proceedings of the 7th Intern. Conference IS'2014, September 24–26, Warsaw, Poland, Vol. 2, Springer, Cham – London, pp.141–152, 2014.

[74] P.G. Zavlangas, S.G. Tzafestas, 'Industrial robot navigation and obstacle avoidance employing fuzzy logic,' in Journal of Intelligent and Robotic Systems, 27(1–2), pp. 85–97, 2000.

[75] D. Simon, H. El-Sherief, 'Real-Time Navigation Using the Global Positioning System,' in IEEE Aerospace and Electronic Systems Magazine, vol. 10, no. 1, pp. 31–37, 1995.

[76] A.V. Palagin, V. N. Opanasenko, 'Reconfigurable-computing technology,' in Cybernetics and Systems Analysis 43, no. 5, pp. 675–686, 2007.

[77] Lin, Mingjie, et al., 'Performance benefits of monolithically stacked 3-D FPGA', in Computer-Aided Design of Integrated Circuits and Systems, IEEE Transactions on 26, no. 2, pp. 216–229, 2007.

[78] A. Drozd, et al. 'Green Experiments with FPGA,' in Green IT Engineering: Components, Networks, and Systems Implementation, V. Kharchenko, et al. (Eds), Vol. 105. Berlin, Heidelberg: Springer International Publishing, pp. 219–239, 2017. DOI: 10.1007/978-3-319-55595-9_11

[79] A. Drozd, et al. 'Checkable FPGA Design: Energy Consumption, Throughput, and Trustworthiness,' in Green IT Engineering: Social, Business and Industrial Applications, Studies in Systems, Decision, and Control, V. Kharchenko, et al. (Eds), Vol. 171. Berlin, Heidelberg:

Springer International Publishing, pp. 73–94, 2018, DOI: 10.1007/978-3-030-00253-4_4.

[80] Y.P. Kondratenko, V.M. Kuntsevich, A.A. Chikrii, V.F. Gubarev (Eds.), 'Advanced Control Systems: Theory and Applications,' Series in Automation, Control and Robotics, River Publishers, Gistrup, 2021.

[81] I. Atamanyuk et al., 'Generalized method for prediction of the electronic devices and information systems' state,' in 14th International Conference on Perspective Technologies and Methods in MEMS Design, MEMSTECH 2018 – Proceedings, Lviv, Ukraine, pp. 18–22, April 2018. DOI: 10.1109/MEMSTECH.2018.8365709

[82] I. Atamanyuk et al., 'Method of Polynomial Predictive Control of Fail-Safe Operation of Technical Systems', in Proc. XIIIth Int. Conf. The Experience of Designing and Application of CAD Systems in Microelectronics, CADSM 2015, 19–23 Febr., 2015, Polyana-Svalyava, Ukraine, pp. 248–251, 2015. DOI: 10.1109/CADSM.2015.7230848

[83] O. Drozd, M. Kuznietsov, O. Martynyuk, et al., 'A method of the hidden faults elimination in FPGA projects for the critical applications,' in Proceedings of 2018 IEEE 9th International Conference on Dependable Systems, Services, and Technologies (DESSERT'2018), 24–27 May, Kyiv, Ukraine, pp. 231–234, 2018. DOI: 10.1109/DESSERT.2018.8409131J.

[84] I. Atamanyuk et al., 'Calculation Methods of the Prognostication of the Computer Systems State under Different Level of Information Uncertainty,' in Proceedings of the 12th Int. Conf. on Information and Communication Technologies in Education, Research, and Industrial Application. Integration, Harmonization, and Knowledge Transfer, V. Ermolayev, et al. (Eds), ICTERI'2016, Kyiv, Ukraine, CEUR-WS, Vol-1614, pp. 292–307, 2016. http://ceur-ws.org/Vol-1614/paper_79.pdf

[85] Y. Kondratenko, P. Khalaf, H. Richter, D. Simon, 'Fuzzy Real-Time Multi-Objective Optimization of a Prosthesis Test Robot Control System,' in Y. Kondratenko, A. Chikrii, V. Gubarev, J. Kacprzyk (Eds) Advanced Control Techniques in Complex Engineering Systems: Theory and Applications, Springer, pp. 165–185, 2019.

[86] Y. Kondratenko, G. Khademi, V. Azimi, D. Ebeigbe, M. Abdelhady, S.A. Fakoorian, T. Barto, A.Y.Roshanineshat, I. Atamanyuk, D. Simon, 'Information, Communication, and Modeling Technologies in Prosthetic Leg and Robotics Research at Cleveland State University,' in Proceedings of the 12th International Conference on Information and Communication Technologies in Education, Research,

and Industrial Application. Integration, Harmonization and Knowledge Transfer, ICTERI'2016, Kyiv, Ukraine, V. Ermolayev, et al. (Eds), CEUR-WS, Vol-1614, pp. 168–183, 2016. http://ceur-ws.org/Vol-1614/paper_34.pdf

[87] A. Piegat, 'Fuzzy Modeling and Control,' Springer, Heidelberg, 2001.

[88] T. Runkler, 'Extended defuzzification methods and their properties', in Fuzzy Systems, 1996., Proceedings of the Fifth IEEE International Conference on, Vol. 1, pp. 694–700, 1996.

[89] T. Runkler, 'Selection of appropriate defuzzification methods using application specific properties,' in IEEE Transactions on Fuzzy Systems, 5.1, pp. 72–79, 1997.

[90] J.M. Merigo, A.M. Gil-Lafuente, R.R. Yager, 'An overview of fuzzy research with bibliometric indicators,' in Applied Soft Computing, No. 27, pp. 420–433, 2015.

[91] A. Kaufmann, M. Gupta, 'Introduction to Fuzzy Arithmetic: Theory and Applications,' in Van Nostrand Reinhold Company, New York, 1985.

[92] Y.P. Kondratenko, N.Y. Kondratenko, 'Soft Computing Analytic Models for Multiplication of Asymmetrical Fuzzy Numbers,' in Studies in Fuzziness and Soft Computing, Volume 393, pp. 201–214, 2021. DOI: 10.1007/978-3-030-47124-8_17

[93] I.P. Atamanyuk, et al. 'The Algorithm of Optimal Polynomial Extrapolation of Random Processes.' In: K.J. Engemann et al. (Eds) Modeling and Simulation in Engineering, Economics, and Management. MS 2012. Lecture Notes in Business Information Processing, Vol. 115. Springer, Berlin, Heidelberg, pp. 78–87, 2012. https://doi.org/10.1007/978-3-642-30433-0_9

[94] H. Richter et al., 'Dynamic Modeling, Parameter Estimation and Control of a Leg Prosthesis Test Robot,' Applied Mathematical Modelling, Vol. 39, no. 2, pp. 559–573, January 2015.

[95] D. Davis, P. Supriya, 'Implementation of Fuzzy-Based Robotic Path Planning.' In: S. Satapathy et al. (Eds) Proceedings of the Second International Conference on Computer and Communication Technologies. Advances in Intelligent Systems and Computing, vol 380. Springer, New Delhi, 2016. https://doi.org/10.1007/978-81-322-2523-2_36

[96] T. Jin, B.-J Choi, 'Obstacle avoidance of mobile robot based on behavior hierarchy by fuzzy logic.' Int. J. Fuzzy Logic Intell. Syst. Vol. 12, No. 3, pp. 245–249, 2012.

[97] R. Zhao, H.-K. Lee, 'Fuzzy-based Path Planning for Multiple Mobile Robots in Unknown Dynamic Environment,' Journal of Electrical Engineering and Technology, vol. 12, no. 2, pp. 918–925, Mar. 2017.

[98] Z. Gomolka, et al., 'From homogeneous network to neural nets with fractional derivative mechanism.' In: Rutkowski, L. et al. (Eds), Int. Conf. on Artificial Intelligence and Soft Computing, ICAISC-2017, Part I, Zakopane, Poland, 11–15 June 2017, LNAI 10245, Springer, Cham, 2017, pp. 52–63. DOI: https://doi.org/10.1007/978-3-319-59063-9_5

[99] G. Khademi, et al. 'Hybrid invasive weed / biogeography-based optimization,' Engineering Applications of Artificial Intelligence, Vol. 64, pp. 213–231, July 2017.

12

Algorithmic Procedures Synthesis of Robust-Optimal Control for Moving Objects

V. L. Timchenko, and D.O. Lebedev

Department of Computerized Control Systems, National Shipbuilding University, Geroev Ukraine ave. 9, 54025, Mykolayiv, Ukraine
E-mail: vl.timchenko58@gmail.com; dns19944@gmail.com

Abstract

The transient process control of a variety of moving objects described by ordinary nonlinear differential equations requires development of robust-optimal systems with variable structure. Using general algorithmic procedure for constructing optimal trajectories, determining switching moments and synthesizing control functions for multidimensional systems are suggested. The control of the mismatch between the physical object trajectory and the optimal model calculated trajectory enables to take the values of optimal control into account and to form a robust subsystem, which provides invariance to incomplete information about the moving object. Simulating quadrotor UAV stabilization and sea vessel maneuvering under conditions of external disturbances and parametric noise demonstrates the required control accuracy in the vicinity of the formed optimal trajectory of the controlled coordinates. The algorithms and circuit solutions for the synthesis of robust-optimal variable-structure systems for multidimensional nonlinear moving objects have been developed as a practical basis of automated procedures of the synthesis of robust-optimal systems and development of software tools for various types of moving object control systems.

Keywords: Algorithmic procedure, robust-optimal control, variable-structure system, marine moving object, quadrotor UAV.

12.1 Introduction

The original classical works of R. Bellman, R. Kalman, N. Krasovsky, A. Feldbaum, L. Pontryagin and other well-known scientists demonstrate the rapid development of the foundations of modern control theory related to the problems of analyzing desired (optimal) trajectories of object movement.

The basic requirements for the tasks of the optimal control synthesis for quadratic criteria of optimality and maximum speed were mathematically formalized, supported by algorithms for the determination of switching moments. It was a breakthrough in control systems, especially in the space sector, where the highest accuracy and efficiency were required for well-defined objects and given properties of the environment [1, 2]. It is well known there are many control problems, e.g. motion under conditions of aerodynamic and/or hydrodynamic resistance of a stochastic medium with incompletely known parameters. However, the deviation of a mathematical model from a moving physical object had so significant impact on control errors that it stimulated the rapid development of methods of adaptive, and later, robust control. Later, summing up the development of these methods, V. Kuntsevich [3] classified them as adaptive-optimal methods requiring identification of parameters and adjustment of control (which is rather difficult to provide in real-time), and robust-optimal methods that create control algorithms with known stable solutions concerning bounded uncertainties. These methods were further developed in subsequent works [4–8].

Other widely used methods based on robust properties of sliding modes (considering the additional power of control for their implementation), but frequent switching of control devices negatively effects their efficiency [8–11]. Robust properties are also inherent in PID controllers, since they use approximate tuning for the parameters of the plant model, respond to an error, and compensate uncertain disturbances [12].

The complexity of the synthesis of multidimensional objects control was revealed and caused the development of a new approach for forming the desired trajectories in solving inverse problems of dynamics. The symmetry of solutions of the motion modeling and control synthesis made it possible to focus on developing the programmed trajectories of moving objects at the initial stage of control synthesis. These methods simplify the synthesis of optimal control since they do not require solving Riccati equations or variational problems. When forming the corresponding requirements for the controlled process and the desired (optimal) trajectories of the moving object, the control is based on solving the inverse problems of dynamics with high dynamic accuracy ensuring the properties of robustness [13–18]. The problems of synthesis of optimal stabilization systems for the output coordinates of various moving objects, including additional qualities of the transient process, are solved on the basis of Lyapunov matrix algebraic equations for the quadratic criterion by the synthesis of optimal controllers in static and dynamic feedbacks even during stabilization of various objects [19–21].

In V. Larin's [22], the important problem of determining the coefficients in the weight matrices of a quadratic performance criterion is solved as an inverse problem for the synthesis of an optimal controller of a linear stationary object, i.e. inversion of the problem of "analytical design of optimal controllers". It should be noted that the computational complexity of the analytical design of suboptimal controllers increases significantly with an increase in the order of the controlled object and does not allow their practical use in the synthesis of control systems [14]. The use of the principles of modal control and Ackerman formula presupposes on heuristic assumptions about the optimality of the transient process and allows the formation of the desired trajectories for various distributions of parameters to ensure the appropriate type of the transient process. During the synthesis of scalar control, the corresponding additional matrices of parameters are rather conditionally assigned. The methods of high computational efficiency based on H optimization have also been developed, primarily for high order moving objects limited mainly by linear systems [23–24].

Some particular characteristics of modern control methods for several classes of moving objects such as marine moving objects and unmanned aerial vehicles, which have been widely applied in modern transport, military and other similar infrastructural systems, are considered in detail. The models of the dynamics of a marine moving object as a rigid body functioning at the interface between air and liquid media have been sufficiently studied [25–28] and are described by multidimensional nonlinear stochastic differential systems of equations. A comprehensive and accurate description of the dynamics of a marine moving object under the influence of random disturbances (currents, wind, and sea waves) seems extremely difficult due to the incomplete definiteness of the dynamic model parameters, as it is impossible to physically form and mathematically describe all the factors that effect the value of parameters or their function and also to measure some components of external disturbances (wind pulsations, irregular waves). The main applicable simplifications for the considered problems of marine moving objects control include of external random stationary disturbances, the horizontal plane of dynamic vibrations of the object, and several other assumptions. A high level of computational complexity limits the use of classical methods of optimal control, considering the multidimensional nonlinear model of the controlled object, the solution of boundary value problems for the synthesis of program control, and the solution of Riccati matrix equations when designing an optimal controller. The criteria of quadratic optimality require the formalization of weighting coefficient matrices based on additional calculation procedures [22] or empirical calculations.

The development of the fundamental feedback principle of control [8] is based on the wide application of static feedbacks for the synthesis of optimal control [21], linear matrix inequalities taking into account the action of arbitrary, bounded perturbations [19], however, they are mainly used for linear or equivalent linear models of controlled systems, incl. marine moving objects [29, 30]. Some alternative approaches, e.g. based on the principles of solving inverse problems of dynamics and structural synthesis [13, 15], do not always enable to optimize the parameters of programmed (assigned) trajectories. The incomplete definiteness of the parameters of both the dynamic model of marine moving object and the environment, as well as the problem of real-time control, leads to the necessity to apply the principles of robust control. For example, the use of variable structure systems [10, 11], that is effective for a certain (reactive) type of propellers, requires the increased control energy consumption and frequent switching, which can lead to a decrease in operability for the controls of a marine moving object.

The implementation of high-tech functional requirements for maneuvering and positioning of marine moving objects can be carried out using applied control systems with a variable structure based on special switching feedbacks. These systems ensure the minimization of energy consumption or maximization of performance for a specific functional task of marine mobile object control (respectively, working technological or transient critical modes) with the required control accuracy and sufficient invariance to the uncertainty of object parameters and the environment [31–36].

The deployment of unmanned aerial vehicles (UAV) enables to significantly expand the use of visual operational monitoring tools of technical objects for various purposes. For example, safe navigation is ensured by the increasing safety requirements for restricted water areas, especially during various technological operations, such as maneuvering, towing, bunkering, loading etc. The analysis of the marine accidents which happened in restricted water areas proves that about 70% of accidents occur due to human factor. It is partly due to the lack of sufficient traffic safety data for operators to make decisions in extreme operational conditions.

The use of UAV (quadrotor UAV) for marine environment monitoring makes it possible to expand the navigation safety information assessment and to increase the control of the marine moving objects improving in this way the marine emergency prevention. It is necessary to improve the quality control systems to ensure the optimal trajectory of stabilization of the quadrotor UAV along all given controlled coordinates (the dynamics are described by a system of six differential equations of the second-order [37, 38]). Quadrotor UAV functions under conditions of limited rotors power and incomplete information of uncontrolled external disturbances (the complexity of a complete description of the physical impact of wind perturbation on a UAV involves the selection of the following components of airflow: high-frequency random pulsations; low-frequency piecewise constant components in horizontal (constant wind) and vertical directions [39]).

The LQR method is widely used for dynamic system stabilization as a modern and effective method to synthesize optimal controllers for UAVs at a minimal energy cost [40–45]. The comparison between LQR and the PID algorithm shows that LQR has better energy efficiency than PID. At the same time, LQR algorithms working well in conditions of noisy control processes do not ensure accurate control in conditions of uncontrolled disturbance compared to the control systems with the sliding modes [46–48]. The use of control synthesis based on quadratic functionals and Lyapunov functions [49–51] is limited to linear systems. Accordingly, the main problem of the QFT method consist in the formation of feedback under the conditions of uncertainty of disturbance parameters [52–54]. H_2/H control synthesis is increasingly used for the synthesis of robust controllers. This optimization-based method is associated with the mathematical description of the expected behavior of a closed system and its stability [55–60]. It has a strictly mathematical basis for the optimization process thus having been applied to both classical optimal and robust control. There are some advantages of the linear approach due to its powerful tools for the design and analysis of controllers. This method considers the resistance to incoming perturbations (aerodynamic moments and forces of wind) as the main source of errors in orientation [61–64]. Neural networks are used to synthesize both robust and optimal controllers [65, 66]. Wind disturbances are compensated by an artificial neural network, the parameters of which are configured in the current mode of operation. The controller is complemented by multilayer neural networks and adaptive control to compensate uncontrolled disturbances. A large volume of online computing required reduces the effectiveness of this approach. The above problems are related to the multidimensionality and nonlinearity of the control object, as well as the uncertainty of wind disturbances significantly complicates the process of synthesis and real-time control.

The proposed approach for the synthesis of robust-optimal systems is based on the use of the system with variable feedbacks and is described for models of marine moving objects [31–36] and quadrotor UAV [67–70]. It is possible to use this system for multidimensional nonlinear non-stationary models of moving objects. The other advantage of this method includes the control synthesis using direct optimality conditions based on energy analysis and robust correction of control processes based on minimizing control errors.

12.2 Problem Review

It is possible to determine how to optimally control dynamic systems under uncertainty conditions for mathematical models of objects and the environment by designing efficient automatic control systems in terms of quality indicators. The criteria of optimality for control processes are primarily determined by the necessity to reduce energy costs under the conditions of limited control resources or maximum speed of transient processes at critical modes of the control object operation. The criteria of optimality that mathematically determine these requirements are opposite in terms of minimax estimates. However, if the quality indicators of specific stages of technological processes are considered as the basis for optimization, then it would be possible to distinguish the stage of the functioning of a dynamic system at the energy consumption minimum and the stage of critical control with the maximum speed of transient processes. Thus, two separate stages of optimization are determined making it possible to apply the corresponding optimality criteria and the procedure for the optimal control synthesis.

The aim of the research is designing a generalized approach to the synthesis of high-precision and robust-optimal control systems that are uncertainties stable and applicable for multidimensional nonlinear moving object. The control of dynamic systems, which include moving objects under the influence of external and parametric disturbances, are considered at transient modes, i.e. generally at stabilization modes of controlled coordinates. The mathematical description (simulation) of dynamic systems in the form of ordinary differential equations represents the structural-parametric model of the control object. The structure of the interaction between the object and the environment is determined by the type of differential equations and is sufficiently definite. The structure can be changed due to the correct simplification of the system of equations, e.g. by replacing the nonlinear model with a linear model. However, the values of the parameters (coefficients) of the model of the controlled object are rather undefined values, and they describe an incompletely informative model. While designing automatic control systems for dynamic objects, the parameters of the mathematical model of the object, which are the basis of synthesis, differ from the calculated ones in real systems due to the impossibility to physically determine and mathematically describe all factors that effect the value or function of the parameter. For example, hydrodynamic drag coefficient and the attached mass of the vessel depend, among other things, on the shape of the vessel hull, which can be changed during operation (including overgrowing with marine organisms), and the density of the marine environment, the value of which is also not constant for different areas of operation of the vessel.

The disturbances under which the controlled object functions can be divided into controlled (measured, physically measurable) and uncontrolled (unmeasured), e.g. wind speed or wind direction that can be accurately measured using an anemometric system. It should be noted that if the control object does not allow the placement of the anemometric system on the hull, e.g. a quadrotor UAV, then, in this case, the wind effect is uncontrollable. Irregular sea waves that determine vessel pitching are uncontrollable disturbances.

The system of equations for the generalized model of the nonlinear nonstationary dynamic system of a moving object is described by Newtonian mechanics in vector-matrix form

$$\ddot{\mathbf{X}}(t) = \mathbf{A_X}\dot{\mathbf{X}}(t) + \mathbf{B_X}\mathbf{U}(t) - \mathbf{g} + \mathbf{C_X}\mathbf{F}(t). \tag{12.1}$$

For optimal control problems, the requirements are formed to ensure the optimal trajectory $\mathbf{X}_{opt}(t)$ of stabilization of the dynamic system along all given controlled coordinates considering the fulfillment of the optimal criterion for the minimum control energy consumption $\mathbf{U}(t)$

$$J = \int_{t_1}^{t_2} Q\left(\mathbf{X}, \mathbf{U}\right) dt = min. \tag{12.2}$$

The maximum speed of transient processes in dynamic systems is described by ensuring the minimum time interval $t_1 \div t_2$

$$J = \int_{t_1}^{t_2} dt = min. \tag{12.3}$$

Thus, the task is to develop practical algorithmic procedures for constructing multidimensional object optimal phase trajectories based on the existing variety of trajectories connecting the initial and final points of the space **R**, satisfying the control constraints and additional requirements–the minimum energy consumption or the minimum time of the transient process (maximum speed).

Suppose that the planning of the trajectory of a nonlinear non-stationary moving object can be described in the form of certain segments with constant values of the corresponding ith (or, respectively, zero $(i + 1)$ - th) derivatives of the state coordinates $\mathbf{X}(t)$ and, thus, described by polynomial dependencies. The type of polynomial dependencies and the required number of trajectory segments are determined by the type of specified optimality criteria and the values of the boundary conditions.

The transition of a moving object from the initial segment of the trajectory to a given segment, considering the control constraints, can be described by the following equations for the state coordinate vector

$$\mathbf{X}_{opt}(t_i^s) = \mathbf{X}(t_{i-1}^s) + ... \pm \overset{(m)}{\mathbf{X}}(t_{i-1}^s)\frac{(t_i^s - t_{i-1}^s)^m}{m!}. \tag{12.4}$$

The solution to the general problem of the dynamic object control must consider the uncertainty of the mathematical model of the object and external perturbations using robust control methods. The robust-optimal approach considered involves the synthesis of a robust correction loop that compensates deviations from the optimal trajectory of the moving object with minimal vector errors $\mathbf{E}(t)$ and its derivatives

$$\ddot{\mathbf{E}}(t) + \mathbf{G}_1\dot{\mathbf{E}}(t) + \mathbf{G}_2\mathbf{E}(t) \rightarrow min \approx 0. \tag{12.5}$$

12.3 Optimal Synthesis

The approach to the synthesis of the optimal system is based on the system with a variable feedback structure. This approach includes the following main steps:

1. Optimal trajectory planning.
2. Determining switching moments of control functions in the feedback loops.

3. Synthesizing optimal control functions in the corresponding feedback loops.

Considering a priori known parameters of the model of a dynamic object, the problem is mathematically exactly solved as an optimal control problem.

12.3.1 Optimal Trajectory Planning

Optimal phase trajectories of the moving object stabilization can be formed using polynomial or exponential representations. Formalizing the trajectory type is determined by the requirements for the optimality of the control process, its type, and values of the boundary conditions, as well as by the constraints on the control actions.

Considering the optimality criterion, one can be guided by the following functional requirements: minimum energy consumption in operating and technological modes of moving object control (taking into account energy consumption for stabilizing the moving object at a given altitude) and maximum speed at critical, extreme modes of operations.

For a multidimensional object of dimension $j = 1,\ldots,n$ with the output vector of phase coordinates $\mathbf{X}(t)$ boundary conditions are defined as some points in the space of variables $\mathbf{R}$ for initial $\overset{(i)}{\mathbf{X}}(0)$ and final $\overset{(i)}{\mathbf{X}}(T)$ boundary conditions. The requirement for physical feasibility of control actions imposes restrictions on their maximum values, and for the polynomial form of trajectories defined by certain dependencies between the values of controls and derivatives of the output coordinates.

When assessing the control energy costs, the optimality criterion is applied, which physically expresses the average control power in the following functional (12.2) in the expanded quadratic form for control power

$$J = \int_0^T \mathbf{U}^T \mathbf{Q} \mathbf{U} dt = \min. \tag{12.6}$$

Thus, the task is to form the algorithmic procedures for constructing optimal phase trajectories that are practically realizable for multidimensional objects based on the existing variety of trajectories connecting the initial and final points of the space $\mathbf{R}$, to satisfy the control constraints and additional requirements–the minimum energy consumption or the minimum time of the transient process (maximum speed).

To simplify the presentation (without loss of generality), consider a linear moving object's model in the horizontal plane without considering perturbations

$$\ddot{\mathbf{X}}(t) = \mathbf{A}\dot{\mathbf{X}}(t) + \mathbf{B}\mathbf{U}(t). \tag{12.7}$$

Work during the movement of the object (in the absence of external perturbations) along the trajectories of stabilization is performed by control and damping (aerohydrodynamic resistance) forces. To estimate the energy losses due to damping forces, the Rayleigh dissipative function must be introduced

$$D(\dot{\mathbf{X}}) = -0.5 \int_0^T \dot{\mathbf{X}}^T \mathbf{A}\dot{\mathbf{X}} dt. \tag{12.8}$$

To optimize the problem of minimum energy consumption, lets consider the Equations (12.7, 12.8), the minimized functional in the form of Lagrange function can be written as

$$L = -0.5\dot{\mathbf{X}}^T(t)\mathbf{A}\dot{\mathbf{X}}(t) + 0.5\mathbf{U}^T(t)\mathbf{Q}\mathbf{U}(t) + \boldsymbol{\lambda}^T[\ddot{\mathbf{X}}(t) - \mathbf{A}\dot{\mathbf{X}}(t) - \mathbf{B}\mathbf{U}(t)]. \tag{12.9}$$

The extremum of functional (12.9) is determined by the following Euler–Lagrange equation considering $\dot{\mathbf{X}}(t) = \mathbf{V}(t)$

$$\frac{\partial L}{\partial \mathbf{V}(t)} - \frac{d}{dt}\frac{\partial L}{\partial \dot{\mathbf{V}}(t)} = \frac{\partial L}{\partial \dot{\mathbf{X}}(t)} - \frac{d}{dt}\frac{\partial L}{\partial \ddot{\mathbf{X}}(t)} = -\mathbf{A}^T\dot{\mathbf{X}}(t) - \mathbf{A}^T\boldsymbol{\lambda}(t) - \dot{\boldsymbol{\lambda}}(t) = 0, \tag{12.10}$$

and for control

$$\frac{\partial L}{\partial \mathbf{U}(t)} - \mathbf{Q}\mathbf{U}(t) - \mathbf{B}^T\boldsymbol{\lambda}(t) - 0. \tag{12.11}$$

From the Equation (12.11), optimal control can be expressed as

$$\mathbf{U}_{opt}(t) = \mathbf{Q}^{-1}\mathbf{B}^T\boldsymbol{\lambda}(t). \tag{12.12}$$

When solving the Equation (12.10), it is assumed that polynomial representation of the moving object trajectories is used. In this case, the object can be at rest or perform a uniform (steady-state relative to the speed $\dot{\mathbf{X}}(t)$ with zero acceleration), or uniformly accelerated (steady-state relative to acceleration $\ddot{\mathbf{X}}(t)$ with zero third derivative), or established for the third derivative $\dddot{\mathbf{X}}(t)$ (with zero fourth derivative) and further, established for mth derivative $\overset{(m)}{\mathbf{X}}(t)$ (with zero $(m+1)$-th derivative), motion.

In general, for $\overset{(m)}{\mathbf{X}}(t) = \overset{(m)}{\mathbf{X}}(0) = \text{const}$ the equation of one segment of the trajectory for phase coordinate vector will take the form

$$\mathbf{X}(t) = \sum_{i=0}^{m-1} \overset{(i)}{\mathbf{X}}(0)\frac{t^i}{i!} \pm \overset{(m)}{\mathbf{X}}(0)\frac{t^m}{m!}. \tag{12.13}$$

Assuming that to achieve a given value we need two segments (acceleration and deceleration), the equations of optimal trajectories will take the following form for joint switching moments t_1^s and T for all state coordinates of a moving object

$$
\begin{array}{ll}
1^{st}\,segment & 2^{nd}\,segment \\
\end{array}
$$

$$
\begin{cases}
\mathbf{X}(t_1^s) = \sum\limits_{i=0}^{m-1} \overset{(i)}{\mathbf{X}}(0)\frac{(t_1^s)^i}{i!} + \overset{(m)}{\mathbf{X}}(0)\frac{(t_1^s)^m}{m!}; \\[2mm]
\dot{\mathbf{X}}(t_1^s) = \sum\limits_{i=0}^{m-1} \overset{(i+1)}{\mathbf{X}}(0)\frac{(t_1^s)^i}{i!}; \\[2mm]
\cdots\cdots \\[1mm]
\overset{(m-1)}{\mathbf{X}}(t_1^s) = \overset{(m-1)}{\mathbf{X}}(0) + \overset{(m)}{\mathbf{X}}(0)t_1^s.
\end{cases}
\qquad
\begin{cases}
\mathbf{X}(T) = \sum\limits_{i=0}^{m-1} \overset{(i)}{\mathbf{X}}(t_1^s)\frac{(T-t_1^s)^i}{i!} - \overset{(m)}{\mathbf{X}}(0)\frac{(T-t_1^s)^m}{m!}; \\[2mm]
\dot{\mathbf{X}}(T) = \sum\limits_{i=0}^{m-1} \overset{(i+1)}{\mathbf{X}}(t_1^s)\frac{(T-t_1^s)^i}{i!} - \overset{(m)}{\mathbf{X}}(0)\frac{(T-t_1^s)^{m-1}}{(m-1)!}; \\[2mm]
\cdots\cdots \\[1mm]
\overset{(m-1)}{\mathbf{X}}(T) = \overset{(m-1)}{\mathbf{X}}(t_1^s) - \overset{(m)}{\mathbf{X}}(0)(T-t_1^s).
\end{cases}
$$

$$(12.14)$$

The Equation (12.10) can be written for the trajectory with a transient time T with a zero third derivative ($m = 2$) in the form

$$\dot{\boldsymbol{\lambda}}(t) + \mathbf{A}^T\boldsymbol{\lambda}t - \mathbf{A}^T(\dot{\mathbf{X}}_0 + \ddot{\mathbf{X}}_0 T) = 0 \qquad (12.15)$$

Together with the equation for the controlled system and control (12.12)

$$\ddot{\mathbf{X}}(t) = \mathbf{A}\dot{\mathbf{X}}(t) = \mathbf{B}\mathbf{Q}^{-1}\mathbf{B}^T\boldsymbol{\lambda}(t) \qquad (12.16)$$

Equation (12.15) forms a boundary value problem for optimal control, the solution of which for a trajectory with zero third and fourth derivative can also be described in the polynomial form

$$\mathbf{U}(t) = \mathbf{U}(t_i^s) + \dot{\mathbf{U}}(t_i^s)t_i^s,$$

$$\mathbf{U}(t) = \mathbf{U}(t_i^s) + \dot{\mathbf{U}}(t_i^s)t_i^s + \ddot{\mathbf{U}}(t_i^s)\frac{(t_i^s)^2}{2}.$$

In general form, for the mth derivative we obtain control in the form

$$\mathbf{U}(t) = \sum\limits_{i=0}^{m-2} \overset{(i)}{\mathbf{U}}(t_i^s)\frac{t_i^s}{i!}, \qquad (12.17)$$

where order m and values $\overset{(i)}{\mathbf{U}}(t_i^s)$ are limited by the physical constraints of the controls.

For the polynomial representation of trajectories, each coordinate is independent of the others (although in the general case considered below, they have different end times of the transient process). It should also be noted that for a zero matrix $\mathbf{A}$, the polynomial and exponential representations of the trajectories coincide. Thus, the polynomial representation of the trajectories, in contrast to the exponential one, makes it possible to construct the moving object stabilization trajectories in advance (before the synthesis of the control) rather simply. The polynomial representation of trajectories is applicable for a complete mathematical description of the motion of the moving object, i.e. considering nonlinearity and non-stationarity. With the exponential representation of the trajectories, it is necessary to numerically solve nonlinear non-stationary differential equations of a higher order which describe the dynamics of the object.

The use of the polynomial form of trajectories makes it possible to estimate the dependence of the optimality functional for the minimum energy consumption on the time of the transient process. Consider the motion of the moving object along the trajectory segments (12.4) for the 3rd and 4th zero derivatives for the coordinate vector $\mathbf{X}(t)$, as well as the final conditions (for simplicity, $\mathbf{X}_0 = \dot{\mathbf{X}}_0 = 0$)

$$\mathbf{X}(t) = 0.5\ddot{\mathbf{X}}_0 t^2; \quad \ddot{\mathbf{X}}(t) = \ddot{\mathbf{X}}_0 = const; \quad \dddot{\mathbf{X}}(t) = 0; \quad \mathbf{X}(T) = \mathbf{X}_k;$$

$$(12.18)$$

$$\mathbf{X}(t) = \frac{1}{2}\ddot{\mathbf{X}}_0 t^2 + \frac{1}{6}\dddot{\mathbf{X}}_0 t^3; \quad \dot{\mathbf{X}}(t) = \ddot{\mathbf{X}}_0 t + \frac{1}{2}\dddot{\mathbf{X}}_0 t^2; \quad \ddot{\mathbf{X}}(t) = \ddot{\mathbf{X}}_0 + \dddot{\mathbf{X}}_0 t;$$

$$\dddot{\mathbf{X}}(t) = \dddot{\mathbf{X}}_0 = const; \quad \overset{(IV)}{\mathbf{X}}(t) = 0; \quad \mathbf{X}(T) = \mathbf{X}_k; \quad \dot{\mathbf{X}}(T) = \dot{\mathbf{X}}_k.$$

$$(12.19)$$

Then it is possible to obtain the necessary initial conditions, respectively, for the 1st

$$\ddot{\mathbf{X}}_0 = 2T^{-2}\mathbf{X}_k,$$

and for the 2nd segment

$$\ddot{\mathbf{X}}_0 = T^{-2}(6\mathbf{X}_k - 2\dot{\mathbf{X}}_k T); \quad \dddot{\mathbf{X}}_0 = T^{-3}(6\dot{\mathbf{X}}_k T - 12\mathbf{X}_k).$$

Further, for the 1st segment, we obtain the equation for the initial conditions for the control

$$\ddot{\mathbf{X}}_0 = \mathbf{A}\dot{\mathbf{X}}_0 + \mathbf{B}\mathbf{U}_0 \Rightarrow \mathbf{U}_0 = \mathbf{B}^{-1}\ddot{\mathbf{X}}_0;$$

$$\dddot{\mathbf{X}}_0 = \mathbf{A}\ddot{\mathbf{X}}_0 + \mathbf{B}\dot{\mathbf{U}}_0 = 0 \Rightarrow \dot{\mathbf{U}}_0 = -\mathbf{B}^{-1}\mathbf{A}\ddot{\mathbf{X}}_0,$$

and the general equation for the control in the form of

$$\mathbf{U}(t) = \mathbf{U}_0 + \dot{\mathbf{U}}_0 t = 2\mathbf{B}^{-1}T^{-2}\mathbf{X}_k - 2\mathbf{B}^{-1}\mathbf{A}T^{-2}\mathbf{X}_k t =$$
$$= 2\mathbf{B}^{-1}T^{-2}\mathbf{X}_k(\mathbf{E} - \mathbf{A}t). \tag{12.20}$$

When carrying out similar transformations for the 2nd segment of the trajectory, it is possible to obtain (considering $\overset{(IV)}{\mathbf{X}_0} = \mathbf{A}\overset{\cdots}{\mathbf{X}}_0 + \mathbf{B}\ddot{\mathbf{U}}_0 = 0$)

$$\mathbf{U}(t) = \mathbf{U}_0 + \dot{\mathbf{U}}_0 t + 0.5\ddot{\mathbf{U}}_0 t^2 = \mathbf{B}^{-1}T^{-2}\{(6\mathbf{X}_k - 2\dot{\mathbf{X}}_k T)+$$
$$+ [T^{-1}(6\dot{\mathbf{X}}_k T - 12\mathbf{X}_k) - \mathbf{A}(6\mathbf{X}_k - 2\dot{\mathbf{X}}_k T)]t-$$
$$- \mathbf{A}T^{-1}(3\dot{\mathbf{X}}_k T - 6\mathbf{X}_k)\}t^2. \tag{12.21}$$

It should be noted that the equations for the boundary condition $\dot{\mathbf{X}}(T)T$ with an increase in the time of the transient process T, as follows from the analysis of the equations for the jth component of the vector $\mathbf{X}(t)$, changes within narrow boundaries, and for small $\ddot{x}_{j0}$ and sufficiently big T_j, almost equal to a constant

$$2x_{jk} = (\ddot{x}_{j0} + \frac{1}{3}\dddot{x}_{j0}\,T_j)T_j{}^2; \quad \dot{x}_{jk} = (\ddot{x}_{j0} + \frac{1}{2}\dddot{x}_{j0}\,T_j)T_j;$$

$$\frac{2x_{jk}}{\dot{x}_{jk}T_j} = \frac{\ddot{x}_{j0} + \frac{1}{3}\dddot{x}_{j0}\,T_j}{\ddot{x}_{j0} + \frac{1}{2}\dddot{x}_{j0}\,T_j} \approx \frac{2}{3} = C.$$

Then, for the elements of the functional (12.6) concerning the time of the transient process T after integration of the form (12.20, 12.21), we obtain the equations proportional to $\sim 1/T$, which signalizes the decrease in the required control power during movement of a moving object along the trajectories of the form (12.3.1), while the time T increases. Thus, we can draw a generalized conclusion that for a given boundary constraint for the coordinate vector, the trajectory with the longest possible transient time is optimal in terms of energy consumption, while the most energy-intensive control is the trajectory with the maximum possible response time considering the physical constraints on the values of the maximum control. Trajectories of the form (12.3.1) of the control object in polynomial form with positive values of the derivatives of the state coordinate vector is optimal relative to the criterion (12.6) when a moving object moves with the minimum possible number of minimum possible values of the derivatives of the coordinate vector

$$\underset{i}{min}\underset{\overset{(i)}{\mathbf{X}}(t_0)}{min}\{\dot{\mathbf{X}}\left[\overset{(i)}{\mathbf{X}}(t_0),t\right]\},$$

fulfilling all specified boundary conditions.

For the tasks of maximum speed, the trajectories are optimal relative to the criterion (12.3) when a moving object moves with the maximum possible number of maximum possible values of the derivatives of the coordinate vector, taking into account the constraints on the control action

$$
\max_i \max_{\overset{(i)}{\mathbf{X}}(t_0)} \{\dot{\mathbf{X}} \left[\overset{(i)}{\mathbf{X}}(t_0), t \right] \},
$$

and the fulfillment of all specified boundary conditions $\overset{(i)}{\mathbf{X}}(t_0)$ = $f[X(t_0), \overset{(i-1)}{\mathbf{U}}{}_{max}(t_0)]$.

12.3.2 Determining Switching Moments of Control Functions in the Feedback Loops

The introduction of constraints on the control action limits the number of possible controlled coordinate derivatives, which effects the form of the optimal trajectory and significantly complicates its calculation. The algorithms of multidimensional system have been developed for the given boundary conditions and the values of the derivatives of the object coordinates vector considering the control constraints. These algorithms are based on the solution of the systems of algebraic equations in the form (12.3.1) and include the formation of leading, sub-leading and driven variables, the formation of some sequence of switching moments of control functions in feedbacks of a controlled object.

To construct an optimal trajectory in terms of the maximum speed (12.3) and determine the switching moments considering the constraints on the control action, the algorithm is formed (Figure 12.1) for the 6*th* order system.

The switching moments for the problem of minimum energy consumption (12.6) are determined on the basis of the system of algebraic equations in the form (12.3.1) for two or more segments by calculating the roots that represent the switching moments t_{ij}^s and T_j that in general case will differ for each jth state coordinates. For this, the corresponding algorithm is applied (Figure 12.2).

12.3.3 Synthesizing Optimal Control Functions in the Corresponding Feedback Loops

When constructing the optimal model of the control system, we assume that the dynamics of a moving object is described by the vector-matrix equation

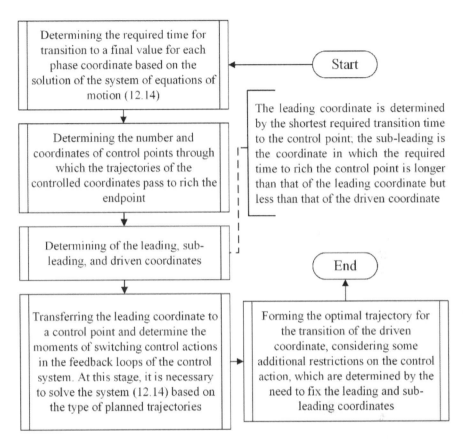

Figure 12.1 Algorithmic procedure *A* for constructing the optimal trajectory in terms of maximum speed

in general form (12.1), the parameters and external disturbances are precisely known and specified.

The equations (12.17) define the representation of control functions in the form of time polynomials, which are switched by certain algorithmic procedures *A* and *B*, and is described by the block diagram shown in Figure 12.3.

An alternative is to represent controls in the form of integrating functions in the feedback loops of a moving object (Figure 12.4). Therefore, we use the differential transformation of the vector-matrix Equation (12.1) considering the above assumptions regarding disturbances, for example, the third and fourth zero derivatives of the coordinate vector. This differential

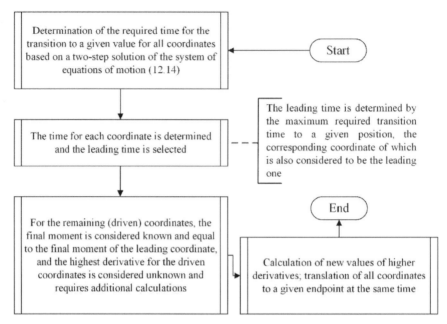

Figure 12.2 Algorithmic procedure *B* for the problem of minimum energy consumption

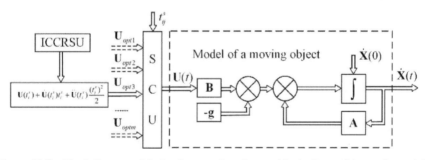

Figure 12.3 Block diagram of the implementation of control in the form of time polynomials ($F(t)$ is not considered; the index X for the matrices A, B is omitted)

transformation is correct due to a clear indication of the initial conditions of coordinates and control on each segment of the optimal trajectory. Thus, we form the corresponding equations for the balance of forces and moments, as well as their derivatives, for a segment of the trajectory with the fulfillment of the condition $\dddot{X}(t) = 0$

$$A_X\ddot{X}(t) + \dot{A}_X\dot{X}(t) + \dot{B}_XU(t) + B_X\dot{U}(t) + \dot{C}_XF(t) + C_X\dot{F}(t) = 0,$$
$$(12.22)$$

and for the condition $\overset{(IV)}{\mathbf{X}}(t) = 0$

$$\mathbf{A_X}\overset{...}{\mathbf{X}}(t) + 2\dot{\mathbf{A}}_{\mathbf{X}}\ddot{\mathbf{X}}(t) + \ddot{\mathbf{A}}_{\mathbf{X}}\dot{\mathbf{X}}(t) + \mathbf{B_X}\ddot{\mathbf{U}}(t) + 2\dot{\mathbf{B}}_{\mathbf{X}}\dot{\mathbf{U}}(t)+$$
$$+\ddot{\mathbf{B}}_{\mathbf{X}}\mathbf{U}(t) + \mathbf{C_X}\ddot{\mathbf{F}}(t) + \dot{\mathbf{C}}_{\mathbf{X}}\dot{\mathbf{F}}(t) + \ddot{\mathbf{C}}_{\mathbf{X}}\mathbf{F}(t) = 0. \quad (12.23)$$

Vector-matrix transformations of the Equations (12.22, 12.23) make it possible to write them in a form that determines the control vector and ensures the movement of a moving object along optimal trajectories, respectively, for the conditions of the third and fourth zero derivatives of the coordinate vector

$$\mathbf{B_X}\dot{\mathbf{U}}(t) + [\mathbf{A_X}\mathbf{B_X} + \dot{\mathbf{B}}_{\mathbf{X}}]\mathbf{U}(t) =$$
$$= -[(\mathbf{A}_{\mathbf{X}}^2 + \dot{\mathbf{A}}_{\mathbf{X}})\dot{\mathbf{X}}(t) + (\mathbf{A_X}\mathbf{C_X} + \dot{\mathbf{C}}_{\mathbf{X}})\mathbf{F}(t)+$$
$$+ \mathbf{C_X}\dot{\mathbf{F}}(t) - \mathbf{A_X}\mathbf{g}]; \quad (12.24)$$
$$\mathbf{B_X}\ddot{\mathbf{U}}(t) + (\mathbf{A_X}\mathbf{B_X} + 2\dot{\mathbf{B}}_{\mathbf{X}})\dot{\mathbf{U}}(t) + (\mathbf{A}_{\mathbf{X}}^2\mathbf{B_X} + 2\dot{\mathbf{A}}_{\mathbf{X}}\mathbf{B_X} + \mathbf{A_X}\dot{\mathbf{B}}_{\mathbf{X}}+$$
$$+ \ddot{\mathbf{B}}_{\mathbf{X}})\mathbf{U}(t) = -[(\mathbf{A}_{\mathbf{X}}^3 + 2\dot{\mathbf{A}}_{\mathbf{X}}\mathbf{A_X} + \mathbf{A_X}\dot{\mathbf{A}}_{\mathbf{X}} + \ddot{\mathbf{A}}_{\mathbf{X}})\dot{\mathbf{X}}(t)+$$
$$+ (\mathbf{A}_{\mathbf{X}}^2\mathbf{C_X} + 2\dot{\mathbf{A}}_{\mathbf{X}}\mathbf{C_X} + \mathbf{A_X}\dot{\mathbf{C}}_{\mathbf{X}} + \ddot{\mathbf{C}}_{\mathbf{X}})\mathbf{F}(t)+$$
$$+ (\mathbf{A_X}\mathbf{C_X} + 2\dot{\mathbf{C}}_{\mathbf{X}})\mathbf{F}(t) + \mathbf{C_X}\ddot{\mathbf{F}}(t) - (2\dot{\mathbf{A}}_{\mathbf{X}} + \mathbf{A}_{\mathbf{X}}^2)\mathbf{g}]. \quad (12.25)$$

For the proposed representations of the control system, the necessary initial conditions of the control functions that ensure the movement of the moving object among the corresponding section of the trajectory are obtained from the following algebraic equations

$$\mathbf{U}(t_i^s) = \mathbf{B}_{\mathbf{X}}^{-1}[\ddot{\mathbf{X}}(t_i^s) - \mathbf{A_X}\dot{\mathbf{X}}(t_i^s) + \mathbf{g} - \mathbf{C_X}\mathbf{F}(t_i^s)],$$
$$\dot{\mathbf{U}}(t_i^s) = \mathbf{B}_{\mathbf{X}}^{-1}[\overset{...}{\mathbf{X}}(t_i^s) - \mathbf{A_X}\ddot{\mathbf{X}}(t_i^s) - \dot{\mathbf{A}}_{\mathbf{X}}\dot{\mathbf{X}}(t_i^s) - \dot{\mathbf{B}}_{\mathbf{X}}\mathbf{U}(t_i^s) - \dot{\mathbf{C}}_{\mathbf{X}}\mathbf{F}(t_i^s)+$$
$$+ \mathbf{C_X}\dot{\mathbf{F}}(t_i^s)]. \quad (12.26)$$

It should be noted that the system is shown in Figure 12.3. is simpler to implement, but the system in Figure 12.4, is implemented by integrating blocks in feedback loops and has additional filtering properties.

12.4 Robust Correction

The solution to the moving objects control problem under conditions of parameter's uncertainty, measurement noise, and external disturbances is based on the use of a variable structure system forming the optimal reference

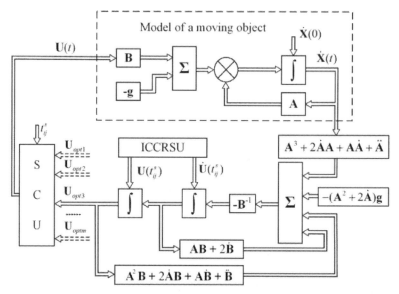

Figure 12.4 Block diagram of the implementation of control in the form of integrating functions in the feedback

model of the moving object motion and the corrective control signal formed by a robust loop based on the mismatch between the optimal and physical signal output. The integral signal of optimal and robust control is transferred to a moving object propellers. The general constraint on the control action should limit the values of the optimal and corrective control

$$|\mathbf{U}_{opt}| + |\mathbf{U}_{cor}| \le |\mathbf{U}_{\max}| \,.$$

The proposed approach for an admissible simplification of the synthesis of robust control includes the linearization of the dynamics equations in the vicinity of the nominal values of the parameters, which makes it possible to apply the superposition principle to analyze control errors. The optimal control and the specified trajectories of the moving object are formed, as described above, taking into account the fact that the nonlinearity of the model, and the discrepancy arising from the application of linearization can be attributed to additional uncertainty, which is compensated by a robust control.

The external disturbances $\mathbf{F}$, some of which $\mathbf{F}_{cont}$ can be measured and controlled, the parametric measurement noise $\boldsymbol{\eta}$, the mismatch between the parameters of the physical moving object, described by the matrix transfer

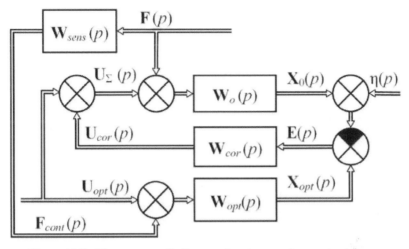

Figure 12.5 The structure of a linear and stationary robust-optimal system

function $\mathbf{W}_o(p)$, and the optimal mathematical model $\mathbf{W}_{opt}(p)$ are considered to be uncertainties that are compensated by the robust loop.

Based on the structural diagram of a robust-optimal system (Figure 12.5), the error vector can be determined as

$$\begin{aligned}
\mathbf{E}(p) &= [\mathbf{I} + \mathbf{W}_o(p)\mathbf{W}_{cor}(p)]^{-1}\{[\mathbf{W}_{opt}(p) - \mathbf{W}_o(p)]\mathbf{U}_{opt}(p) + \\
&\quad + \mathbf{W}_{opt}(p)\mathbf{F}_{cont}(p) - \mathbf{W}_o(p)\mathbf{F}(p) - \mathbf{\eta}(p)\} = \mathbf{E}_\mathbf{U}(p) + \\
&\quad + \mathbf{E}_\mathbf{F}(p) + \mathbf{E}_\mathbf{\eta}(p).
\end{aligned} \tag{12.27}$$

In a more simplified form, the error vector can be represented in an approximate form due to the unmeasurable part of the perturbations $\Delta\mathbf{F}(p)$

$$\mathbf{E}_\mathbf{F}(p) \approx [\mathbf{I} + \mathbf{W}_o(p)\mathbf{W}_{cor}(p)]^{-1}\mathbf{W}_{opt}(p)\Delta\mathbf{F}(p).$$

The requirements for the structure and parameters of a robust correction in the form of the minimum H_2 norm [71] of the matrices of sensitivity functions can be written for:

- input programmed optimal control based on the mismatch between the parameters of the optimal model and the physical object

$$\mathbf{S}_\mathbf{U} = [\mathbf{I} + \mathbf{W}_o(j\omega)\mathbf{W}_{cor}(j\omega)]^{-1}[\mathbf{W}_{opt}(j\omega) - \mathbf{W}_o(j\omega)];$$

- the difference between the values of the vector of external disturbances $\mathbf{F}(t)$ and their controlled part $\mathbf{F}_{cont}$

$$\mathbf{S}_\mathbf{F} = [\mathbf{I} + \mathbf{W}_o(j\omega)\mathbf{W}_{cor}(j\omega)]^{-1}\mathbf{W}_{opt}(j\omega);$$

- values of the parametric noise vector η

$$\mathbf{S_\eta} = [\mathbf{I} + \mathbf{W}_o(j\omega)\mathbf{W}_{cor}(j\omega)]^{-1}.$$

The robustness of the control system is limited when the condition for the H_2 norm of the sensitivity matrix is met

$$\left\|[\mathbf{I} + \mathbf{W}_o(j\omega)\mathbf{W}_{cor}(j\omega)]^{-1}[\mathbf{W}_{opt}(j\omega) - \mathbf{W}_o(j\omega)]\right\| \leq 1;$$

$$\left\|[\mathbf{I} + \mathbf{W}_o(j\omega)\mathbf{W}_{cor}(j\omega)]^{-1}\mathbf{W}_{opt}(j\omega)\right\| \leq 1; \qquad (12.28)$$

$$\left\|[\mathbf{I} + \mathbf{W}_o(j\omega)\mathbf{W}_{cor}(j\omega)]^{-1}\right\| \leq 1.$$

The analysis of the equations obtained (12.28) shows that together with the small mismatch between model and the physical moving object's matrix norm $\|\mathbf{W}_{opt}(j\omega) - \mathbf{W}_o(j\omega)\| \leq 1$ amplification is applied in the straight part of the robust loop, which does not have a negative effect on stability but instead increases the accuracy of the control system.

The structure and parameters of a robust loop can be obtained from the condition of the minimization of the error vector and its derivatives in the form (12.5). The equation of the physical moving object (12.1) (considering only the controlled disturbance $\mathbf{F}_{cont}$ and the use of a robust loop) takes the following nonstationary form (considering linearized matrices $\bar{\mathbf{A}}(t), \bar{\mathbf{B}}(t), \bar{\mathbf{C}}(t)$)

$$\ddot{\mathbf{X}}(t) = \bar{\mathbf{A}}(t)\dot{\mathbf{X}}(t) + \bar{\mathbf{B}}(t)[\mathbf{U}_{opt}(t) + \mathbf{U}_{cor}(t)] + \bar{\mathbf{C}}(t)\mathbf{F}_{cont}(t) - \mathbf{g}. \quad (12.29)$$

The Equation (12.1) for the optimal moving object model can be written as

$$\ddot{\mathbf{X}}_m(t) = \bar{\mathbf{A}}_m(t)\dot{\mathbf{X}}_m(t) + \bar{\mathbf{B}}_m(t)\mathbf{U}_{opt}(t) + \bar{\mathbf{C}}_m(t)\mathbf{F}_m(t) - \mathbf{g}. \quad (12.30)$$

To determine the control signal using The Equations (12.29) and (12.30), the approximate equation for the vector of control errors can be written in the form

$$\ddot{\mathbf{E}}(t) \approx -\mathbf{G}_1\dot{\mathbf{E}}(t) - \mathbf{G}_2\mathbf{E}(t) \approx \bar{\mathbf{A}}_m(t)\dot{\mathbf{E}}(t) - \bar{\mathbf{B}}_m(t)\mathbf{U}_{cor}(t). \quad (12.31)$$

Using (12.31), the dependence for the robust correction loop vector can be obtained

$$\mathbf{U}_{cor}(t) = \bar{\mathbf{B}}_m^{-1}(t)[\bar{\mathbf{A}}_m(t)\dot{\mathbf{E}}(t) + \mathbf{G}_1\bar{\mathbf{A}}_m(t) + \mathbf{G}_2\mathbf{E}(t)]. \quad (12.32)$$

The structural implementation of a robust-optimal control system of variable structure using robust correction is shown in Figure 12.6.

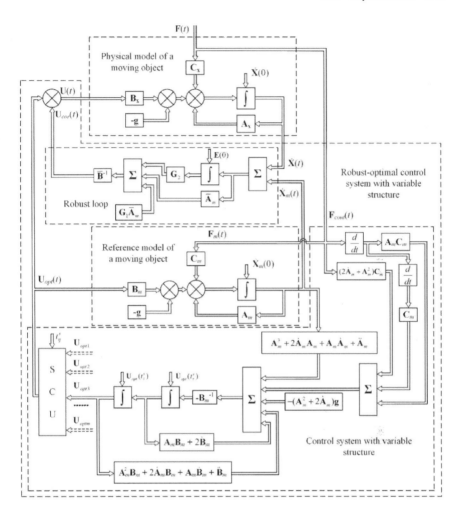

Figure 12.6 The structural implementation of a robust-optimal control system of variable structure

12.5 Experiments

12.5.1 Sea Vessel Maneuvering

To simulate the process of stabilizing the course angle $\varphi(t) = x_1(t)$ and angular speed $\omega(t) = x_2(t)$ of the vessel, controlled by changing the angle of the rudder blade $\alpha(t)$, considering the values of the reduced aerohydro-dynamic coefficients and external disturbance $f(t)$ [27,31], let's consider the

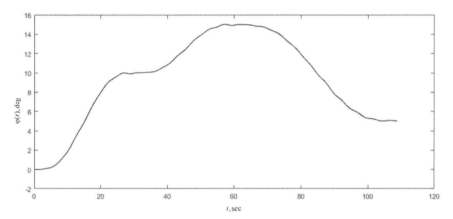

Figure 12.7 Course angle over time

problem of sea vessel maneuvering at maximum speed (12.3)

$$\dot{\mathbf{X}}(t) = \mathbf{A_X}\mathbf{X}(t) + \mathbf{B}\mathbf{U}(t) + \mathbf{C}f(t),$$

where $\mathbf{X}(t) = (x_1(t) \ x_2(t) \ x_3(t))^\mathrm{T}$; $\mathbf{U}(t) = (\alpha(t) \ \dot{\alpha}(t))^\mathrm{T}$;

$$\mathbf{A}_X = \begin{pmatrix} 0 & 1 & 0 \\ 0 & 0 & 1 \\ 0 & -0.03\,|x_2(t)| & -0.084 \end{pmatrix}; \mathbf{B} = \begin{pmatrix} 0 & 0 \\ 0 & 0 \\ 0.0002 & 0.0063 \end{pmatrix};$$

$$\mathbf{C} = (0 \ 0 \ 0.00034)^\mathrm{T}.$$

Simulating the process of vessel stabilization is considered taking into account the impact of uncontrolled irregular waves [27] $W_f(p)$ and the mismatch of the parameters of the mathematical model and the physical vessel, set within 20%, as well as the influence of parametric noise of a given intensity.

The simulation graphs include time trajectories for course angle $\varphi(t)$ (Figure 12.7), rudder angle $\alpha(t)$ (Figure 12.8) and errors $\varepsilon_\varphi(t)$ along the course angle (Figure 12.9).

The simulation of the vessel stabilization under the influence of irregular sea waves, set by the shaping filter for a given wave intensity of four points, was considered taking into account the noise of measuring the output coordinates. The specified maneuvering trajectory includes three sections at different course angles, which is typical when sailing in limited water areas (sea channels, straits, etc.), as well as, for example when laying underwater pipelines. The simulation results demonstrate that the control accuracy is ensured with errors of less than 1%.

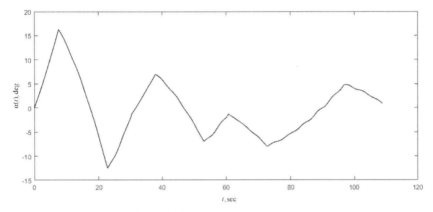

Figure 12.8 Rudder angle over time

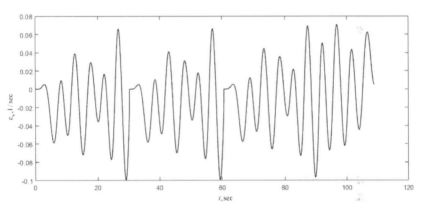

Figure 12.9 Course angle error over time

12.5.2 Quadrotor UAV Stabilization

The motion dynamics of a quadrotor UAV are described by the system of six second-order differential Equation (12.1) considering the effects of wind disturbances. The complexity of a complete description of the physical effect of wind disturbance on a quadrotor UAV suggests [39] the selection of the following components of the airflow: high-frequency random pulsations; low-frequency piecewise constant components in horizontal (constant wind) and vertical directions (ascending or descending flows). Thus, the quadrotor UAV motion is described by the following system of differential equations [37, 41, 68]

$$\dot{\mathbf{X}}(t) = \mathbf{A}(\mathbf{X})\mathbf{X}(t) + \mathbf{B}(\mathbf{X})\mathbf{U}(t) + \mathbf{CF}(t) - \mathbf{g},$$

where $\mathbf{X}(t) = (v_x\ v_y\ v_z\ \omega_\theta\ \omega_\psi\ \omega_\varphi)^{\mathrm{T}}$; $(\mathbf{F}(t) = f_x\ f_y\ f_z\ f_\theta\ f_\psi\ f_\varphi)^{\mathrm{T}}$;

$$\mathbf{g} = \begin{pmatrix} 0 & 0 & 9.81 & 0 & 0 & 0 \end{pmatrix}^{\mathrm{T}};$$

$$\mathbf{A}\,(\mathbf{X}) = -\mathrm{diag}\,(0.004v_x, 0.004v_y, 0.01v_z, 0.8\omega_\theta,\ 0.8\omega_\psi, 0.8\omega_\varphi)\,;$$

$$\mathbf{B}\,(\mathbf{X}) = \begin{pmatrix} cos\varphi sin\theta cos\psi + sin\varphi sin\psi & 0 & 0 & 0 \\ sin\varphi sin\theta cos\psi + cos\varphi sin\psi & 0 & 0 & 0 \\ cos\theta cos\psi & 0 & 0 & 0 \\ 0 & 1 & 0 & 0 \\ 0 & 0 & 1 & 0 \\ 0 & 0 & 0 & 0.5 \end{pmatrix};$$

$$\mathbf{U}\,(t) = \begin{pmatrix} U_1 \\ U_2 \\ U_3 \\ U_4 \end{pmatrix} = \begin{pmatrix} N_1 + N_2 + N_3 + N_4 \\ -N_1 - N_2 + N_3 + N_4 \\ -N_1 + N_2 + N_3 - N_4 \\ N_1 - N_2 + N_3 - N_4 \end{pmatrix};$$

$$\mathbf{C} = \mathrm{diag}\,(0.08,\ 0.036,\ 1.14,\ 0.008,\ 0.008,\ 0.004)\,.$$

The functional task of visual monitoring of marine traffic determines the high-precision stabilization of the quadrotor UAV in the horizontal plane and the movement along a given trajectory followed by a "soft" landing. The minor controlled coordinates of the yaw angles φ, roll ψ; and pitch θ are controlled by three optimized PID controllers [38]. The trajectories concerning the higher coordinates x, y, z are formed for the condition of the third zero derivatives of the vector of coordinates. For the given stabilization trajectory of the quadrotor UAV, the switching moments of the control functions $T_1, T_2,$ T are determined.

The block diagram of the robust-optimal quadrotor UAV control system (Figure 12.10) includes the following blocks: *a*) CAVCU is a block of synthesizing the specified values of the orientation angles of a quadrotor UAV; *b*) CSFU is a block of synthesizing the corrective signal of a robust loop; *c*) SCU is a block of switching control actions; *d*) SW is a switching key; *e*) $\mathbf{W}_{cor}$ is a transfer matrix function that can be obtained from (12.32) considering linearization and stationarity of matrices $\mathbf{A}$ and $\mathbf{B}$. The physical and reference models are set with a mismatch condition within $\pm$ 15%. A piecewise constant component (average hourly wind speed $V_w = 10$ m/s) and a high-frequency component formed by "white noise" of a given intensity cause the wind impact.

The results of the simulation of movement along a given quadrotor UAV trajectory (Figures 12.11–12.13: error and velocity graphs for the higher

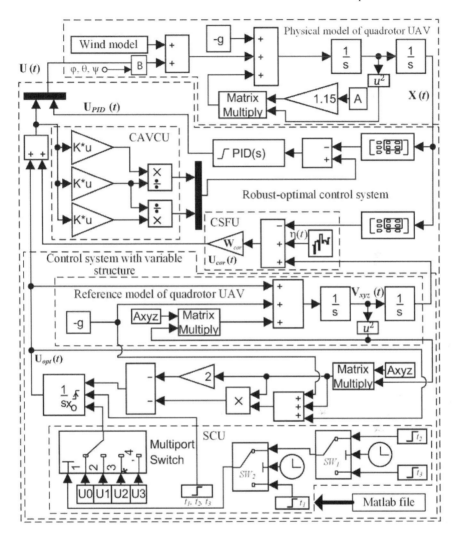

Figure 12.10 Simulink-based robust-optimal stabilization scheme for a quadrotor UAV on a given trajectory

coordinates, as well as lower coordinates of angular orientation, respectively) demonstrate sufficiently small values of control errors (less than 1%).

For comparative analysis, the control accuracy and energy consumption for the control system under consideration with the optimization of the trajectory of three higher coordinates of the motion and three PID controllers (Figures 12.14, 12.15) are compared; the graphs of the higher coordinates

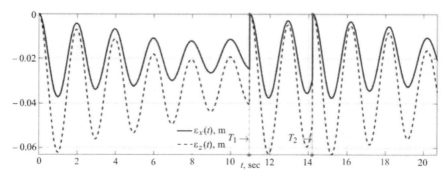

Figure 12.11   Errors of the higher coordinates over time

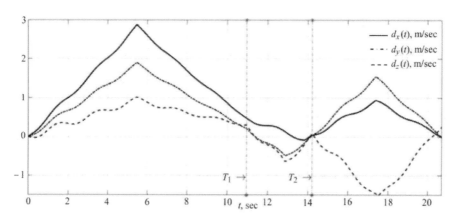

Figure 12.12   Velocity of the higher coordinates over time

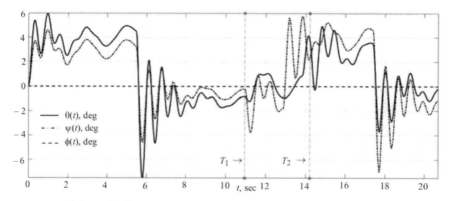

Figure 12.13   Lower coordinates of angular orientation over time

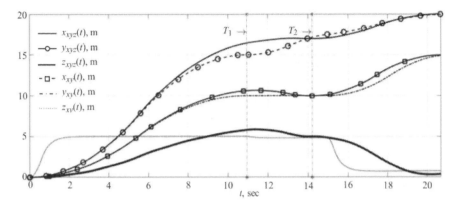

Figure 12.14 Comparison of control accuracy between optimal control with variable structure and PID control over time

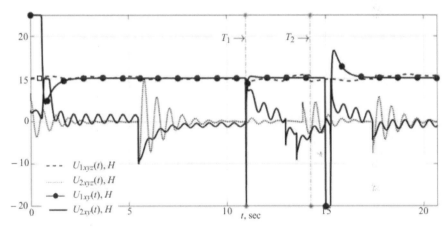

Figure 12.15 Comparison of energy consumption between optimal control with variable structure and PID control over time

and control, respectively, are shown by means of the index *xyz*), and system with optimization by two higher coordinates *x*, *y* and a block of four PID controllers for the remaining four coordinates (the graphs shown by mean of the index *xy*), are given. The simulation results demonstrate a decrease in energy consumption (values are given in arbitrary units) of control for the proposed control system.

The simulation results (Figures 12.16, 12.17, the red line inside the "tube" shows the optimal trajectory) demonstrates the errors change in the form of ñtubesż for different values of external uncontrolled wind disturbance.

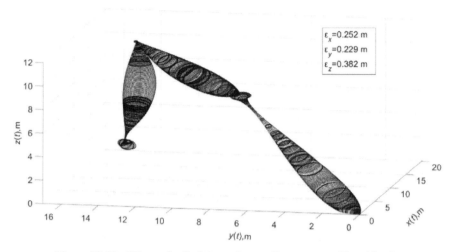

Figure 12.16   "Error tubes" of the main coordinates x, y, z; $V_w = 10$ m/s

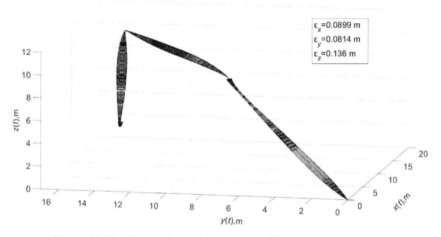

Figure 12.17   "Error tubes" of the main coordinates x, y, z; $V_w = 3$ m/s

For example, the problem of high-precision control of a quadrotor UAV under the conditions of wind disturbance arises during navigation and environmental monitoring of the marine environment to expand the information base for assessing the safety of navigation. The simulation results show the robust stability of the control system of the quadrotor UAV to external disturbances, the uncertainty of the mathematical model and the minimum deviations of the current trajectory from the formed optimal trajectory of stabilization. For the circuitry implementation, a Simulink-oriented scheme of a robust-optimal control system for a quadrotor UAV was formed.

12.6 Conclusion

Optimal control systems provide moving objects with the highest control quality indicators in terms of speed, accuracy, and energy consumption. In the case of multidimensional nonlinear systems, there are significant restrictions on the applicability of classical optimal control methods associated with the computational difficulties due to boundary value problems complex solutions, the correctness of the formation of weight matrices of optimality criteria and others.

To optimize nonlinear systems up to the sixth order, the proposed algorithmic procedure of the synthesis of variable structure system for different types of moving objects based on preliminary construction of optimal trajectories of stabilization of dynamic processes has been applied. The generated minimax optimality conditions for various criteria allow to consider the constraints on the control actions and solve the synthesis problem for the corresponding functional control problems: maximum speed (for critical modes) and minimum energy consumption (for technological modes). The uncertainty of the external environment and the dynamic object requires robust correction of the optimal system. The generated optimal trajectories and control determine the formation of the additional robust control loop to compensate incomplete a priori definiteness of the mathematical model and uncontrolled (unmeasured) disturbances and noises.

The proposed procedure for the synthesis of robust-optimal systems has been tested for several models of a nonlinear moving object and demonstrated the versatility and engineering practicality, as well as efficiency under the influence of external limited disturbances in terms of accuracy and control energy consumption. A number of algorithmic procedures for optimal trajectories construction, switching moments determination, and synthesis of control functions for multidimensional systems have been developed. Controlling the mismatch between the trajectory of a physical object and the calculated optimal trajectory of the reference model allows to form a robust subsystem that provides invariance to incomplete information about the dynamic system taking into account the values of the optimal control from the reference system. The presented results of the simulation of quadrotor UAV stabilization and maneuvering of a vessel under the conditions of external disturbance and parametric noise demonstrate the provision of the necessary control accuracy in the vicinity of the formed optimal trajectory of controlled coordinates.

References

[1] Krasovsky, N.N.: Some problems of the theory of stability of motion, Phizmathlit, Moscow (1959) (in Russian)

[2] Feldbaum, A.A.: Theory Of Optimal Systems, Phizmathlit, Moscow (1963) (in Russian)

[3] Kuntsevich, V.M.: Synthesis of Robust Optimal Adaptive Control Systems for Nonstationary Objects under Bounded Disturbances. In: Journal of Automation and Information Sciences, NY, Begell house inc., vol. 36, issue 3, pp.14–24 (2004)

[4] Kuntsevich, V.M., et al (Eds): Control Systems: Theory and Applications. Series in Automation, Control and Robotics, River Publishers, Gistrup, Delft, (2018)

[5] Kondratenko, Y.P., Kuntsevich, V.M., Chikrii, A.A., Gubarev, V.F. (Eds): Advanced Control Systems: Theory and Applications. Series in Automation, Control and Robotics, River Publishers, Gistrup, (2021)

[6] Kondratenko, Y.P., Chikrii, A.A., Gubarev, V.F., Kacprzyk, J. (Eds): Advanced Control Techniques in Complex Engineering Systems: Theory and Applications. Dedicated to Professor Vsevolod M. Kuntsevich. Studies in Systems, Decision and Control, Vol. 203. Cham: Springer Nature Switzerland AG (2019)

[7] Kuntsevich, V.M.: Estimation of Impact of Bounded Perturbations on Nonlinear Descrete Systems. In: Kuntsevich, V.M. et al. (Eds). Control Systems: Theory and Applications. Series in Automation, Control and Robotics, River Publishers, Gistrup, Delft, pp. 3–15 (2018)

[8] Emelynov, S.V., Korovin, S.K.: New types of feedback, Nauka, Phizmathlit, Moscow, (1997) (in Russian)

[9] Emelynov, S.V.: Automatic control systems of variable structure: synthesis of scalar and vector systems by state and by output. In: Nonlinear dynamics and control, Phizmathlit, Moscow, vol. 5, pp. 5–24 (2007) (in Russian)

[10] Utkin, V.I.: Sliding Modes in Control and Optimization, Springer-Verlag, (1992)

[11] Shtessel, Y.B., Moreno, J.A., Fridman, L.M.: Twisting sliding mode control with adaptation: Lyapunov design, methodology and application. In: Automatica, vol. 75, pp. 229–235 (2017)

[12] Dorf, R.,. Bishop, R.: Modern Control Systems, LBZ, Moscow (2002) (in Russian)

[13] Boychuk, L.M.: Method of structural synthesis of nonlinear automatic control systems, Energia, Moscow (1971) (in Russian)

[14] Galiulin, A.S.: Methods for solving inverse problems of dynamics, Moscow, Nauka, (1986) (in Russian)

[15] Krutko, P.D.: Robust Stable Structures of Controlled Systems of High Dynamic Accuracy. Algorithms and dynamics of motion control of model objects. In: Control theory and systems, no. 2, pp. 120–140 (2005) (in Russian)

[16] Krutko, P.D.: Inverse problems of dynamics in the theory of automatic control, Mashinostroenie, Moscow, (2004) (in Russian)

[17] Kondratenko, Y.P., Timchenko,V.L.: Optimal feedback switching method for linear control system. In: Intern. System and Networks: Mathematical theory and applications, Berlin, Academie Verlag, Vol. 2, pp. 291–292 (1994)

[18] Timchenko, V.L., Lebedev, D.O., Robust-optimal stabilization of nonlinear dynamic systems. In: Radio Electronics, Computer Science, Control, no. 3., pp. 185–197 (2019)

[19] Gabasov, R., Kirillova, F.M., Ruzhitskay, E.A.: Implementation of Bounded Feedback in a Nonlinear ontrol Problem. In: Cybernetics and Systems Analysis, no. 1, pp. 108–116 (2009) (in Russian)

[20] Balashevich, N.V., Gabasov, R., Kalinin, A.I. and Kirillova, F.M.: Optimal Control of Nonlinear Systems. In: Computational Mathematics and Mathematical Physics, vol. 42, no. 7, pp. 931–956 (2002)

[21] Horowitz, I.M.: Survey of quantitative feedback theory (QFT). In: Int. Journal of Robust and Non-Linear Control, vol. 11., no. 10., pp. 887–921 (2001)

[22] Larin, V.B.: On the reversal of the problem of analytical design of controllers. In: Control and Information Promblems, no. 1, pp. 17–25 (2004) (in Russian)

[23] Yaesh, I., Boyarski, S., Shaked U.: Probability-guaranteed robust H∞ performance analysis and state-feedback design. In: Systems & Control Letters, vol, 48, issue 5, pp. 351–364 (2003)

[24] Chesnov, V.N.: Synthesis of H_{∞} - controllers for Multidimensional Systems of Given Accuracy and Stability. In: Autom. Remote Control, vol. 72, no. 10, pp. 2161–2175 (2011)

[25] Roberts, G.N.: Trends Marine Control Systems. Im: IFAC, CAMS'07, Zagreb, Croatia (2007)

[26] Ornerdi, c E., Robert,s G.N.: A fuzzy track-keeping autopilot for ship steering. In: Journal of Marine Engineering and technology, London, no. A2, pp. 23–35 (2003)

[27] Lukomsky, Y.K., Chugunov, V.S.: Marine mobile objects control systems, L., Sudostroenie, (1988) (in Russian)

[28] Fossen, T.I., Perez, T.: Kinematic models for maneuvering and seakeeping of marine vessels. In. Journal of modeling, identification and control, vol. 28, issue 1, pp. 19–30 (2007)

[29] Johansen, T., Fossen, T.I.: Control allocation—A survey. In. Automatica, vol.49, pp. 1087–1103 (2013)

[30] Comasòlivas, R., Escobet, T. and Quevedo, J.: Automatic design of robust PID controllers based on QFT specifications. In: Proceeding of IFAC Conference on Advances in PID Control, pp. 715–720 (2012)

[31] Timchenko, V.L., Kondratenko, Y.P.: Robust stabilization of marine mobile objects on the basis of systems with variable structure of feedbacks. In: Intern. Journal of Automation and Information Sciences, NY, Begell house inc., vol. 43, issue 6, pp. 16–29 (2011)

[32] Timchenko, V.L.: Synthesis of variable structure systems for stabilization of ships at incomplete controllability. In: Intern. Journal of Automation and Information Sciences, NY, Begell house inc., vol. 44, issue 6, pp. 8–19 (2012)

[33] Timchenko, V.L., Ukhin,O.A.: Optimization of Stabilization Processes of Marine Mobile Object in Dynamic Positioning Mode. In: Intern. Journal of Automation and Information Sciences, NY, Begell house inc., vol. 46, issue 7, pp. 40–52 (2014)

[34] Timchenko, V.L., Ukhin, O.A.: Variable Structure Robust-Optimal Systems for Control of Marine Vehicles. In: Proceeding of the IEEE 4th International Conference MSNMC, Kyiv, Ukraine, pp. 151–155 (2016)

[35] Timchenko, V.L., Lebedev, D.O.: Automated Algorithmic Procedure of Robust-Optimal Control System's Synthesis for Marine Vehicles. In: Proceeding of the IEEE 4th International Conference MSNMC, Kyiv, Ukraine,, pp. 98–101 (2018)

[36] Timchenko, V.L., Ukhin, O.A. and Lebedev, D.O.: Optimization of nonlinear systems of variable structure for control of marine moving vehicles. In: Intern. Journal of Automation and Information Sciences, NY, Begell house inc., vol. 49, issue 7, pp. 33–47 (2017)

[37] Altug, E., Ostrowski, J.P. and Taylor. C.J.: Control of a quadrotor helicopter using dual cameravisual feedback. In: The International Journal of Robotics Research, vol. 24, no. 5, pp. 329–341 (2005)

[38] Pyrkin, A.A., Maltseva, T.A., Labadin, D.V., Surov, M.O. and Bobtsov, A.A.: Sintez sistemy upravleniya kvadrokopterom s ispol'zovaniyem uproshchennoy matematicheskoy modeli. In: Izv. Vuzov. Priborostroyeniye, RF, SPb, vol. 56, no. 4, pp. 47–51 (2012) (in Russian).

[39] Andreev, M.A., Miller, B., Miller, B.M. and Stepanyan, K.V.: Path planning for unmanned aerial vehicle under complicated conditions and hazards. In: International Journal of Computer and Systems Sciences, vol. 51, no. 2, pp. 328–338 (2012)

[40] Younes, Y.A., Drak, A., Noura, H. et al.: Robust Model-Free Control Applied to a Quadrotor. In: J. Intell. Robotics Syst., 84, pp. 37–52 (2016)

[41] Larin, V.B., Tunik, A.A.: Synthesis of the quad-rotor flight control system. In: Proceeding of the IEEE 4th International Conference MSNMC, Kyiv, Ukraine, pp. 12–17 (2016)

[42] Wang, D., Liu, D., Li, H., et al.: An Approximate Optimal Control Approach for Robust Stabilization of a Class of Discrete-Time Nonlinear Systems With Uncertainties. In: IEEE Transactions on Systems, Man, and Cybernetics: Systems, vol. 46, no. 5, pp. 713–717 (2016)

[43] Lu, T., Wen, P.: Time Optimal and Robust Control of Twin Rotor System. In: Proceeding of the IEEE International Conference on Control and Automation, Guangzhou, China, pp. 862–866 (2007)

[44] Faruque, I.A., Muijres, F.T., Macfarlane, K.M. et al.: Identification of optimal feedback control rules from micro-quadrotor and insect flight trajectories. In: Biol Cybern., 112, pp. 165–179 (2018)

[45] Kamran, J.M., Ahmed, S.F., Bakar, M.I., Ali, A.: Horizontal Motion Control of Underactuated Quadrotor Under Disturbed and Noisy Circumoyostances, Information and Communication Technology. In: Advances in Intelligent Systems and Computing, Springer, vol. 625., pp. 63–79 (2018)

[46] Jayakrishnan, H.J.Z: Position and Attitude control of a Quadrotor UAV using Super Twisting Sliding Mode. In: IFAC-PapersOnLine, vol. 49, issue 1, pp. 284–289 (2016)

[47] Moreno, J.A., Osorio, M.: A Lyapunov approach to second-order sliding mode controller and observers. In: Proceeding of the 47th Conference on Decision and Control, pp. 2856–2861 (2008)

[48] Matveev, A.S., Teimoori, H., Savkin, A.V.: A method for guidance and control of an autonomous vehicle in problems of border patrolling and obstacle avoidance. In: Automatica, vol. 47, issue 3, pp. 515–524 (2011)

[49] Leira, F.S., Johansen, T.A. and Fossen,T.I.: Automatic detection, classification and tracking of objects in the ocean surface from UAVs using a thermal camera. In: Proceeding of Aerospace Conf., pp. 1–10(2015)

[50] Liu, Y., Chen, C., Wu, H. et al.: Structural stability analysis and optimization of the quadrotor unmanned aerial vehicles via the concept of Lyapunov exponents. In: Int J Adv Manuf Technol 94, pp. 3217–3227 (2018)

[51] Liu, Y., Li X., Wang, T., Zhang, Y., Mei, P.: The Stability Analysis of Quadrotor Unmanned Aerial Vechicles. In: Wearable Sensors and Robots. Lecture Notes in Electrical Engineering, Springer, vol. 399, pp. 383–394 (2017)

[52] Liu, Y., Li, X., Wang, T., et al.: Quantitative stability of quadrotor unmanned aerial vehicles. In: Nonlinear Dyn, pp. 1819–1833 (2017)

[53] Gharib, M. R., and Moavenian, M.: Full dynamics and control of a quadrotor using quantitative feedback theory. In:Int. J. Numer. Model, 29, pp. 501–519 (2017)

[54] Xu, Y., Tong, Ch.: Quantitative feedback control of a quadrotor. In: Proceedings of the IEEE International Symposium on Industrial Electronics, Hangzhou, China, pp. 1309–1314 (2012)

[55] Raffo, G. V., Ortega M. G., Rubio F. R.: An integral predictive/nonlinear H∞ control structure for a quadrotor helicopter. In: Automatica, vol. 46, issue 1, pp. 29–39 (2010)

[56] Viegas, D., Batista, P., Oliveir,a.P., Silvestre, C.: Decentralized H$_2$ observers for position and velocity estimation in vehicle formations with fixed topologies. In: Systems & Control Letters, vol. 61, issue 3, pp. 443–453 (2012)

[57] Falkenberg, O., Witt, J., Pilz, U., Weltin, U., Werner, H.: Model Identication and H$_\infty$ Attitude Control for Quadrotor MAV's. In: ICIRA, Part II, pp. 460–471 (2012)

[58] Sushchenko, O.A., Shyrokyi, O.V.: *H$_2$/H*-Optimization of system for stabilization and control by line-of-sight orientation of devices operated at UAV. In: Proceedings of the IEEE 3rd International Conference APUAVD, pp. 235–238 (2015)

[59] Zhang, X., Kamgarpou,r M., et al.: Robust optimal control with adjustable uncertainty sets. In: Automatica, vol. 75, pp. 249–259 (2017)

[60] Sushchenko, O.A.: Robust control of angular motion of platform with payload based on H$_\infty$-synthesis. In: Journal of Automation and Information Sciences, vol. 48, issue 12, pp. 13–26 (2016)

[61] Satici, A.C., Poonawala, H., Spong, M.W.: Robust Optimal Control of Quadrotor UAVs. In: IEEE Access, vol. 1, pp. 79–93 (2013)

[62] Vossen, G., Maurer, H.: On L1-minimization in optimal control and applications to robotics. In: Optim. Control Appl. Meth., 27, pp. 301–321 (2006)

[63] Chen, Y., Han, J., & Zhao, X., Three-dimensional path planning for unmanned aerial vehicle based on linear programming. In: Robotica, 30(5), pp. 773–781 (2012).

[64] Jiang, B., Bishop, A.N., Anderson, B. D.., Drake S.P.: Optimal path planning and sensor placement for mobile target detection. In: Automatica, Vol. 60, pp. 127–139 (2015)

[65] Lewis, F.L., Vrabie, D., Vamvoudakis, K. G.: Reinforcement Learning and Feedback Control: Using Natural Decision Methods to Design. Optimal Adaptive Controllers. In: IEEE Control Systems Magazine, vol. 32, no. 6, pp. 76–105 (2012)

[66] Bisheban, M., Lee T.: Geometric Adaptive Control With Neural Networks for a Quadrotor in Wind Fields. In: Proceeding of the IEEE Transactions on Control Systems Technology, pp. 1–19 (2019)

[67] Timchenko, V.L., Lebedev, D.O., Kuklina, K.A. and Timchenko, I.V.: Robust-optimal control system of quad for maritime traffic's monitoring. In: Proceeding of the IEEE 4th international conference APUAVD, Kyiv, Ukraine,, pp. 192–196 (2017)

[68] Timchenko, V.L., Lebedev, D.O., Optimization of Processes of Robust Control of Quadcopter for Monitoring of Sea Waters. In: Journal of Automation and Information Sciences, NY., Begell house, vol. 51, issue. 2, pp. 1–10 (2019)

[69] Timchenko, V.L., Lebedev, D.O.: Robust control systems with suboptimal models for stabilization of UAV. In: Proceedings of the 2019 IEEE 5th International Conference APUAVD, Kyiv, Ukraine, pp. 124–129 (2019)

[70] Timchenko, V.L., Lebedev, D.O.: Synthesis of Optimal Phase Trajectory for Quadrotor UAV Stabilization. In: Proceedings of the IEEE 6th International Conference MSNMC, Kyiv, Ukraine, pp. 20–24 (2020)

[71] Sumacheva, V.A., Kharitonov, V.L.: Calculation of the H_2 Norm of the Transfer Matrix of a System of Neutral Type. In: Differential Equations And Control Processes, vol. 4, pp. 22–32 (2014)

Part III
Recent Developments in Collaborative Automation

13

Modeling of Cyber-Physical Systems

Illya Holovatenko, and Andrii Pysarenko

Department of Information Systems and Technologies, National Technical University of Ukraine "Igor Sikorsky Kyiv Polytechnic Institute", Kyiv, Ukraine
E-mail: illyaholovatenko@gmail.com; andrew.pisarenko@gmail.com

Abstract

The purpose is to review current research in the field of constructing cyber–physical systems. The structure of the chapter is as follows: a review of the most effective approaches and methods of constructing cyber–physical systems; using these methods—design cyber–physical–logistical system; perform the review of the obtained results and come up with a plan for further researches based on the reviewed methods. For validation of the designed system we came up with a live scenario from the logistical systems: the autonomous object delivers goods from one warehouse to another. We also will model obstacles along the way. The system should proactively react to this and recalculate the optimal path. This will lead to the timely delivery of goods to the destination.

Keywords: Cyber–physical system, logistics, optimal control, route planning.

13.1 Introduction

Cyber–physical systems are the integration of computational and physical processes. Controllers and computer networks control physical processes, usually through feedback, which is also affected by physical processes and vice versa.

327

The economic and social potential of such systems is much greater than that of their counterparts, which is why the main global investment goes to the development of this technology. Nevertheless, there are significant problems, especially because the physical components of such systems require certain security and reliability requirements that are qualitatively different from those for general purpose computing.

Logistics is a set of entities involved in the delivery of goods and services to the end customer. Logistics management is the effective end-to-end process management, which includes: 1) planning the optimal delivery route; 2) optimal appointment of "performers" (autonomous objects that carry out delivery); 3) coordination of system subjects among themselves; 4) risk forecasting; 5) timely response to emergency situations. The logistics system faces problems that can potentially lead to inefficiencies and waste in supply chains, such as delayed deliveries, rising fuel costs, inconsistent suppliers, and ever-increasing end-user expectations.

13.2 Review of Modeling Methods for Cyber–Physical Systems

One of the approaches to modeling cyber–physical systems is their representation in the form of templates based on FCA (formal concept analysis) to optimize the interoperability of cyber–physical systems in Industry 4.0 [1]. Interoperability can be defined as the ability of two or more systems to exchange, understand and consume information [10].

A cyber–physical system is presented in the form of a certain metamodel of the interaction of the nth number of cyber–physical systems. The components of a cyber–physical system are sets of physical and cyber components. A cyber–physical system is a structural agglomeration of these elements, which can also include other cyber–physical subsystems, forming a new cyber–physical system.

There are two relationships of different nature between the components of a cyber–physical system:

1) The relationship between the physical components of cyber–physical subsystems, implying the transfer of any kind of physical object.
2) The relationship of communication between cyber components, which means the presence of information and/or control channel between the components.

The components of the system perform certain functions depending on their role in the system, and following this, they have some input and output signals that capture information flows between this element and the elements with which it is associated with the help of the above relations.

The definition of the system, in this case, will be as follows. Let there be a system

$$\{\mathbf{P}, \mathbf{C}, \mathbf{CPS}, \mathbf{R}^\mathbf{P}, \mathbf{R}^\mathbf{c}\} \qquad (13.1)$$

where **P** is the set of physical components that are related by the relationship $\mathbf{R}^\mathbf{P}$;

1) **C**—cyber components that are related by the relationship $\mathbf{R}^\mathbf{c}$;
2) **CPS**—a cyber–physical system, including as a child, having the same properties as the parent.

Figure 13.2 shows an example of building a composite cyber–physical system. It includes three child cyber–physical systems (CPS1, CPS2, CPS3) as part of one parent (CPS4); each of the systems has at least one physical (P) and one cyber component (C); communication occurs only between components of the same type (physical component physical component, cyber component ⇔ cyber component).

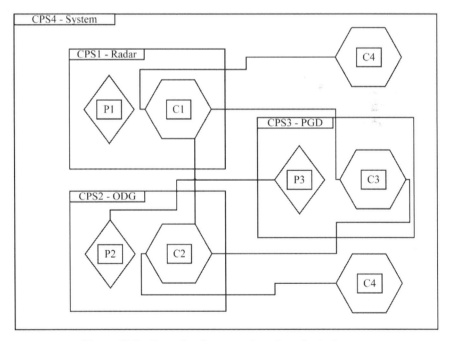

Figure 13.2   Example of a composite cyber–physical system

The resulting metamodel of a cyber–physical system is further described in UML notation as shown in Figure 13.3.

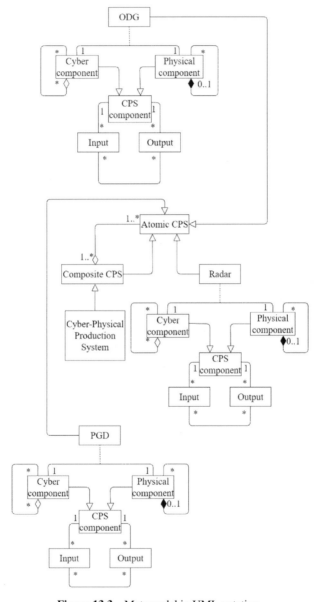

Figure 13.3   Metamodel in UML notation

The main focus is on the "form a single CPS" relationship between the "Cyber Component" and "Physical Component" classes, which relate to each other as "one to one" (one cyber component to one physical component).

As stated in the metamodel, the "Cyber Component" and "Physical Component" classes are subclasses of the "CPS Component" base class, which contains the fundamental properties. Each "CPS Component" can either act as a communicator between two components, or be an independent unit and communicate with other homogeneous components. Communication is done via input and output interfaces.

The next approach to the description of cyber–physical systems is the use of Business Process Modeling Notation (BPMN 2.0), taking into account the specifics of cyber–physical systems. Classifying new metaclasses requires a deep understanding of BPMN metaclasses [2, 11–14].

For this, new abstract classes of cyber–physical systems are introduced as a set of BPMN subclasses. This concept will allow the developer of a cyber–physical system to describe the logic of internal processes, various types of actions inside and real objects.

In a cyber–physical system, business processes must be defined according to a centralized approach to execute business logic. BPMN defines the behavior of different participants using different pools.

The first pool is for physical processes. The second pool is for cyber processes. The third pool is a central controller that directly coordinates the interaction between cyber and physical processes. Therefore, the BPMN4CPS process must consist of at least three pools: physical, controller, and cyber.

The separation into the processing logic, executed at the physical level, the controller and the cyber part, forces the modeler to explicitly indicate the interaction between them using messages. Data objects provide information about what physical actions need to be performed and/or what information they generate. Data objects are the temporarily stored data of a running process instance.

Figure 13.4 shows an example illustrating the execution of a closed loop of a cyber–physical task in the context of three pools: physical, controller, and cyber process pools.

In cyber–physical systems, there are three types of tasks: cyber, manual, and physical. These types of actions are also covered in the BPMN metamodel.

Cybertask is an extension of the standard BPMN service task. This task is performed at the cyber level and, in essence, is the support for the controller in the formation of the control action. For example, solving optimization

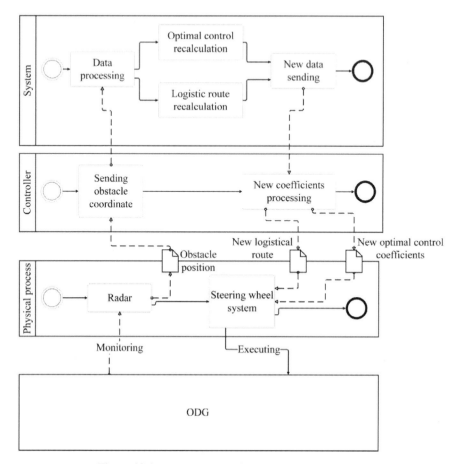

Figure 13.4   A closed-loop of a cyber–physical system

problems, the cyber level finds a solution to a certain problem, which it sends to the controller.

A physical task is an atomic activity within a process. The physical process can be divided into two parts:

1) retrieve sensor data and send them to the controller level;

2) receipt by executive devices of control actions from the controller.

In this model, the controller plays the role of a kind of "mediator" and "aggregator". The role of the "mediator" is that the controller receives data from the sensors and the cyber layer and sends it to the physical layer.

The role of the "aggregator" is that having stabilized the system in the event of a disturbance and having received data from the cyber level, the controller brings this data into a single form that is understandable for executive devices.

The third approach to modeling cyber–physical systems is the model-based design method for constructing dynamic systems [3, 15–17].

Designing a complex cyber–physical system, especially with heterogeneous subsystems distributed over networks, is challenging. Commonly used design methods are complex and include mathematical modeling of physical systems, formal computation models, modeling of heterogeneous systems, software synthesis, verification, validation, and testing.

The proposed algorithm consists of 10 steps:

1) problem statement;
2) description of physical processes by equations;
3) description of minimum/maximum parameters and/or restrictions;
4) synthesis of control;
5) selection of computing software;
6) choice of hardware;
7) modeling of the process;
8) design;
9) software development;
10) performance check and testing.

Let's briefly describe each step of the algorithm.

Problem statement.

A preparatory step that requires maximum attention. In this step, a basic understanding of the problem is formed. For the most part, this step involves a written description of the problem, without the use of complex mathematics.

Description of physical processes by equations.

In this step, an understanding of the relevant physical systems is formed, such as the environment in which the cyber–physical system is located, or the physical processes that need to be controlled.

Models of physical processes are simplified representations of real systems and usually take the form of systems of differential equations.

Description of minimum/maximum parameters and/or restrictions;

At this stage, need to highlight the constants, adjustable parameters, and variables that need to be controlled.

Also need to determine the values that characterize the system and/or physical processes, such as minimum and maximum values, initial states, final states.

Synthesis of control.

This step requires determining the conditions under which the physical processes are controllable and synthesizing the control law. It is also necessary to define the parameters of time delays, sampling rate, and quantization so that the physical dynamics of interest can be accurately measured and appropriately controlled.

Selection of computing software.

The formal computation model defines semantics that often leads to greater analyzability and the ability to model cyber–physical systems using heterogeneous modeling tools.

Models described by formal computation models are easier to analyze in terms of determinism, runtime, state reachability, and memory usage.

Choice of hardware.

When having a basic understanding of all the components of the system, it is necessary to select the components from which the production model will be built.

The selection of equipment should be based on the fact that the finished model should adequately interact with the simulated physical systems and implement the control law in a high-precision measure.

For each component, consider its input and output bandwidth, input-to-output latency, power consumption, resolution, and measurement speed, as well as mechanical parameters such as form factor, electrical noise rejection, durability, and lifespan.

Modeling of the process.

This step allows us to verify the adequacy of the developed model of the cyber–physical system. It is necessary to collect quantitative and qualitative parameters to make sure that the optimal control law takes the system from the initial state to the final one. Should also check that all limits and ranges for each module are met.

Design.

This involves assembling the production model itself according to the specifications. It is best to build a model so that tests can be run at each step to ensure that the individual components and subsystems are performing as planned. This, in turn, facilitates the transition from simulation to testing.

Software development.

The creation of the controller software is a critical step in this process. The resulting model of the system in the previous steps, in some modeling environments, can be easily converted into executable code.

If the synthesis of the code is not possible the handwritten code must carefully follow the chosen computation models. Code execution must be verified on a real object because code generators and compilers can introduce software artifacts.

Performance check and testing.

Computing systems can be isolated from physical systems through hardware testing, while programmable hardware, such as controllers or FPGAs, simulates feedback from physical or other computational processes. Runtime and latency measurements can be used to refine previous models, and test results can indicate errors in modeling or implementation.

Formal validation and validation provide insight into the behavior of an algorithm over some or all combinations of its inputs or over time.

13.3 Problem Formulation

The concept of logistics includes several components: goods; objects, that deliver goods (trucks, ships, planes, trains); and places where goods are delivered (end consumer, intermediate warehouse, loading/unloading point, ports, railway junctions).

Let's designate each of the elements. Objects that deliver goods (ODG), places where goods are delivered (PGD). Now the logistics management systems receive information on the status of the cargo from PGD. These systems do not reflect what happens between individual PGDs. In the event of an emergency on the ODG route (the ship got into a storm, the truck got stuck in a traffic jam), the system will not display this information and will not generate the appropriate control action to optimize the delivery of the cargo.

Let us introduce a system of connections within this system:

1) communication "ODG" – "PGD";
2) communication "ODG" – "ODG";
3) communication "PGD" – "PGD";

Figure 13.5   An example of inter-component interaction within a cyber–physical system

4) Communication "System" – "PGD";
5) Communication "System" – "ODG".

Consider the following cyber–physical system:

1) ODG received a logistic route for the delivery of cargo from PGD1 to PGD2 and an algorithm for optimal, in terms of energy consumption, control from a centralized cloud server;
2) ODG begins to move, in parallel tracking external disturbance on the route (for example, using a radar);
3) In the event of a situation in which move along the planned route is impossible, ODG sends the relevant information to the centralized cloud;
4) The cloud, in turn, recalculates the logistic route and the algorithm for the optimal control in terms of energy consumption;
5) Having received the updated data, ODG continues its journey to PGD2.

Such a dynamic response to changes in the external environment will minimize the problem of untimely delivery of cargo. Figure 13.5 shows an example of how ODG communicates with the cloud.

13.4 Building Models of a Logistic Cyber–Physical System

We use the cyber–physical system metamodel method. Figure 13.6 shows the inter-component interaction of a cyber–physical system, proposed in the formulation of the problem.

An instance of a cyber–physical system class is denoted as a compound ellipse, where the solid half indicates the physical component and the dotted

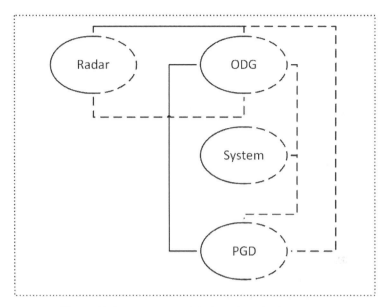

Figure 13.6 The metamodel of a cyber–physical system

half indicates the cyber component. Accordingly, the "form a single CPS" relation is represented by combining the half ellipses into a whole ellipse.

Some properties of a cyber–physical system can be directly deduced from Figure 13.6 by the type of connection between cyber–physical components. For example, a cyber–physical component that has no physical inputs and outputs and only interacts in the cybernetic dimension can be thought of as a computing node. In this notation, "System" is a computational node.

In addition, a component that receives only physical input and provides only cyber output can be called a sensor. A radar that interacts with ODG through physical signals and provides a cyber response on the state of the road. In the metamodel, it can be referred to as sensors.

Figure 13.7 shows the implementation of a model of a logistics cyber–physical system in BPMN4CPS terminology.

Here, the extended BPMN notation divides the processing logic into three parts: the cyber part, the controller, and the physical part.

These parts interact with each other through the exchange of messages. The set of components here is as follows

1) ODG;
2) radar installed on the ODG;

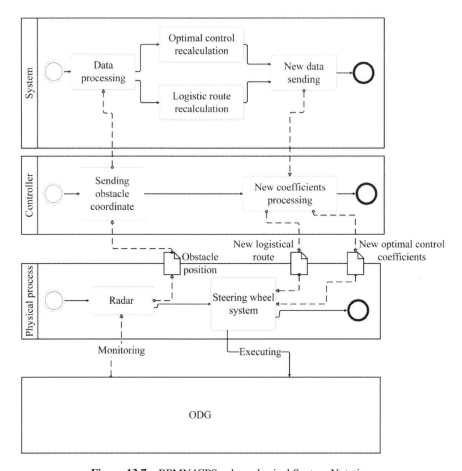

Figure 13.7 BPMN4CPS cyber–physical System Notation

3) controller;

4) the cyber component of the system responsible for data processing (cloud computing resources).

This notation clearly shows the relationship between components. ODG is a basic component.

The level of the physical process is a radar as a sensor device and a steering wheel system as an executive component. The radar tracks objects in the environment and the distance to them. Having received information about the next object from the radar, the controller sends data to the cloud computing center. The result of data processing by the cloud computing center is sent to the controller, which transmits control actions to the input of the steering wheel control system to change the direction and/or speed of movement, according to the current situation.

The level of controller interaction is a set of hardware. The main task of this layer is to reconcile data between the physical and cyber component layers. Having received data from the radar, the controller must bring them into an acceptable format for a computing resource. The controller performs a similar function, having received data from a computing resource, with only one difference, which brings the data into a format that is acceptable for an executive device at the physical level.

And finally, the level of cyber processes. The main component of this layer is the cloud-computing resource. Receiving data for analysis, namely the state of the environment received from the radar, a decision is made whether it is necessary to recalculate the logistic route and/or the optimal control coefficients for the further following to the PGD. If necessary, the corresponding algorithms perform the calculation of the refined route and the synthesis of the optimal traffic control. Subsequently, this data is sent to the controller level for subsequent transmission to the actuators.

Now we turn to the construction of the model by the method proposed in [3].

Problem statement.

Consider the following problem:

1) ODG received a logistic route for cargo delivery from PGD_1 to PGD_2 and an algorithm for optimal, in terms of energy consumption, control from a centralized cloud server;
2) ODG begins to move, in parallel tracking external influences on the route (for example, using a radar);
3) In the event of a situation in which move along the planned route is impossible, ODG sends the relevant information to the centralized cloud;
4) The cloud, in turn, recalculates the logistic route and the algorithm for the optimal control in terms of energy consumption;
5) Having received the updated data, ODG continues its journey to PGD_2.

Description of physical processes by equations.

Table 13.1 describes the main physical definitions and constants of ODG.

Table 13.1   Physical definitions of ODG

Symbol	Notation
L	ODG width
r_r	Right wheel radius
r_l	Left wheel radius
ω_r	Right wheel angular velocity
ω_l	Left wheel angular velocity
V_r	Right wheel velocity
V_l	Left wheel velocity
ω	Overall angular velocity
θ	The rotation angle of the ODG
φ	The rotation angle of the wheel

Let φ be the angle of rotation of the wheel. It is known from kinematics that the angular velocity is expressed by the following equation:

$$\omega = \frac{d\varphi}{dt} \tag{13.9}$$

For the right wheel, the angular velocity is:

$$\omega_r = \frac{d\varphi_r}{dt}. \tag{13.10}$$

The angular velocity of the left wheel is:

$$\omega_l = \frac{d\varphi_l}{dt}. \tag{13.11}$$

Having both components of the angular velocity of the wheels, it is possible to express the total angular velocity of the ODG, ω_{ODG}:

$$\omega_{ODG} = r\frac{\omega_r - \omega_l}{L}. \tag{13.12}$$

From the angular velocity it is possible to determine the overall linear velocity of the ODG, V_{ODG}:

$$V_{ODG} = r\frac{\omega_r + \omega_l}{2}. \tag{13.13}$$

Then the projections of the velocity can be defined as:

$$V_x = V\cos(\omega) = \dot{x} \tag{13.14}$$

$$V_y = V\sin(\omega) = \dot{y} \tag{13.15}$$

Description of minimum/maximum parameters and/or restrictions.

The main problem of logistics is the late delivery of goods. Based on this, the first limitation is introduced: the final time of cargo delivery from PGD_1 to PGD_2. It should also be mentioned that it is not possible to achieve full compliance with the time frame. The obvious proof of this is that proactively responding to environmental changes and recalculating the logistics route also takes a certain amount of time. So, it is rational to introduce admissible boundary values for the final delivery time of the cargo.

The movement of the ODG along the route is the main task, but one should also keep in mind that significant financial costs of logistics are the cost of fuel or electricity consumed by the ODG. This leads to the second criterion—the optimality of following the ODG in terms of fuel or electricity consumption.

Synthesis of control.

Control synthesis is carried out by the approach described in [4, 18]. Table 13.2 lists the conventions and their descriptions that will be used below.

Consider a linear model [5, 6]:

$$\begin{cases} \dot{\mathbf{x}}(t) = \mathbf{A}(t)\mathbf{x}(t) + \mathbf{B}(t)\mathbf{u}(t), \\ \mathbf{y}(t) = \mathbf{C}(t)\mathbf{x}(t). \end{cases} \qquad (13.17)$$

Table 13.2 Optimal tracking system notation

Symbol	Notation
A,B,C,D	State-space matrices
e	Difference between desirable and real outputs
F	The terminal constant of criterion
H	Hamiltonian
J	Optimal criterion
K	Solution of the Riccati equation
p	Additional vector function
Q,R	Weighting matrices of criterion
u	Control law
x	Vector of states of the state-space model
y	Vector of outputs of the state-space model
z	Desired output

And the energy-optimal consumption criterion:

$$J = \frac{1}{2}e^T(T)\mathbf{Fe}(T) + \frac{1}{2}\int_{t_0}^{T}[e^T(t)\mathbf{Q}(t)e(t) + u^T(t)\mathbf{R}(t)\mathbf{u}(t)]dt \quad (13.18)$$

where

$$e(t) = \mathbf{z}(t) - \mathbf{y}(t) \quad (13.19)$$

$e(t)$ – difference between desirable $\mathbf{z}(t)$ and real $\mathbf{y}(t)$ outputs. Rewrite $e(t)$ as a function of $\mathbf{z}(t)$ and $\mathbf{x}(t)$:

$$e(t) = \mathbf{z}(t) - \mathbf{C}(t)\mathbf{x}(t) \quad (13.20)$$

Substitute the error (13.19) into the optimality criterion (13.18) and find the Hamiltonian H of the system:

$$H = \frac{1}{2}[\mathbf{z}(t) - \mathbf{C}(t)\mathbf{x}(t)]^T\mathbf{Q}(t)[\mathbf{z}(t) - \mathbf{C}(t)\mathbf{x}(t)] + \frac{1}{2}\mathbf{u}^T(t)\mathbf{R}(t)\mathbf{u}(t)p_0(t)$$
$$+ \mathbf{A}(t)^T\mathbf{x}(t)^T\mathbf{p}(t) + \mathbf{B}(t)^T\mathbf{u}(t)^T\mathbf{p}(t). \quad (13.21)$$

The additional vector function $\mathbf{p}(t)$ is described by:

$$\mathbf{p}(t) - \begin{bmatrix} p_0(t) \\ p_1(t) \\ ... \\ p_{n+1}(t) \end{bmatrix} \quad (13.22)$$

so,

$$\mathbf{p}(T) = \begin{bmatrix} -1 & 0 & ... & 0 \end{bmatrix}^T \quad (13.23)$$

and can be found from the following equation:

$$\dot{\mathbf{p}}(t) = -\frac{\partial H}{\partial \mathbf{x}(t)} \quad (13.26)$$

Since there are no control restrictions:

$$\max_u H : \frac{\partial H}{\partial \mathbf{u}(t)} = 0 \quad (13.27)$$

so,

$$\frac{\partial H}{\partial \mathbf{u}(t)} = \mathbf{R}(t)\mathbf{u}(t) + \mathbf{B}^T(t)\mathbf{p}(t) = 0 \quad (13.28)$$

where optimal control is written as:

$$\mathbf{u}(t) = -\mathbf{R}^{-1}(t)\mathbf{B}^T(t)\mathbf{p}(t) \tag{13.29}$$

Differentiate (13.26) and obtain:

$$\dot{\mathbf{p}}(t) = -\mathbf{C}^T(t)\mathbf{Q}(t)\mathbf{C}(t)\mathbf{x}(t) - \mathbf{A}^T(t)\mathbf{p}(t) + \mathbf{C}^T(t)\mathbf{Q}(t)\mathbf{z}(t) \tag{13.30}$$

Substitute the resulting control (13.29) into the state Equation (13.17):

$$\dot{\mathbf{x}}(t) = \mathbf{A}(t)\mathbf{x}(t) - \mathbf{B}(t)\mathbf{R}^{-1}(t)\mathbf{B}^T(t)\mathbf{p}(t) \tag{13.31}$$

Let the relationship between $\mathbf{x}(t)$ and $\mathbf{p}(t)$ be written by the following equation:

$$\mathbf{p}(t) = \mathbf{K}(t)\mathbf{x}(t) - \mathbf{g}(t). \tag{13.32}$$

Therefore, state Equation (13.31) can be rewritten as:

$$\dot{\mathbf{x}}(t) = [\mathbf{A}(t) - \mathbf{S}(t)\mathbf{K}(t)]\mathbf{x}(t) + \mathbf{S}(t)\mathbf{g}(t) \tag{13.33}$$

where,

$$\mathbf{S}(t) = \mathbf{B}(t)\mathbf{R}^{-1}(t)\mathbf{B}^T(t) \tag{13.34}$$

Differentiate the vector function $\mathbf{p}(t)$ in (13.32):

$$\dot{\mathbf{p}}(t) = \dot{\mathbf{K}}(t)\mathbf{x}(t) + \mathbf{K}(t)\dot{\mathbf{x}}(t) - \dot{\mathbf{g}}(t) \tag{13.35}$$

and substitute (13.33) in the (13.35):

$$\dot{\mathbf{p}}(t) = [\dot{\mathbf{K}}(t) + \mathbf{K}(t)\mathbf{A}(t) - \mathbf{K}(t)\mathbf{S}(t)\mathbf{K}(t)]\mathbf{x}(t) + \mathbf{K}(t)\mathbf{S}(t)\mathbf{g}(t) - \dot{\mathbf{g}}(t) \tag{13.36}$$

Substitute (13.32) into the equation of optimal control (13.29) and obtain:

$$\mathbf{u}(t) = -\mathbf{R}^{-1}(t)\mathbf{B}^T(t)[\mathbf{K}(t)\mathbf{x}(t) - \mathbf{g}(t)] \tag{13.37}$$

Where $\mathbf{K}(t)$ – is a real, symmetric, positively defined matrix of dimension $n \times n$, which is the solution of the obtained Riccati equation:

$$\mathbf{K}(t) = -\mathbf{K}(t)\mathbf{A}(t) - \mathbf{A}^T(t)\mathbf{K}(t) + \mathbf{K}(t)\mathbf{B}(t)\mathbf{B}^{-1}(t)\mathbf{B}^T(t)\mathbf{K}(t)$$
$$- \mathbf{C}^T(t)\mathbf{Q}(t)\mathbf{C}(t) \tag{13.38}$$

with a boundary condition:

$$\mathbf{K}(T) = \mathbf{C}^T(T)\mathbf{F}\mathbf{C}(T) \tag{13.39}$$

The vector $\mathbf{g}(t)$ (with n components) is the result of the solution of the differential equation:

$$\dot{\mathbf{g}}(t) = -[\mathbf{A}(t) - \mathbf{B}(t)\mathbf{R}^{-1}(t)\mathbf{B}^T(t)\mathbf{K}(t)]^T\mathbf{g}(t) - \mathbf{C}^T(t)\mathbf{Q}(t)\mathbf{z}(t) \quad (13.40)$$

with an appropriate boundary condition:

$$\mathbf{g}(T) = \mathbf{C}^T(T)\mathbf{F}\mathbf{z}(T) \quad (13.41)$$

Therefore, by solving the Riccati Equation (13.38) find the $\mathbf{K}(t)$ matrix also, solving the differential Equation (13.40) in the reverse time will give the value of the entire vector $g(t)$. Substituting the obtained values of $\mathbf{K}(t)$ and $\mathbf{g}(t)$ into Equation (13.37), obtain the optimal input control of the system, which minimizes criterion (13.18) and solves the tracking problem.

Selection of computing software.

Modeling of the physical component of the system will be performed in the MATLAB package using the Automated Driving Toolbox [7] add-on. The Automated Driving Toolbox provides algorithms and tools for the design, simulation, and testing of advanced driver-assistance systems (ADAS) and autonomous driving. It is also possible to design and test computer vision systems, lidar/radar sensing systems, in addition to path planning capabilities and the development of vehicle controllers.

Choice of hardware.

Based on the task at hand, a cyber–physical system must have a certain computing center, which can be represented as a cloud server. The cloud provider can be Microsoft Azure. For the functioning of a cyber-component system, it is enough to have a server with the following parameters:

1) Linux operating system;
2) 8 GB of RAM;
3) 4 virtual processor cores;
4) 64 GB of storage.

These parameters correspond to the virtual machine of the F4s v2 series [8] optimized for computing.

Modeling of the process.

MATLAB is the main tool for modeling a scenario that suits the task at hand. First, using the Automated Driving Toolbox is easier with MATLAB. Secondly, modeling of cyber–physical interaction through a computing "cloud" can be achieved through MATLAB Instrument Control Toolbox [9]. Based on this, using the MATLAB ecosystem for modeling is preferable.

The subsequent steps "Design", "Software development" and "Performance check and testing" are associated with the real implementation of the system and require a real self-driving ODG traveling from PGD_1 to PGD_2. For research purposes, these steps will be skipped and the results will be collected from the "Modeling of the process" step.

13.5 Simulation Results

Figure 13.8 shows the initial logistic route from PGD_1 to PGD_2 as received by the ODG from the centralized cloud.

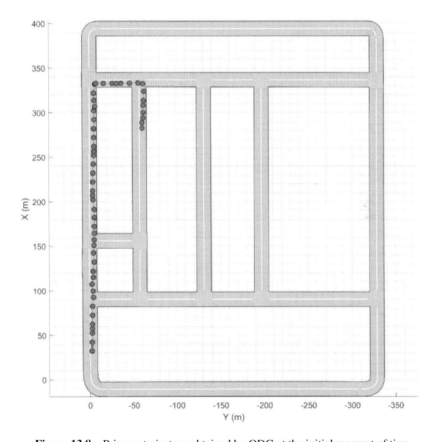

Figure 13.8 Primary trajectory obtained by ODG at the initial moment of time

```
1    [scenario, axes] = plotDrivingScenario('map.png');
2
3    occupancyGrid = createOccupancyGrid('map.png', 225);
4
5    ccConfig = inflationCollisionChecker('InflationRadius', 1);
6    costmap = vehicleCostmap(occupancyGrid, 'CollisionChecker', ccConfig);
7    plot(costmap);
8    xy = handleStartAndEndPointsFromGrid();
9    close 2
10   planner = pathPlannerRRT(costmap, 'ConnectionDistance', 25, ...
11       'MinTurningRadius', 1);
12   path = plan(planner, xy(1,:), xy(2,:));
13   lengths = 0:10:path.Length;
```

Figure 13.9   Script for constructing the primary trajectory

The algorithm for constructing the primary trajectory written in MAT-LAB is shown in Figure 13.9.

The script consists of three parts:

1) building a road model using the Automated Driving Toolbox;
2) converting the road model into the so-called cost map;
3) the last step is to find the trajectory between two points using the RRT (Rapidly-exploring random tree) algorithm.

Then the ODG began following the logistic route. While driving, external influences on the roads, such as traffic jams or obstacles, are continuously monitored.

Figure 13.10 shows the state of the cyber–physical system at the time of finding an obstacle on the path of the ODG. A MATLAB script that emulates the operation of the radar and the movement of the ODG is shown in Figure 13.11.

As soon as the radar detects an obstacle, the data is sent by the controller to the cloud. The data received from the radar is an array of GPS coordinates of the obstacle. The computing server calculates a new trajectory, taking into account the found obstacle, and sends a new route and optimal control coefficients to the ODG controller. Figure 13.12 shows a new logistics route after actively responding to an obstacle in the way.

The obstacle tracking algorithm is shown in Figure 13.13.

Figure 13.14 shows the state of the cyber–physical system after obtaining a new logistics route and optimal control coefficients. ODG follows PGD_2.

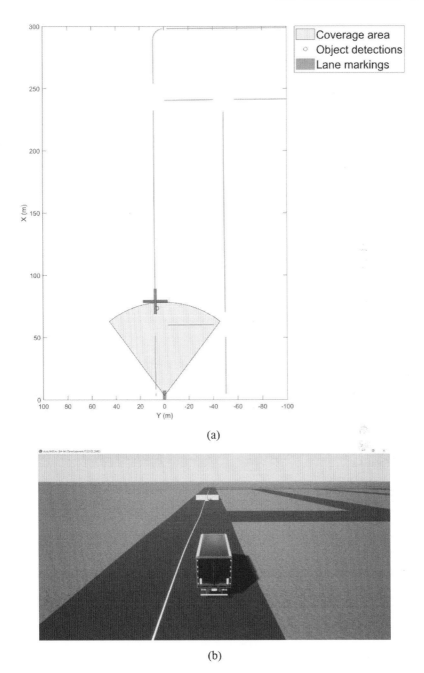

(a)

(b)

Figure 13.10 The state of the cyber–physical system at the moment the radar detects an obstacle (the state of the radar (a), the state of the ODG (b)).

```
1    [radar, lmPlotter, olPlotter, detPlotter] = addRadar();
2
3    runScenario(scenario, truck, radar, ...
4        lmPlotter, olPlotter, detPlotter, ...
5        planner);
```

Figure 13.11   MATLAB script responsible for the operation of the radar and movement of the ODG.

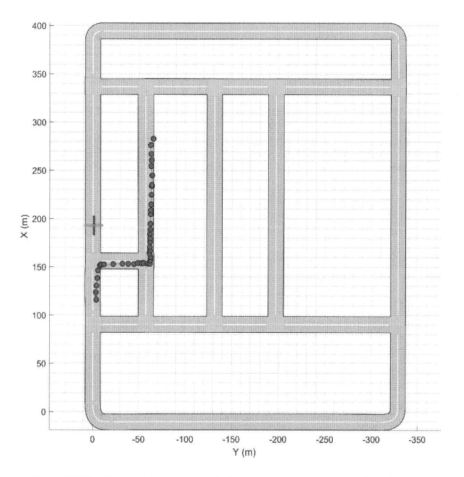

Figure 13.12   New logistics route recalculated due to the occurrence of an obstacle.

```
 1   function [radar, lmPlotter, olPlotter, detPlotter] = addRadar()
 2       radar = radarDetectionGenerator('DetectionCoordinates', ...
 3           'Sensor Cartesian', 'MaxRange', 75, ...
 4           'RangeResolution', 10, 'AzimuthResolution', 10, ...
 5           'FieldOfView', [75 15], 'UpdateInterval', 0.01, ...
 6           'HasRangeRate', false);
 7
 8       bep = birdsEyePlot('XLim', [0 300], 'YLim', [-100 100]);
 9
10       caPlotter = coverageAreaPlotter(bep, 'DisplayName', ...
11           'Coverage area', ...
12           'FaceColor', 'blue');
13       detPlotter = detectionPlotter(bep, 'DisplayName', 'Object detections');
14       lmPlotter = laneMarkingPlotter(bep, 'DisplayName', 'Lane markings');
15       olPlotter = outlinePlotter(bep);
16
17       plotCoverageArea(caPlotter, radar.SensorLocation, ...
18           radar.MaxRange, radar.Yaw, radar.FieldOfView(1));
19   end
20
21   function [] = runScenario(scenario, truck, radar, ...
22       lmPlotter, olPlotter, detPlotter)
23       while advance(scenario)
24           [lmv ,lmf] = laneMarkingVertices(truck);
25           plotLaneMarking(lmPlotter, lmv, lmf)
26           tgtpose = targetPoses(truck);
27           [obdets, numOfDetections, obValid] = radar(tgtpose, ...
28               scenario.SimulationTime);
29           [objposition, objyaw, ...
30               objlength, objwidth, ...
31               objoriginOffset, color] = targetOutlines(truck);
32           plotOutline(olPlotter, objposition, objyaw, ...
33               objlength, objwidth, ...
34               'OriginOffset', objoriginOffset, ...
35               'Color', color)
36           if obValid
37               detPos = cellfun(@(d)d.Measurement(1:2), ...
38                   obdets, 'UniformOutput', false);
39               detPos = vertcat(zeros(0,2), cell2mat(detPos')');
40               plotDetection(detPlotter, detPos)
41           end
42           if numOfDetections > 0
43               sendDataToCloud();
44           end
45       end
46   end
```

Figure 13.13 MATLAB script showing the characteristics of the radar and the obstacle tracking algorithm.

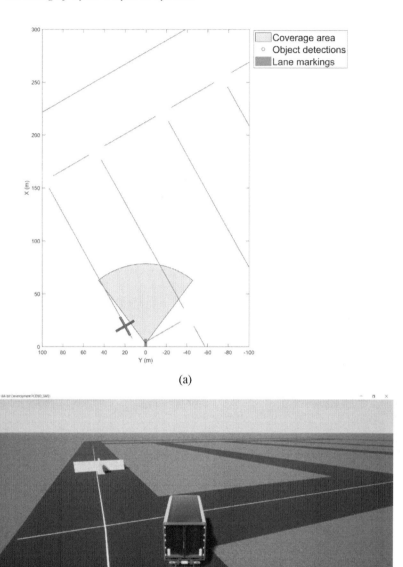

(a)

(b)

Figure 13.14 The state of the cyber–physical system after receiving a new logistic route (the state of the radar (a), the state of the ODG (b).

13.6 Conclusion

The chapter discusses the most promising in terms of completeness of description approaches to modeling cyber–physical systems. Not only the analysis of modeling approaches to cyber–physical systems is carried out but also the results of building a model for the selected object are presented.

The FCA metamodel approach is excellent for building a high-level view of the cyber–physical system under development. The result gives a clear understanding of what systems/functional blocks the cyber–physical system consists of and how these nodes are interconnected (using cyber or physical links). Based on the types of incoming and/or outgoing links, you can determine the type of object. For example, a cyber–physical component that has no physical inputs and outputs and only interacts in the cybernetic dimension can be thought of as a computing node. In addition, a component that receives only physical input and provides only cyber output can be called a sensor.

The BPMN approach is great for the next design phase—describing the internal interactions between systems/functional blocks. If the previous approach showed which components in the system are physical and which are cyber, then this approach illustrates in detail how these components interact with each other. What is the nature of the original physical signal from the steering wheel system to the ODG? Or, what type of cyber signal comes from the radar to the level of controller interaction. The advantage of this approach is precisely in detailing the "hidden" interactions between components and how these interactions form a closed loop of data flow from cyber elements to physical ones and vice versa.

The last so-called model-based approach, perhaps, in part, is a continuation of the previous one. However, it should be performed most effectively in conjunction with it. There are many model-based steps worth completing before starting to use the BPMN-based method. For example, "problem statement," "description of physical processes by equations," "Description of minimum/maximum parameters and/or restrictions." Completing these steps will adequately describe the internal interactions between the components. The main disadvantage of this approach is the absence of any graphical representations of intermediate results in the process of building a cyber–physical system. In turn, its great advantage is algorithmic. Performing each step, you can get the entire set of data, characteristics, details of technical implementation for adequate testing of the performance of the developed cyber–physical system.

Each approach has its advantages and disadvantages. The authors see the sense in some symbiosis of these approaches. The first approach is ideal for the stage of initial prototyping of the problem that has arisen. Following the first steps of the third approach "description of physical processes by equations" and "Description of minimum/maximum parameters and/or restrictions", it is possible to collect the necessary set of values/data to describe the internal interactions between the components of a cyber–physical system following the second approach. Further, the development of a cyber–physical system should be continued according to the model-based approach algorithm.

References

[1] D. Morozov, et al. 'Multi-paradigm modeling of cyber–physical Systems', IFAC-PapersOnLine, 51(11), 2018, pp. 1385–1390. DOI: https://doi.org/10.1016/j.ifacol.2018.08.334

[2] Graja, et al. 'BPMN4CPS: A BPMN Extension for Modeling cyber–physical Systems', 2016 IEEE 25th International Conference on Enabling Technologies: Infrastructure for Collaborative Enterprises (WETICE), 2016, DOI: https://doi.org/10.1109/wetice.2016.41

[3] J. C. Jensen, et al. 'A model-based design methodology for cyber–physical systems', 2011 7th International Wireless Communications and Mobile Computing Conference, 2011, DOI: https://doi.org/10.1109/iwcmc.2011.5982785

[4] A. Holovatenko, et al. 'Energy-Efficient Path-Following Control System of Automated Guided Vehicles', J Control Autom Electr Syst 32, 2021, pp. 390–403. DOI: https://doi.org/10.1007/s40313-020-00668-8

[5] A. Chikrii. Control of Moving Objects in Condition of Conflict. In: Kuntsevich, V.M. et al. (Eds). Control Systems: Theory and Applications. Series in Automation, Control and Robotics, River Publishers, Gistrup, Delft, 2018, pp. 17–43.

[6] V. F. Gubarev, et al. Model Predictive Control for Discrete MIMO Linear Systems. In: Kondratenko, Y.P., Chikrii, A.A., Gubarev, V.F., Kacprzyk, J. (Eds) Advanced Control Techniques in Complex Engineering Systems: Theory and Applications. Dedicated to Prof. V.M. Kuntsevich. Studies in Systems, Decision and Control, Vol. 203. Cham: Springer Nature Switzerland AG, 2019, pp. 63-81. DOI: https://doi.org/10.1007/978-3-030-21927-7_4

[7] MathWorks, Automated Driving Toolbox. https://www.mathworks.co m/products/automated-driving.html, n.d.

[8] Microsoft Azure, Linux Virtual Machines Pricing. https://azure.micros oft.com/en-us/pricing/details/virtual-machines/linux, n.d.

[9] MathWorks, Instrument Control Toolbox. https://www.mathworks.com/ products/instrument.html, n.d.

[10] IEEE: Standard Computer Dictionary, 1990. A Compilation of IEEE Standard Computer Glossaries. In NY. 610-1990. ISBN: 1559370793.

[11] S. Meyer, et al. Internet of things architecture iot-a project deliverable d2.2 concepts for modelling iot-aware processes.

[12] S. Meyer, et al. "Internet of things-aware process modeling: Integrating iot devices as business process resources," in 25th International Conference on Advanced Information Systems Engineering, 2013, pp. 84–98.

[13] S. Meyer, et al., "The things of the internet of things in BPMN," in Advanced Information Systems Engineering Workshops, 2015, pp. 285–297.

[14] K. Sperner, et al., "Introducing entity-based concepts to business process modeling," in Business Process Model and Notation - Third International Workshop, BPMN 2011, Lucerne, Switzerland, November 21-22, 2011. Proceedings, 2011, pp. 166–171.

[15] C. Brooks, et al., "Model Engineering Using Multimodeling", in 1st International Workshop on Model Co-Evolution and Consistency Management (MCCM 08), September 2008.

[16] K. Balasubramanian, et al., "Developing Applications Using Model-Driven Design Environments," IEEE Computer, vol. 39, no. 2, pp. 33, February 2006.

[17] G. Karsai, et al., "Model-Integrated Development of Embedded Software," Proceedings of the IEEE, vol. 91, no. 1, January, 2003.

[18] J. Varela-Aldás, et al. (2021) Application for the Cooperative Control of Mobile Robots with Energy Optimization. In: Kurosu M. (eds) Human-Computer Interaction. Interaction Techniques and Novel Applications. HCII 2021. Lecture Notes in Computer Science, vol. 12763. Springer, Cham. DOI: https://doi.org/10.1007/978-3-030-78465-2_25.

14

Reliability Control of Technical Systems based on Canonical Decomposition of Random Sequences

I. Atamanyuk[1,2], Y. Kondratenko[3], and M. Solesvik[4]

[1]Warsaw University of Life Sciences, Nowoursynowska 166,
02-787 Warsaw, Poland
[2]Mykolayiv National Agrarian University, Georgy Gongadze street,
Mykolaiv, 9, 54020, Ukraine
[3]Petro Mohyla Black Sea National University, 68 Desantnykiv street,
10, Mykolaiv, 54003, Ukraine
[4]Western Norway University of Applied Sciences, Inndalsveien 28,
5063 Bergen, Norway
E-mail: atamanyuk@mnau.edu.ua; y_kondrat2002@yahoo.com;
marina.solesvik@hvl.no

Abstract

A method for determining the suitability of a technical object for further operation based on the information about the current state and the history of its functioning during an arbitrary period of time is obtained. The check is carried out on the basis of the analysis of the state of the object under study at future points in time and the value of the posterior probability of no-failure operation. The method for predicting the individual reliability of objects is based on the canonical decomposition of a random sequence describing the change in the value of the controlled parameter over time. As an estimate of the future state, the conditional mathematical expectation is used; to estimate the probability of no-failure operation, statistical modeling of a random sequence beyond the observation area is performed.

The method can also be used to control the reliability of the objects that are characterized by many parameters (in this case, the vector canonical decomposition is used). The proposed approach does not impose any restrictions on the random sequence of change of the controlled parameters and, thus, makes it possible to fully take into account the peculiarities of the functioning of the object under study.

The work presents block diagrams that characterize the features of using the proposed method. Expressions for assessing the quality of predicting the state of the controlled object are obtained.

Keywords: Random sequences, canonical decomposition, reliability control, extrapolation of random sequences.

14.1 Introduction

The main features of the development of modern technology [1] are increasing the degree of automation [2, 3]; increasing loads [4, 5], speeds, temperatures, pressures; reduction in size and weight; increasing requirements for accuracy and efficiency of functioning [6, 7]; combining individual units into systems with a single control [8, 9], etc. The increase in complexity and requirements for performance indicators inevitably leads to the need to increase requirements for the reliability and durability of equipment (especially of critical infrastructure systems [10–12]).

The problem of individual prediction [13–16] of state and reliability is of particular interest. Its solution allows not only to obtain an estimate of the reliability of each specific sample of products, but also, in the presence of developed diagnostic support, to move from maintenance by time or resource to planning operation based on the actual state.

Technical objects are operated under conditions of continuous influence of different kinds of disturbances, the impact of which has random nature from the point of view of the intensity, duration and moment of occurrence. Accordingly, the resulting changes in the state of the operated objects also turn out to be random [17–19] and form a certain random process in time. Thus, the problem of predicting the future state of an object from the known past and present is a problem of a probabilistic-statistical nature, and its solution is always associated with the study of random processes occurring in the devices, the state of which is to be predicted. Such studies are constantly carried out in practice, both in the process of testing products for reliability, and directly in the process of their operation. This information constitutes the necessary information basis of reliability control of complex technical objects.

14.2 Problem Statement

Let us introduce the parameter S as a state of the technical control object and the random sequence $\{S\} = S(i), i = \overline{1,I}$ as the parameter S changes in a discrete diapason of time points t_i, $i = \overline{1,I}$. The values $S(i)$, $i = \overline{1,I}$ of the parameter S must satisfy the condition

$$S(i) \in [c,d], \quad i = \overline{1,I}. \tag{14.1}$$

If the parameter S crosses the boundaries of the admissible region $[c,d]$, a refusal is recorded. Based on the measurements $s(\mu)$, $\mu = \overline{1,k}$, $k < I$ at discrete moments in time t_μ, $\mu = \overline{1,k}$, it is required to draw a conclusion about the suitability of an object for operation at future moments in time t_i, $i = \overline{k+1,I}$.

14.3 Solution

14.3.1 Forecasting the State of Control Objects

The most general approach for solving the problem of assessing the future state of the control object is based on the Kolmogorov-Gabor polynomial [20–22]. However, finding its parameters for a large number of known values and the used order of nonlinear relation is a very complicated and time-consuming procedure (for example, for 11 known values and order of nonlinearity 4, it is necessary to obtain and solve 1819 partial differential equations of the mean square of the extrapolation error). In this regard, when forming the forecast algorithms realized in practice, various simplifications and restrictions on the properties of a random sequence are used. For example, Pugachev [23, 24] proposed a number of suboptimal methods of nonlinear extrapolation with a limited order of stochastic relation based on the approximation of the posterior probability density of the estimated vector by the Hermite orthogonal polynomial expansion or in the form of an Edgeworth series. Solution of the non-stationary equation of Kolmogorov [25] (a particular case of Stratanovich [26] differential equation for describing Markov processes) is obtained under the conditions that (a) the diffusion coefficient is a constant value, and (b) the drift coefficient can be presented as linear dependence from the state. There are exhaustive solutions of optimal linear extrapolation problem for different types of random sequences and different levels of data support in considered forecasting task. Among examples are (a) Kolmogorov's equation for random sequences with stationary character which are measured without errors; Kalman's technique [27] for noisy random sequences with Markov properties; Wiener-Hopf's filter-extrapolator which is efficient for noisy sequences with stationary character; Kudritskiy's algorithms of optimal linear extrapolation [28–30] synthesized using the Pugachev's canonical expansion, and others. Nevertheless, we should remember that only for Gaussian random sequences it is possible to achieve the highest forecast accuracy using methods of linear extrapolation.

The universal mathematical model (taking into account the restrictions, which are imposed on the random sequence under consideration) is the canonical decomposition of such type [31–33]:

$$S(i) = M[S(i)] + \sum_{\nu=1}^{i} \sum_{\lambda=1}^{N} L_{\nu}^{(\lambda)} \rho(\lambda, \nu; 1, i), \quad i = \overline{1, I}, \tag{14.2}$$

where the elements of the canonical expansion $L_{\nu}^{(\lambda)}$, $\rho(\lambda, \nu; h, i)$ are determined by the recurrent relations:

$$L_{\nu}^{(\lambda)} = S^{\lambda}(\nu) - M[S^{\lambda}(\nu)] - \sum_{\mu=1}^{\nu-1} \sum_{j=1}^{N} L_{\mu}^{(j)} \rho(j, \mu; \lambda, \nu) -$$
$$- \sum_{j=1}^{\lambda-1} L_{\nu}^{(j)} \rho(j, \nu; \lambda, \nu), \quad \lambda = \overline{1, N}, \nu = \overline{1, I}; \tag{14.3}$$

$$\rho(\lambda, \nu; h, i) \beta_{h\nu}^{(\lambda)}(i) = \frac{M\left[L_{\nu}^{(\lambda)}\left(S^{h}(i) - M[S^{h}(i)]\right)\right]}{M\left[\left\{L_{\nu}^{(\lambda)}\right\}^{2}\right]}$$

$$= \frac{1}{D_{\lambda}(\nu)}\{M\left[S^{\lambda}(\nu)S^{h}(i)\right] - M\left[S^{\lambda}(\nu)\right]M\left[S^{h}(i)\right] -$$

$$- \sum_{\mu=1}^{\nu-1} \sum_{j=1}^{N} D_{j}(\mu) \rho(j, \mu; \lambda, \nu) \rho(j, \mu; h, i) - \sum_{j=1}^{\lambda-1} D_{j}(\nu) \rho$$

$$(j, \nu; \lambda, \nu) \rho(j, \nu; h, i)\}, \quad \lambda = \overline{1, h}, \ \nu = \overline{1, i}, \ h = \overline{1, N}, \ i = \overline{1, I}; \tag{14.4}$$

$$D_\lambda\left(\nu\right) = M\left[\left\{L_\nu^{(\lambda)}\right\}^2\right] = M\left[S^{2\lambda}\left(\nu\right)\right] - M^2\left[S^\lambda\left(\nu\right)\right] -$$

$$-\sum_{\mu=1}^{\nu-1}\sum_{j=1}^{N} D_j\left(\mu\right)\left\{\rho\left(j,\mu;\lambda,\nu\right)\right\}^2 - \sum_{j=1}^{\lambda-1} D_j\left(\nu\right)\left\{\rho\left(j,\nu;\lambda,\nu\right)\right\}^2,$$

$$\lambda = \overline{1,N}, \nu = \overline{1,I}. \tag{14.5}$$

The coordinate functions $\rho\left(\lambda,\nu;h,i\right)$, $\nu = \overline{1,i}$; $\lambda,h = \overline{1,N}$; $i = \overline{1,I}$ can be presented as

$$\rho\left(\lambda,\nu;\,h,i\right) = \begin{cases} 1, & \text{if } (h = \lambda) \wedge (\nu = i)\,; \\ 0, & \text{if } (i < \nu) \vee ((h < \lambda) \wedge (\nu = i))\,. \end{cases}$$

The block-diagram of the calculation algorithm for the parameters $D_\lambda\left(\nu\right)$, $\lambda = \overline{1,N}$, $\nu = \overline{1,I}$ and $\rho\left(\lambda,\nu;h,i\right)$, $\nu = \overline{1,i}$, $\lambda = \overline{1,N}$, $h = \overline{1,N}$, $i = \overline{1,I}$ of the canonical expansion (14.2) is shown in Figure 14.1. Nonlinear model (14.2) of the random sequence $\{S\} = S\left(i\right)$, $i = \overline{1,I}$ includes N arrays $\left\{L^{(\lambda)}\right\}$, $\lambda = \overline{1,N}$, where $L_i^{(\lambda)}$, $\lambda = \overline{1,N}$, $i = \overline{1,I}$ are centered uncorrelated random coefficients. Such coefficients contains data concerning the corresponding components $S^\lambda(i)$, $\lambda = \overline{1,N}$, $i = \overline{1,I}$. The coordinate functions $\rho\left(\lambda,\nu;h,i\right)$, $\nu = \overline{1,i}$, $\lambda,h = \overline{1,N}$, $i = \overline{1,I}$ correspond to probabilistic relations of the order $\lambda + h$ between the sections t_ν and t_i, $\left(\nu,i = \overline{1,I}\right)$. The expression (14.2) is also true if some stochastic relations of the random sequence $\{S\} = S\left(i\right)$, $i = \overline{1,I}$ are absent. In this case, the corresponding coordinate functions take the values zero and the data of the relation are automatically excluded from the canonical expansion.

Suppose that as a result of the measurement, the first value $s(1)$ of the sequence $\{S\}$ at the point t_1 is known. Therefore, the values of the coefficients L_1^λ, $\lambda = \overline{1,N}$ are known:

$$l_1^\lambda = s^\lambda(1) - M\left[S^\lambda(1)\right] - \sum_{j=1}^{\lambda-1} l_1^j \rho(j,\lambda;1,1), \lambda = \overline{1,I}. \tag{14.6}$$

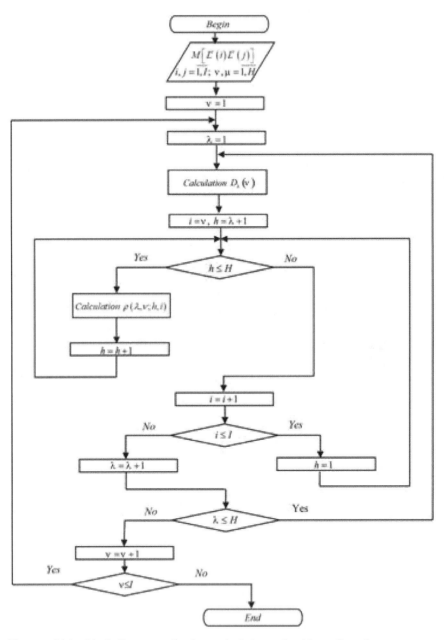

Figure 14.1 Block-diagram of the calculation algorithm for the parameters $D_\lambda(\nu)$, $\rho(\lambda, \nu; h, i)$ of the polynomial power of canonical expansion (14.3)

Substitution of $l_1^{(1)}$ into (14.2) allows to obtain a polynomial canonical expansion of the posterior random sequence $\{S^{(1,1)}\} = S(i/S_1(1))$:

$$S^{(1,1)}(i) = S(i/s(1)) = M[S(i)] + (S(1) - M[S(1)])\rho(1,1;1,i,) +$$
$$+ \sum_{\lambda=2}^{N} L_1^{(\lambda)} \rho(\lambda,1;1,i) + \sum_{\nu=1}^{i} \sum_{\lambda=1}^{N} L_\nu^{(\lambda)} \rho(\lambda,\nu,1,i), \ \ i = \overline{1, I}. \quad (14.7)$$

Applying the operation of mathematical expectation to (14.7) gives an estimate of the future values of the sequence that is optimal by the criterion of the minimum of the mean-square error of extrapolation, provided that one value is used to determine this estimate:

$$m_x^{(1,1)}(1,i) = M[S(i/s(1))] = M[S(i)] + (s(1) - M[S(1)])$$
$$\rho(1,1;1,i), \ \ i = \overline{1, I}. \quad (14.8)$$

Taking into account that the coordinate functions $\rho(\lambda, \nu, h, i), \lambda, h = \overline{1, N}, \ \nu, i = \overline{1, I}$ are determined from the condition of the minimum of the mean-square error of approximation in the intervals between arbitrary values $S^\lambda(\nu)$ and $S^h(i)$, the expression (14.8) can be generalized to the case of forecasting $S^h(i), h = \overline{1, N}, i = \overline{2, I}$:

$$m_x^{(1,1)}(h,i) = M[S^h(i/x(1))] = M[S^h(i)] +$$
$$+ (s(1) - M[S(1)])\rho(1,1;h,i), \ \ i = \overline{1, I}, \quad (14.9)$$

where $m_s^{(1,1)}(h,i)$ is the optimal estimate of the future value $S^h(i)$, provided that the value $S(1)$ is used for the forecast.

Concretization of the second value $l_1^{(2)}$ in (14.7) gives the canonical expansion of the posterior sequence $\{S^{(1,2)}\} = S(i/s_1(1), s_1(1))^2$

$$S^{(1,2)}(i) = S(i/s(1), s(1)^2) = M[S(i)] + (s(i) - M[s(1)])\rho(1,1;1,i) +$$
$$+ [s^2(1) - (s(1) - M[S(1)])\rho(1,1;2,1)]\rho(2,1;1,i)$$
$$+ \sum_{\lambda=3}^{N} L_i^{(\lambda)} \rho(\lambda,1;1,i) + \sum_{\nu=2}^{i} \sum_{\lambda=1}^{N} L_\nu^{(\lambda)} \rho(\lambda,\nu,1,i), i = \overline{1, I}.$$

$$(14.10)$$

Application of the mathematical expectation operation to (14.10) allows to obtain an extrapolation algorithm by two values $s(1), s(1)^2$ using the expression (14.9):

$$m_s^{(1,2)}(h,i) = M[S^h(i/s(1),s(1)^2)] =$$
$$= m_s^{(1,1)}(h,i) + [s^2(1) - m_s^{(1,1)}(2,1)]\rho(2,1;h,i), \quad i = \overline{1,I}.$$
(14.11)

The component $S^N(1)$ should be used as a posteriori data (for the first section t_1 in the last recurrent cycle)

$$m_s^{(1,N)}(h,i) = M[S^h(i/s(1),s(1)^2 \ldots s(1)^N)] =$$
$$= m_s^{(i,N-1)}(h,i) + [s^N(1) - m_s^{(1,N-1)}]\rho(N,1;h,i), \quad i = \overline{1,I}.$$
(14.12)

and then the forecast is specified by applying the value $s(2)$

$$m_s^{(2,1)}(h,i) = M[S^h(i/s(1),s(1)^2 \ldots s(1)^N, s(2))] =$$
$$= m_s^{(1,N)}(h,i) + [s(2) - m_x^{(1,N)}(1,2)]$$
$$\rho(1,2;h,i), \quad i = \overline{1,I}.$$
(14.13)

The last recurrent cycle can be presented as the next calculation operation (for the section t_2),

$$m_s^{(2,N)}(h,i) = M[S^h(i/s(1),\ldots,s(1)^N,s(2),\ldots,s(2)^N)] =$$
$$= m_s^{(2,N-1)}(h,i) + [s^N(2) - m_s^{(2,N-1)}(N,2)]$$
$$\rho(N,2;h,i), \quad i = \overline{1,I}.$$

Generalization of the obtained regularity allows writing a forecast procedure for an arbitrary number of known values [34–36]:

$$m_s^{(\mu,l)}(h,i) = \begin{cases} M[S^h(i)] \text{ for } \mu = 0; \\ m_s^{(\mu,l-1)}(h,1) + (s^l(\mu) - m_s^{(\mu,l-1)}(l,\mu))\rho(l,\mu;h,i) \\ \qquad \text{for } l \neq 1; \\ m_s^{(\mu,l-1)}(h,1) + (s^l(\mu) - m_s^{(\mu-l,N)}(l,\mu))\rho(l,\mu;h,i) \\ \qquad \text{if } l = 1. \end{cases}$$
(14.14)

The expression $m_s^{(\mu,l)}(h,i) = M[S^h(i)/s^\nu(j), j = \overline{1,\mu-l}, \nu = \overline{1,N}; s^\nu(\mu), \nu = \overline{1,l}]$ for $h = l, l = N, \mu = k$ is the optimal estimate

$m^{(k,N)}(1,i)$ of the future value $s(i), i = \overline{k+,1,I}$, provided that the values $s^{\nu}(j), \nu = \overline{l,N}, j = \overline{1,k}$ are used to calculate this estimate, i.e. the results of measurements of the sequence $\{S\}$ at points $t_j, j = \overline{1,k}$ are known.

The expression for the desired estimate $m_s^{(k,N)}(1,i)$ can be written in the following explicit form:

$$m_s^{(k,N)}(1,i) = M[S(i)] + \sum_{j=1}^{k} \sum_{\nu=1}^{N} (s^{\nu}(j) - M[S^{\nu}(j)]) P_{((j-1)N+\nu)}^{(kN)}$$

$$((i-1)N + 1), \tag{14.15}$$

where

$$P_{\lambda}^{(\alpha)}(\xi) = \begin{cases} P_{\lambda}^{(\alpha-1)}(\xi) - P_{\lambda}^{(\alpha-1)}(\alpha)\gamma_k(i), & \text{if } \lambda \leq \alpha - 1 \\ \gamma_{\alpha}(\xi), & \text{for } \lambda = \alpha \end{cases} \tag{14.16}$$

$$\gamma_{\alpha}(\xi) = \begin{cases} \rho(\text{mod}_N(\alpha), [\alpha/N]+1; 1, [\alpha/N]+1), & \text{for } \xi \leq kN; \\ \rho(\text{mod}_N(\alpha), [\alpha/N]+1; 1, i), & \text{if } \xi = (i-1)N + 1. \end{cases} \tag{14.17}$$

[.] is a rounding operation in the expression (14.17).

If the first k values are fixed, then the canonical expansion of a posterior random sequence $\{S\}$ can be written in the following way

$$S^{(k,N)}(i) = S(i/s^{\nu}(j), \nu = \overline{1,N}, j = \overline{1,k}) = m^{(k,N)}(i,j) +$$

$$+ \sum_{\mu=k+1}^{i} \sum_{\lambda=1}^{N} L_{\mu}^{(\lambda)} \rho(\lambda,\mu; 1, i), \quad i = \overline{k+1,I}. \tag{14.18}$$

The single extrapolation error (for known values $s^{\nu}(j), \nu = \overline{1,N}, j = \overline{1,k}$) can be calculated in such way

$$\delta[i/s^{\nu}(j), \nu = \overline{1,N}, j = \overline{1,k}] = m_s^{(k,N)}(1,i) - s^{(k,N)}(i), i - \overline{k+1,I}, \tag{14.19}$$

where $s^{(k,N)}(i), i - \overline{k+1,I}$ are the future values of the extrapolated realization.

The realization $s^{(k,N)}(i), i - \overline{k+1,I}$ in the area of forecasting develops in a random way, so its exact values cannot be specified. However, the expression (14.18) is an exact model of the stochastic evolution of the realization

$s^{(k,N)}(i), i - \overline{k+1, I}$, which allows to reduce the expression (14.19) to the form

$$\Delta[i/s^{\nu}(j), \nu = \overline{1, N}, j = \overline{1, k}] = m_x^{(k,N)}(i, j) - S^{(k,N)}(i), \quad i = \overline{k+1, I}. \tag{14.20}$$

Thus, the error of a single extrapolation has a stochastic nature that deals with the random character of the investigated processes.

Implementation of the mathematical expectation procedure to (14.20), taking into account the expression (14.18), give us possibility to calculate the systematic component of the single extrapolation's error

$$\varepsilon[i/s^{\nu}(j), \nu = \overline{1, N}, j = \overline{1, k}] = M[\Delta(i/s^{\nu}(j), \nu = \overline{1, N}, j = \overline{1, k})] =$$

$$= \sum_{\mu=k=1}^{i} \sum_{\lambda=1}^{N} M[L_{\mu}^{(\lambda)}] \rho(\lambda, \mu, 1, i) = 0, i = \overline{k+1, I}. \tag{14.21}$$

The latter Equation (14.21) illustrates that the procedures (14.14), (14.15) provides calculation (for each predicted realization) of unbiased estimates of future values.

This also underlines that the proposed algorithm gives possibility to provide the unbiasedness for the average

$$\varepsilon^{(k,N)}(i) = M[\varepsilon(i/s^{\nu}(j), \nu = \overline{1, N}, j = \overline{1, k})] = 0, \quad i = \overline{k+1, I}. \tag{14.22}$$

The variance help us to estimate the significance of the random component of the error for single extrapolation

$$D_{\Delta}[i/s^{\nu}(j), \nu = \overline{1, N}, j = \overline{1, k}] = M[(\Delta[i/s^{\nu}(j), \nu = \overline{1, N}, j = \overline{1, k}] -$$

$$- \varepsilon[i/\chi^{\nu}(j), \nu = \overline{1, N}, j = \overline{1, k}])^2] = M\left[\left(\sum_{\mu=k+1}^{l} \sum_{\lambda=1}^{N} L_{\mu}^{(\lambda)} \rho(\lambda, \mu; 1, i)\right)^2\right]$$

$$= \sum_{\mu=k+1}^{i} \sum_{\lambda=1}^{N} D_{\lambda}(\mu) \{\rho(\lambda, \mu; 1, i)\}^2, i = \overline{k+1, I}. \tag{14.23}$$

Since the specific values of the extrapolated realization do not impact to the variance $D_{\Delta}[i/s^{\nu}(j), \nu = \overline{1, N}, j = \overline{1, k}]$ it is possible to form an expression for the variance of the posterior random sequence

$$D_\Delta^{(k,N)}(i) = M[D_\Delta[i/X^\nu(j), \nu = \overline{1,N}, j = \overline{1,k}]]$$

$$= \sum_{\mu=k+1}^{i} \sum_{\lambda=1}^{N} D_\lambda(\mu)\rho(\lambda, \mu; 1, i)\beta_{1\mu}^{(\lambda)}(i). \qquad (14.24)$$

Taking into account (14.22), (14.24), the expression for the mean square of the forecast error can be written as

$$E^{(k,N)}(i) = [\varepsilon^{(k,N)}(i)]^2 + D_\Delta^{(k,N)}(i) =$$

$$D_\Delta^{(k,N)}(i) = \sum_{\mu=k+1}^{i} \sum_{\lambda=1}^{N} D_\lambda(\mu)\rho(\lambda, \mu; 1, i), i = \overline{k+1, I}. \qquad (14.25)$$

Thus, the mean square of the forecast error based on the procedures (14.14), (14.15) is equal to the variance of the posterior random sequence.

It is necessary to note that the parameters of the procedure (14.14), calculated by the size of the statistical sample of realizations, can be specified using new a posteriori information $s_{L+1}(\nu), \nu = \overline{1,I}$ in real time in the process of solving the forecasting problem:

$$M_{(L+1)}[S^n] = \frac{\sum_{l=1}^{L+1} s_i^n}{L+1} = \frac{\sum_{l=1}^{L} s_i^n L}{(L+1)L} = \frac{s_{L+1}^n L}{L+1} = \frac{M_{(L+1)}[S^n]L + s_{L+1}^n}{(L+1)};$$

$$D_{n(L+1)}(\nu) = \frac{\sum_{l=1}^{L+1} (l_{\nu,1}^{(n)})^2}{L} = \frac{\sum_{l=1}^{L} (l_{\nu,1}^{(n)})^2 (L-1)}{L(L-1)} + \frac{(l_{\nu,1}^{(n)})^2}{(L)}$$

$$\frac{D_{n(L+1)}(\nu) + (S_{(L+1)}^n(\nu) - m^{(n-1,\nu)}(n,\nu))}{L};$$

$$\rho_{L+1}(n, \nu; \lambda, i) = \frac{\sum_{l=1}^{L+1} l_{\nu,l}^{(n)}(S_l^\lambda - M_{(L+1)}[S^\lambda(i)])}{LD_{n,(L+1)}(\nu)}$$

$$= \frac{\rho_{L+1}(n, \nu; \lambda, i)(L-1) + (s_{L+1}^n(\nu) - m^{(n-1,\nu)}(n,\nu))}{(s_{L+1}^\lambda(i) - M_{(L+1)}[S^\lambda(i)])}$$

Thus, the results of extrapolation can be used for significant (up to several arithmetic operations) simplification of the recalculating procedure for the algorithm (14.14) parameters in the case if new statistical data of the investigated random sequence come. It allows to use the updated parameters for the current solution of the forecast problem.

By analogy with the procedure of forming the algorithm (14.14), the method of extrapolation of the realization of the vector random sequence $\{\bar{S}\}$, taking into account the stochastic relations $M[S_j^l(\mu)S_h^q]$, $\mu, i = \overline{1, I}$; $j, h = \overline{1, H}$; $l, q = \overline{1, N}$ between the components, can be obtained as following:

$$
m_{s;j,h}^{(\mu,l)}(q,i) = \begin{cases}
M[S_h(i)], & \text{if } \mu = 0; \\[2mm]
m_{s;j,h}^{(\mu,l-1)}(q,i) + (s_j^l(\mu) - m_{x;j,h}^{(\mu,l-1)}(l,\mu))\beta_{j,i}^{(h,q)}(\mu,i), \\
\quad \text{for } l > 1, j > 1; \\[2mm]
m_{s;j-1,h}^{(\mu,N)}(q,i) + (s_j^l(\mu) - m_{x;j-1,j}^{(\mu,N)}(l,\mu))\beta_{j-1,N}^{(h,q)}(\mu,i), \\
\quad \text{for } l = 1, j > 1; \\[2mm]
m_{s;H,h}^{(\mu-1,N)}(q,i) + (s_1^l(\mu) - m_{s;H,1}^{(\mu-1,N)}(l,\mu))\beta_{1,1}^{(h,s)}(\mu+1,i), \\
\quad \text{if } l = 1, j = 1;
\end{cases}
$$

$$(14.26)$$

where $m_{s;j,h}^{(\mu,l)}(1,i) = M[S_h(i)/s_\lambda^n(\nu)]$, $\lambda = \overline{1,H}, n = \overline{1,N}, \nu = \overline{1,\mu-1}$; $s_\lambda^n(\mu)$, $\lambda = \overline{1,j}$] is the optimal, in the mean-square sense, estimate of the future values of the investigated random sequence, provided that a posteriori information $s_\lambda^n(\nu), \lambda = \overline{1,H}, n = \overline{1,N}$ is used for the forecast.

The algorithm is based on the canonical model

$$
S_h(i) = M[S_h(i)] + \sum_{\nu=1}^{i-1}\sum_{l=1}^{H}\sum_{\lambda=1}^{N} L_{\nu l}^{(\lambda)}\beta_{l\lambda}^{(h,1)}(\nu,i)
$$

$$
+ \sum_{l=1}^{h-1}\sum_{\lambda=1}^{N} L_{il}^{(\lambda)}\beta_{l\lambda}^{(h,1)}(i,i) + W_{ih}^{(1)}, \quad l = \overline{1,I}; \tag{14.27}
$$

$$
L_{\nu l}^{(\lambda)} = S_l^{\lambda}(\nu) - M[S_l^{\lambda}(\nu)] - \sum_{\mu=1}^{\nu=1}\sum_{m=1}^{H}\sum_{j=1}^{N} L_{\mu m}^{(l)}\beta_{mj}^{(l,\lambda)}(\mu,\nu) -
$$

$$
- \sum_{m=1}^{l-1}\sum_{j=1}^{N} L_{\nu m}^{(j)}(\nu,\nu) - \sum_{j=1}^{\lambda-1} L_{\nu l}^{(j)}\beta_{lj}^{(l,\lambda)}(\nu,\nu), \quad \nu = \overline{1,I}; \tag{14.28}
$$

$$D_{l,\lambda}(v) = M\left[\left\{L_{vl}^{(\lambda)}\right\}^2\right] = M\left[S_l^{2\lambda}(v)\right] - M^2\left[S_l^\lambda(v)\right]$$

$$-\sum_{\mu=1}^{v-1}\sum_{m=1}^{H}\sum_{j=1}^{N}D_{mj}(\mu)\left\{\beta_{mj}^{(l,\lambda)}(\mu,v)\right\}^2 - \sum_{m=1}^{l-1}\sum_{j=1}^{N}D_{mj}(v)\left\{\beta_{mj}^{(l,\lambda)}(v,v)\right\}^2$$

$$-\sum_{j=1}^{\lambda-1}D_{lj}(v)\left\{\beta_{lj}^{(l,\lambda)}(v,v)\right\}^2, v=\overline{1,I}; \tag{14.29}$$

$$\beta_{l,\lambda}^{(h,q)}(v,i) = \frac{M\left[L_{vl}^{(\lambda)}\left(S_h^q(i) - M\left[L_{vl}^{(\lambda)}S_h^q(i)\right]\right)\right]}{M\left[\left\{W_{vl}^{(\lambda)}\right\}^2\right]}$$

$$= \frac{1}{D_{l,\lambda}(v)}M\left[S_l^\lambda(v)S_h^q(i)\right] - M\left[S_l^\lambda(v)\right]M\left[S_h^q(i)\right]$$

$$-\sum_{\mu=1}^{v-1}\sum_{m=1}^{H}\sum_{j=1}^{N}D_{mj}(\mu)\beta_{mj}^{(l,\lambda)}(\mu,v)\beta_{mj}^{(h,s)}(\mu,i)$$

$$-\sum_{m=1}^{l-1}\sum_{j=1}^{N}D_{mj}(v)\beta_{mj}^{(l,\lambda)}(v,v)\beta_{mj}^{(h,s)}(v,i)$$

$$-\sum_{j=1}^{\lambda-1}D_{lj}(v)\beta_{mj}^{(l,\lambda)}(v,v)\beta_{mj}^{(h,s)}(v,i), \lambda=\overline{1,h}v=\overline{1,i}. \tag{14.30}$$

The difference with (14.14) lies in that that after taking full account of the posterior information $s_j(\mu), s_j^2(\mu)...s_j^N(\mu)$ at the fixed section t_μ, the transition to the accumulation is carried out to predict the posterior information of the next component S_{j+1}, in the case j=N the transition to the next section $t_{\mu+1}$ is carried out. Diagram in Figure 14.2 reflects the peculiarities of the computational process when using the extrapolator (14.26).

The expression for the mean-square error of extrapolation using the algorithm (14.26) by the known values $s_j^n(\mu), \mu=\overline{1,k}; j=\overline{1,H}; n=\overline{1,N}$ has the form

$$E_h^{(k,N)}(i) = M\left[S_h^2(i)\right] - M^2\left[S_h(i)\right] -$$

$$-\sum_{\mu=1}^{k}\sum_{j=1}^{H}\sum_{n=1}^{N}D_{jn}(\mu)\left\{\beta_{jn}^{(h,1)}(\mu,i)\right\}^2, i=\overline{k+1,I}. \tag{14.31}$$

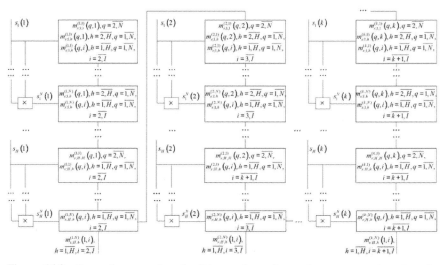

Figure 14.2 Generative procedure for future values of a random sequence based on the algorithm (14.26)

The mean-square error of extrapolation $E_h^{(K,N)}(i)$ is equal to the variance of the posterior random sequence

$$S_h^{(k,N)}(i) = S\left(i/s_l^\nu(j),\ \nu = \overline{1,N}, j = \overline{1,k}, l = \overline{1,H}\right) = m_{H,h}^{(k,N)}(1,i)$$

$$+ \sum_{\nu=k+1}^{i-1}\sum_{l=1}^{H}\sum_{\lambda=1}^{N} L_{\nu l}^{(\lambda)}\beta_{l\lambda}^{(h,1)}(\nu,i) + \sum_{l=1}^{h-1}\sum_{\lambda=1}^{N} L_{il}^{(\lambda)}\beta_{l\lambda}^{(h,1)}(i,i) + L_{ih}^{(1)},$$

$$i = \overline{k+1, I}. \tag{14.32}$$

If the vector random sequence has only linear stochastic relations $(M[S_j(\mu)S_h(i)] \neq 0,\ .M[S_j^l(\mu)S_h^s(i)] = 0, l + s >)$ then the extrapolation algorithm takes the form [37–39]:

$$m_{x;h}^{(\mu,l)}(i) = \begin{cases} M\left[X_h(i)\right], if \mu = 0; \\ m_{x;h}^{(\mu,l-1)}(i) + \left[x_l(\mu) - m_{x;h}^{(\mu,l-1)}(\mu)\right]\varphi_{h\mu}^{(l)}(i), if l \neq 1; \\ m_{x,h}^{(\mu,H)}(i) + \left[x_1(\mu) - m_{x;h}^{(\mu-1,H)}(\mu)\right]\varphi_{h\mu}^{(1)}(i), if l = 1; \end{cases}$$

$$\tag{14.33}$$

where $m_{x;h}^{(\mu,l)}(i) = M[X_h(i)/X_\lambda(v\ ), \lambda = \overline{1,H}, v = \overline{1,\mu-1}; x_j(\mu)$, $j = 1, l]h = \overline{1,H}, i = \overline{k,I}$ is the linear optimal by the criterion of the minimum of a mean-square error of forecasting estimate of the future values of the investigated sequence, provided that the values $s_\lambda(v), \lambda = \overline{1,H}, v = \overline{1,\mu-1}; s_j(\mu), j = \overline{1,l}$ are known.

The parameters of the algorithm are the elements of the linear vector canonical decomposition

$$S_h(i) = M[S_h(i)] + \sum_{v=1}^{i}\sum_{\lambda=1}^{H} v_v^{(\lambda)}\varphi_{hv}^{(\lambda)}(i), i = \overline{1,I}, \tag{14.34}$$

where

$$V_v^{(\lambda)} = S_\lambda(v) - M[S_\lambda(v)] - \sum_{\mu=1}^{v-i}\sum_{j=1}^{H} v_\mu^{(j)}\varphi_{\lambda\mu}^{(j)}(v)$$

$$- \sum_{j=1}^{\lambda-1} V_v^{(j)}\varphi_{\lambda\mu}^{(j)}(v), v = \overline{1,I}; \tag{14.35}$$

$$\varphi_{hv}^{(\lambda)}(i) = \frac{M[V_v^{(\lambda)}(S_h(i) - M[S_h(i)])]}{M\left[\{V_v^{(\lambda)}\}^2\right]} = \frac{1}{D_{\lambda(v)}}(M[S_\lambda(v)S_h(i)]$$

$$- M[S_\lambda(v)]M[S_h(i)] - \sum_{\mu=1}^{v-1}\sum_{j=1}^{H} D_j(\mu)\varphi_{\lambda\mu}^{(j)}(v)\varphi_{h\mu}^{(j)}(i)$$

$$- \sum_{j=1}^{\lambda-1} D_j\varphi_{\lambda v}^{(j)}(v)\varphi_{hv}^{(j)}(i), \lambda = \overline{1,I}, v = \overline{1,i}; \tag{14.36}$$

$$D_\lambda(v) = M\left[\{V_v^{(\lambda)}\}^2\right] = M\left[\{X_\lambda(v)\}^2\right] - M^2[X_\lambda(v)]$$

$$- \sum_{\mu=1}^{v-1}\sum_{j=1}^{H} D_j(\mu)\left\{\varphi_{\lambda\mu}^{(j)}(v)\right\}^2 - \sum_{j=1}^{\lambda-1} D_j(v)\left\{\varphi_{\lambda v}^{(j)}(v)\right\}^2, v = \overline{1,I}. \tag{14.37}$$

The expression for the mean-square error of extrapolation using the algorithm (14.33) is written as

$$E_h^{(\mu,l)}(i) = D_h(i) - \sum_{v=1}^{\mu-1}\sum_{j=1}^{H} D_j(v)\left\{\varphi_{hv}^{(j)}(i)\right\}^2 - \sum_{j=1}^{l-1} D_j(\mu)\left\{\varphi_{h\mu}^{(j)}(i)\right\}^2,$$
$$i = \overline{\mu+1, I}.$$
(14.38)

14.3.2 Identification of Random Sequences Model's Parameters based on Statistical Goodness-of-fit Tests

The coordinate functions allow taking into account (a) the mechanism of after effect and (b) estimation of the stochastic relation in the canonical expansion (14.2):

$$\rho(\lambda, v; 1, i) = \frac{M[L_v^{(\lambda)}(S(i) - M[S(i)])]}{D_\lambda(v)} = \frac{M[L_v^{(\lambda)}\overset{o}{X}(i)]}{D_\lambda(v)}, i > v.$$
(14.39)

and these functions determine the level (degree) of stochastic relationship between subsequent sections of the random sequence $S(i)$, $i > v$ and coefficients $L_v^{(\lambda)}$. Thus, we can assume that the coefficients $L_v^{(\lambda)}$ in the section t_v have zero's influence to the subsequent values of the random sequence, if (starting from any $i_{k_v} > v$), $\rho(\lambda, v; 1, i) \equiv 0$ is true for the arbitrary value λ. Accordingly, the after effect interval k_v from the sampling point t_v is determined in this case as

$$k_v = i_{k_v} - v.$$
(14.40)

In the situation under consideration, it is possible to form the components of the canonical expansion using finite sample of the size L. Thus, the process of k_v determination can be transformed to solving the next problem: 1) according to the sample data, estimates of the coordinate functions $\rho_L(\lambda, v; 1, i), \lambda = \overline{1, N}, i = \overline{v, I}$ can be obtained; (b) it is necessary to find such value $i_{k_v}, v < i_k$ for which the identity $\rho(\lambda, v; 1, i) \equiv 0, i > v$ is satisfied with a set degree of confidence.

Analyzing the model (14.2) and its linearization mechanism it is possible to transform the problem of after effect estimation and determination of the nonlinearity order of a random sequence to the problem of the linear relation between $L_v^{(\lambda)}$ and $S(i)$, $v < i$, the standard quantitative characteristic of which is the normalized correlation coefficient

$$r^{(\lambda)}(v,i) = \frac{M\left[L_v^{(\lambda)}S(i)\right]}{\sqrt{D_\lambda(v)}\sqrt{D_x(i)}}, \lambda = \overline{1,N}, v = \overline{1,I}, i = \overline{v,I}. \quad (14.41)$$

The expression (14.41), using (14.39), can be transformed to the following form

$$r^{(\lambda)}(v,i) = \frac{\sqrt{D_\lambda(v)}\rho(\lambda,v;1,i)}{\sqrt{D_x(i)}}, i > v, \quad (14.42)$$

where the components $D_\lambda(v)$, $\rho(\lambda,v;1,i)$ of the canonical expansion are used for calculation of the correlation coefficient $r^{(\lambda)}(v,i)$. Since the estimates of all these components can be obtained during the initial statistical data processing, it is possible to estimate the normalized correlation coefficient $r^{(\lambda)(L)}(v,i)$ for any values v,λ and i. From this point of view, we can reformulate the task of significance estimation for the correlation between the coefficient $L_v^{(\lambda)}$ and the ith section of the investigated random sequence to the problem of testing the statistical hypothesis

$$r^{(\lambda)}(v,i) = 0 \quad (14.43)$$

and the alternative hypothesis $r^{(\lambda)}(v,i) \neq 0$.

The random variable

$$\frac{1}{2}\ln\left[\frac{1 + r^{(\lambda)(L)}(v,i)}{1 - r^{(\lambda)(L)}(v,i)}\right]$$

should be considered distributed normally, with the mathematical expectation [40]

$$m = \frac{1}{2}\ln\left[\frac{1 + r^{(\lambda)(L)}(v,i)}{1 - r^{(\lambda)(L)}(v,i)}\right] + \frac{r^{(\lambda)}(v,i)}{2(L-1)}$$

and the variance $D = \frac{1}{L-3}$, therefore the value

$$a^{(\lambda)}(v,i) = \sqrt{L-3}\left[\frac{1}{2}\ln\left[\frac{1 + r^{(\lambda)(L)}(v,i)}{1 - r^{(\lambda)(L)}(v,i)}\right]\right.$$
$$\left. - \left(\frac{1}{2}\ln\left[\frac{1 + r^{(\lambda)(L)}(v,i)}{1 - r^{(\lambda)(L)}(v,i)}\right] + \frac{r^{(\lambda)}(v,i)}{2(L-1)}\right)\right]$$

has the standard normal distribution (0,1).

Thus, the data can be consistent with the hypothetical value $r^{(\lambda)}(v, i)$ (with the significance level), if the value

$$\frac{1}{2} \ln \left[\frac{1 + r^{(\lambda)(L)}(v, i)}{1 - r^{(\lambda)(L)}(v, i)} \right] + \frac{r^{(\lambda)}(v, i)}{2(L - 1)}$$

lies within

$$\left[\frac{1}{2} \ln \left[\frac{1 + r^{(\lambda)(L)}(v, i)}{1 - r^{(\lambda)(L)}(v, i)} \right] - \frac{z_\alpha}{\sqrt{L - 3}}, \frac{1}{2} \ln \left[\frac{1 + r^{(\lambda)(L)}(v, i)}{1 - r^{(\lambda)(L)}(v, i)} \right] + \frac{z_\alpha}{\sqrt{L - 3}} \right]$$

where z_α is the value of the standard normal deviate which corresponds to the level of confidence α. In other cases, the hypothetical value $r^{(\lambda)}(v, i)$ does not correspond to the statistics.

Because the condition $(r^{(\lambda)}(v, i) = 0)$ is checking, the random variable $a^{(\lambda)}(v, i)$ takes the form [41, 42]

$$a^{(\lambda)}(v, i) = \sqrt{L - 3} \ln \left[\frac{1 + r^{(\lambda)(L)}(v, i)}{1 - r^{(\lambda)(L)}(v, i)} \right]$$

and the hypothesis (14.43) can be accepted if the following condition is fulfilled

$$- z_\alpha < a^{(\lambda)}(v, i) < z_\alpha. \tag{14.44}$$

Hypothesis (14.44) is multi-tested many by increasing the parameter λ until the certain boundary value $N^{(v,i)}$ at which condition (14.44) is true ($N^{(v,i)}$ is a higher order of the nonlinear relation between the sections t_v and) t_v After that, the number of interval increases ($i = i + 1$) and the search procedure for the higher order of nonlinearity is repeated for a new interval.

If for some i_{k_v} of the area $i > v$ and the arbitrary λ (as a rule, it is sufficient to check at $\lambda = 1$: nonlinear relations decay faster than linear ones) the statement $r^{(\lambda)}(v, i) = 0, i > i_{k_v}$ turns out to be true, then this means that the after effect interval for the sampling point t_v is equal to $i_{k_v} - 1$ and for all $i > i_{k_v}$ the value of the coordinate function $\rho(\lambda, v; 1, i)$ should be taken to be zero.

By checking the after effect for all sampling points $t_v, v = \overline{1, I^*}$ (in which the random sequence S is investigated) it is possible to determine the parameters N and I of the canonical expansion (14.2) as

$$I = \max_v \left(i_{k_v} - v \right),$$

$$N = \max_{v,i} N^{(v,i)}.$$

14.3.3 Method of Reliability Control of Technical Objects

Taking into account that the values of the controlled parameter S change in the forecast area in a random manner, the comprehensive characteristic of the reliability of the investigated technical object is the probability of failure-free operation

$$P^{(k)}(I) = P\left\{c < S^{(k)}(i) < d, i = \overline{k+1, I}/x(\mu), \mu = \overline{1, k}\right\}. \quad (14.45)$$

Thus, the problem is reduced to determining the probability that the realization of the a posteriori random sequence $S^{(k)}\left(i/s(\mu), \mu = \overline{1, k}\right), i = \overline{k+1, I}$ will not go beyond the boundaries of the admissible region $[A; d]$. The probability estimate $P^{(k)}(I)$ can be obtained as a result of a statistical experiment using the expression

$$P^{*(k)}(I) = \frac{m}{L} \quad (14.46)$$

where m – is the number of realizations that satisfy the condition (14.1);

L – the total number of realizations.

Thus, on the basis of the above, the method of the predictive control of the failure-free operation of technical objects consists of the realization of the following stages:

- collection of statistical data on the functioning of the same type of objects (for objects with one parameter and for objects with many parameters);
- estimation of discretized moment functions;
- calculation of the parameters of the mathematical model (14.2)— for objects with one parameter; (14.27)—for objects with many parameters);
- calculation of the state of the control object at future moments in time (algorithm (14.14) or (14.26));
- verification of the condition (14.1) (in case of non-fulfillment—the operation of the object is stopped and the subsequent analysis is stopped);
- multiple modeling of the values of random coefficients $L_i^{(\lambda)}$, $i = \overline{k+1, I}$, $\lambda = \overline{1, N}$ or $L_{\nu l}^{(\lambda)}$ (for objects with many parameters) with

a given distribution law and with the subsequent formation, using the expression (14.18) or (14.32), of a set of possible continuations of the realization of the investigated random sequence in the forecast area $[t_{k+1}...t_I]$;

- verification of the condition of non-intersection of the boundaries of the admissible region of change of the controlled parameters by the obtained trajectories and determination of an estimate of the probability of failure-free operation of a technical object;
- making a decision on the subsequent operation of the object based on the obtained estimate of the probability of failure-free operation.

14.4 Conclusion

Thus, a method of reliability control of technical objects, the parameters of which are of a stochastic nature, is obtained in the work. An individual reliability estimation based on the information about the functioning of an object during operation is a peculiarity of the proposed approach. Making decision on suitability at future points in time is carried out in two stages: at the first stage, the belonging of the predicted parameters to the permissible region is checked, and at the second stage, the compliance of the probability value of no-failure operation with the required level is checked. The mathematical model based on the canonical decomposition does not imply any restrictions on the random sequence of change of the parameters of the control object, which makes it possible to take into account the features of its functioning as much as possible. Expressions for the mean-square error of extrapolation are obtained, which allow us to estimate the accuracy of solving the forecasting problem. The resulting method can be used to control the reliability of arbitrary technical objects and systems [9, 10, 43–45], as well as their components.

References

[1] A. Adarsh, R. Mangey, 'System reliabilty: solutions and technologies', Taylor & Francis Group, 2019.
[2] V. Varakuta, T. Hlimancov, S. Starodubcev, S. Ptashka, 'Determination of risk indicators of loss of negative situations on technological dangerous objects', Science and Technology of the Air Force of Ukraine, Issue 1(34), 2019, pp. 107–116.

[3] P.M. Frank, 'Fault diagnosis in dynamic systems using analytical and knowledge-based redundancy: A survey and some new results', J. Automatica 3, 459–474, 1990.

[4] R.J. Patton, P.M. Frank, R.N. Clarke., 'Fault diagnosis in dynamic systems: theory and application', Prentice-Hall Inc., 1989.

[5] R.J. Patton, R.N. Clark, P.M. Frank, 'Issues of fault diagnosis for dynamic systems', Springer Science & Business Media, 2000.

[6] P.M. Frank, X.-C. Ding, 'Survey of robust residual generation and evaluation methods in observer-based fault detection systems', Journal of process control 7.6, 403–424, 1997.

[7] S.X. Ding, 'A unified approach to the optimization of fault detection systems', International journal of adaptive control and signal processing 14.7, 725–745, 2000.

[8] N. Silva., M. Vieira, 'Towards Making Safety-Critical Systems Safer: Learning from Mistakes', In: ISSREW, 2014, 2014 IEEE International Symposium on Software Reliability Engineering Workshops (ISSREW), 162–167, 2014.

[9] V.M. Kuntsevich, et al (Eds), 'Control Systems: Theory and Applications', Series in Automation, Control and Robotics, River Publishers, Gistrup, Delft, 2018.

[10] Y.P. Kondratenko, V.M. Kuntsevich, A.A., Chikrii V.F. Gubarev, (Eds), 'Advanced Control Systems: Theory and Applications', Series in Automation, Control and Robotics, River Publishers, Gistrup, 2021.

[11] V. Varakuta, T. Hlimancov, S. Starodubcev, S. Ptashka, 'Determination of risk indicators of loss of negative situations on technological dangerous objects', J. Science and Technology of the Air Force of Ukraine, No. 1(34), pp. 107–116. https://doi.org/10.30748/nitps.2019.34.15.

[12] A.V. Boyarchuk, etc., 'Safety of Critical Infrastructures: Mathematical Analysis and Engineering Methods of Analysis and Ensuring', National Aerospace University named after N.E.Zhukovsky "KhAI", Kharkiv, 2011.

[13] M. Nafria, 'Circuit reliability prediction: challenges and solutions for the device time-dependent variability characterization roadblock', IEEE Latin America Electron Devices Conference (LAEDC), Mexico, 19–21 April 2021.

[14] H. Jiang, 'Time Dependent Variability in Advanced FinFET Technology for End-of-Lifetime Reliability', IEEE International Reliability Physics Symposium (IRPS) Monterey, CA, USA, 21–25 March 2021.

[15] H. Yu, Y. Ran, G.B. Zhang, G. Ying, 'A dynamic time-varying reliability model for linear guides considering wear degradation', Nonlinear Dynamics, Volume 103, pp. 699–714, 2021.

[16] A. Kremer, L. Dücsö, B. Bertsche, 'Reliability Prediction using Design of Experiments', Proceedings of the 30th European Safety and Reliability Conference and the 15th Probabilistic Safety Assessment and Management Conference, Venice, Italy, 1–5 November 2020.

[17] A.V. Palagin, V.N. Opanasenko, 'Design and application of the PLD-based reconfigurable devices.' In: Design of Digital Systems and Devices: Adamski M., Barkalov A., Wegrzyn M. (Eds.), Lecture Notes in Electrical Engineering. Verlag, Berlin, Heidelberg: Springer, Volume 79, pp. 59–91, 2011.

[18] Sokolov Y.N., Kharchenko V.S., Illyushko V.M., et al, 'Applications of Computer Technologies for Software-Hardware Complexes Reliability and Safety Assessment Systems', National Aerospace University named after N.E.Zhukovsky "KhAI"s, Kharkiv, 2013.

[19] Odarushchenko O.N., Ponochovny Y.L., Kharchenko V.S., et al, 'High Availability Systems and Technologies', National Aerospace University named after N.E.Zhukovsky "KhAI", Kharkiv, 2013.

[20] P.C. Jack, 'Statistical techniques in simulation', Publishing house Kleijnen, 1978.

[21] C. Chatfield, 'Time series forecasting', London: Chapman, and Hall, 2000.

[22] E. Wentzel, 'Theory of Probability and its Engineering Applications', Nauka, Moscow, 1998.

[23] V. Pugachev, 'Theory of Random Functions: And Its Application to Control Problems', Pergamon Press, London, 2013.

[24] V. Pugachev, I. Sinitsyn, 'Stochastic systems: theory and applications', Word Scientific Publishing, London, 2002.

[25] B.S. Everitt (Ed.), 'The Cambridge dictionary of statistics', Cambridge University press, New York, 2006.

[26] R.S. Tsay, 'Nonlinear time series models: testing and applications. Course in Time Series Analysis', New York: Wiley, 2001.

[27] D.C. Montgomery, G.C. Runger, 'Applied Statistics and Probability for Engineers', Wiley Custom, 2012.

[28] V.D. Kudritskii, 'Generalized optimal linear extrapolation algorithm for realizations of random sequences', Engineering Simulation, 12(3), pp. 339–348, 1995.

[29] V.D. Kudritskii, 'Optimal linear filtering and extrapolation of realizations of a vector random function observed with errors', Cybernetics and Systems Analysis, 36(4), pp. 579–584, 2000.

[30] V.D. Kudritskii, 'Algorithm of Wiener filtering and extrapolation for nonstationary random processes observed with correlated errors', Cybernetics and Systems Analysis, 35(3), pp. 406–412, 1999.

[31] I. Atamanyuk, Y. Kondratenko, 'Canonical mathematical model and information technology for cardio-vascular diseases diagnostics', Proceedings of the 14th International Conference The Experience of Designing and Application of CAD Systems in Microelectronics, CADSM 2017, Lviv – Polyana, Ukraine, pp. 438–440 May 2017. DOI: 10.1109/CADSM.2017.7916170

[32] Y. Kondratenko, et al, 'Robotics and Prosthetics at Cleveland State University: Modern Information, Communication, and Modeling Technologies.' Ginige, A. et al. (Eds): Information and Communication Technologies in Education, Research, and Industrial Applications. ICTERI 2016. Communications in Computer and Information Science, vol. 783. Springer, Cham, pp. 133–155, 2017. DOI: https://doi.org/10.1007/978-3-319-69965-3_8

[33] Y. Kondratenko, et al, 'University Curricula Modification Based on Advancements in Information and Communication Technologies.' Proceedings of the 12th International Conference on Information and Communication Technologies in Education, Research, and Industrial Application. Integration, Harmonization and Knowledge Transfer, 21–24 June, 2016, Kyiv, Ukraine, Ermolayev, V. et al. (Eds), ICTERI'2016, CEUR-WS, Vol-1614, pp. 184–199, 2016.

[34] V. Shebanin, et al, 'Application of fuzzy predicates and quantifiers by matrix presentation in informational resources modeling', Proceedings of XII International Conference "MEMSTECH 2016", 22-24 April 2016, Lviv-Poljana, pp. 146–149.

[35] I.P. Atamanyuk, 'Algorithm of extrapolation of a nonlinear random process on the basis of its canonical decomposition', J. Cybernetics and Systems Analysis, 41(2), pp. 267–273, 2005.

[36] I.P. Atamanyuk, 'Optimal Polynomial Extrapolation of Realization of a Random Process with a Filtration of Measurement Errors', Journal of Automation and Information Sciences, Volume 41, Issue 8, Begell House, USA, pp. 38–48, 2009. DOI: 10.1615/JAutomatInfScien.v41.i8.40

[37] V. Shebanin, et al., 'Simulation of vector random sequences based on polynomial degree canonical decomposition', Eastern-European Journal of Enterprise Technologies, Vol. 5, No. 4 (83), 2016, pp. 4–12, DOI: http://dx.doi.org/10.15587/1729-4061.2016.80786

[38] I. Atamanyuk, Y. Kondratenko, N. Sirenko, 'Forecasting Economic Indices of Agricultural Enterprises Based on Vector Polynomial Canonical Expansion of Random Sequences', Proceedings of the 12-th Int. Conference ICTERI'2016, CEUR-WS, 21-24 June, 2016, Kyiv, Ukraine, Vol-1614, pp. 458–468, 2016.

[39] I. Atamanyuk, Y. Kondratenko, N. Sirenko, 'Management System for Agricultural Enterprise on the Basis of Its Economic State Forecasting', Complex Systems: Solutions and Challenges in Economics, Management and Engineering, Vol. 125, pp. 453–470, 2018. DOI: https://doi.org/10.1007/978-3-319-69989-9_27

[40] E. Lehmann, J. Romano, 'Testing Statistical Hypotheses', Springer-Verlag New York, 2005, p. 786.

[41] C. Walck, 'Handbook on statistical distributions for experimentalists,' Particle Phys. Group, Fysikum, Univ. Stockholm, Stockholm, Sweden, Tech. Rep. SUF-PFY/96-01, 2007.

[42] G.E.P. Box, G.M. Jenkins, 'Time–series Analysis, Forecasting and Control', Holden–Day, San Francisco, 1970.

[43] A.N. Tkachenko, et al. 'Evolutionary adaptation of control processes in robots operating in non-stationary environments.' J. Mechanism and Machine Theory. – Printed in placecountry-regionGreat Britain, 1983. – Vol. 18, No. 4. – pp. 275–278. DOI: 10.1016/0094-114X(83)90118-0

[44] I.P. Atamanyuk, et al, 'The Algorithm of Optimal Polynomial Extrapolation of Random Processes.' In: Modeling and Simulation in Engineering, Economics and Management, K.J. Engemann, et al. (Eds.), International Conference MS 2012, New Rochelle, NY, USA (May 30 – June 1, 2012), Proceedings. Lecture Notes in Business Information Processing, Volume 115, Springer, 2012, pp. 78–87. DOI: 10.1007/978-3-642-30433-0_9

[45] M. Solesvik, et al, 'Joint Digital Simulation Platforms for Safety and Preparedness.' In: Y. Luo (Ed), Cooperative Design, Visualization, and Engineering. CDVE 2018. Lecture Notes in Computer Science, vol. 11151. Springer, Cham, pp. 118–125, 2018.

15

Petunin Ellipsoids in Automatic Control Systems Design

D.A. Klyushin, S.I. Lyashko, and A.A. Tymoshenko

Taras Shevchenko National University of Kyiv, Prospekt Glushkova 4D, 03680, Kyiv, Ukraine
E-mail: dokmed5@gmail.com; lyashko.serg@gmail.com; inna-andry@ukr.net

Abstract

Mathematical modeling of modern automatic control systems requires storing and processing large amounts of data. Complex systems are described by numerous parameters that have different significance and obey certain restrictions. These constraints often take the form of a system of linear inequalities. This paper proposes a new way of describing the optimization domain with linear constraints in the design space using Petunin ellipsoids. The statistical properties of these ellipsoids are given, making them an effective tool for describing data in the design space. The problem of construction of the optimal Petunin ellipsoids in design space with linear constrains is considered in the paper. The algorithms of construction of the Petunin ellipsoids without constraints and with linear constraints are described. The issues of the complexity and statistical features are analyzed.

The second part is devoted to a new distribution-free approach to computing ellipsoidal conformal prediction set based on the Hill's assumption and the Petunin ellipsoids. The resulting prediction set is a Petunin ellipsoid with precise confidence level depending on the number of points only. We show that our method works for classical distributions (uniform and Gaussian) with high performance. In addition, our method allows determining the median of the random points set (the deepest point), outliers (outer points) and the statistical depth for every point of the set.

Keywords: Petunin ellipsoids, minimum volume ellipsoid, minimum trace ellipsoid, constrained optimization, design space, data mining, prediction set, outlier detection.

15.1 Introduction

Modern computer-aided design of automatic control systems is impossible without data analysis that describes the parameters of objects and the results of experiments in the form of multidimensional vectors in the design space. The complexity of the problem is due to the fact that the number of such parameters is very large and therefore it is desirable to single out among them those that are important from a certain point of view (for example, statistically significant) and discard points that can be interpreted as outliers. Moreover, each parameter, as a rule, is subject to certain restrictions. It is either limited to the minimum and maximum values, or is related to other parameters by linear relationships.

The problems of designing technical systems are often reduced to solving variational problems of minimizing functionals subject to the fulfillment of constraints in the form of a system of linear inequalities. This leads to the need to determine the optimization area in the design space, in which the search for the optimal solution that satisfies the given constraints is performed. A large scatter of parameters in the design space can complicate mathematical modeling and worsen the properties of numerical optimization algorithms, so the optimization area should be compact and contain only statistically significant points. In addition, many efficient optimization algorithms make significant use of the geometric properties of the optimization domain (for example, convexity). It is desirable that the optimization domain be convex and described by simple equations. These requirements are met by ellipsoids covering the given points. Ellisoids of minimum volume are often used as such ellipsoids, ellipsoids with minimal traces, Dikin ellipsoids and other modifications. The choice of these ellipsoids is dictated by geometric considerations and does not take into account the statistical nature of the parameters of the designed technical systems. demands without considering the data volume.

The statistical nature of the source data dictates the choice of the appropriate tool. In particular, similar problems were considered in [1], where a comparative analysis of several methods for constructing optimal descriptions of design data was carried out, in particular, an ellipsoid of minimum volume, an ellipsoid with a minimum trace, and the Monte Carlo method. Despite the effectiveness of the methods demonstrated, they have a big drawback-they do not take into account the statistical nature of the parameters, focusing on geometric properties. The Petunin ellipsoid described in [3–8] allows us to elegantly solve this problem.

The second topic discussed in the paper is construction of prediction sets, which was investigated in many papers. They are closely tied with the notions of the statistical depth, the multivariate ordering and the quintile contour construction. In particular, using B.M. Hill's work [9] and Yu. Petunin Ellipses [10, 11], statistical properties for several distributions are tested.

Other popular topics about prediction sets include classification of the methods of multivariate ordering was proposed by Barnett [12]. The concept of the statistical depth was developed in the papers of Tukey [13], Oja [14], Koshevoy and Mosler [15], Liu [16], Zuo and Serfling [17], Cascos [18], Lange and Mozharovsky [19] etc.).

The methods of the construction of tolerance intervals and prediction sets were considered by Wilks [20], Wald [21], Tukey [22], Fraser and Guttman [23], Guttman [24], Aichison and Dunsmore [25], Bairamov and Petunin [26], Shafer and Vovk [27], Lei et al. [28], Ndiaye and Takeuchi [29] etc.

The purpose of this section is to describe new method of construction of ellipsoidal prediction set for a cloud of the random points using exchangeability of the points.

For more information about advanced control systems, their design and applications we may refer to [30–34]. Modern research in theory and applications of control systems is represented there, based on contributions by researchers from more than 16 different that report on recent developments and new directions in advanced control systems. Also new theoretical findings, industrial applications and case studies on complex engineering systems are represented.

15.2 Petunin Ellipses and Their Statistical Properties

15.2.1 Petunin Ellipse and Ellipsoids

Let $X = x_1, ..., x_n$ be a set of random points, where $x_i \in \mathbb{R}^d$, and $Ax \leq b$ is a system of linear inequalities, where $A = \{a_{ij}\}_{i,j=1}^{s,d}$. The problem is to find the optimal ellipsoid containing at least given number of points, having minimal volume among the similar ellipsoids and laying within the polygon corresponding to the linear constraints $Ax \leq b$.

To describe the algorithm of the construction of the Petunin ellipsoids let us consider two cases: $d = 2$ and $d > 2$ without constraints. Essentially,

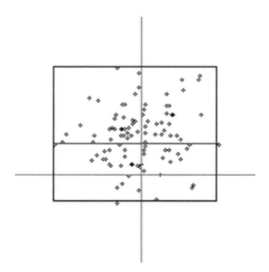

Figure 15.1   Petunin rectangle Π

this is the same algorithm that has geometrical features dependent from the space dimension. Case $d = 2$ is more descriptive and allow simple geometric concepts to be used for illustration.

Case $d = 2$. Construct a convex hull of $X = \{(x_1, y_1), ..., (x_n, y_n)\}$ Find a diameter of the convex hull and its ends (x_k, y_k) and (x_l, y_l). Construct a segment L connecting these ends. Find points laying at the most distance from L: (x_r, y_r) and (x_q, y_q).

Construct segments L_1 and L_2 that are parallel to L, pass through (x_r, y_r) and (x_q, y_q), and have the length equals to the length of L. Construct segments L_3 and L_4 that are orthogonal to L and pass through (x_k, y_k) and (x_l, y_l). Segments L_1, L_2, L_3 and L_4 form a rectangle Π. Denote a short side by a and a long side by b (Figure 15.1).

Make translation, rotation and shrinking with a coefficient $\alpha = \frac{a}{b}$ so that Π transforms to a square Π' with a center (x_0', y_0'). The random points are transformed to points $(x_1', y_1'), (x_2', y_2'), ...(x_n', y_n') \in \Pi'$. Find distances $r_1, r_2, ..., r_n$ from (x_0', y_0') to $(x_1', y_1'), (x_2', y_2'), ...(x_n', y_n') \in \Pi'$. Let $R = \max(r_1, r_2, ..., r_n)$. Construct a circle C with the center (x_0', y_0') and radius R. Thus, $(x_1', y_1'), (x_2', y_2'), ..., (x_n', y_n') \in C$ (Figure 15.2).

Performing inverse transformations of this circle we obtain an ellipse E (Figure 15.3). The computational complexity of this algorithm is equal to computational complexity of the construction of a convex hull $O(nlgn)$.

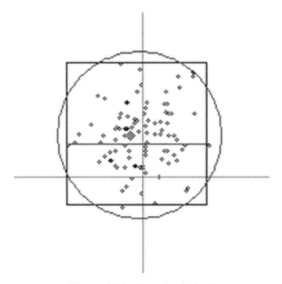

Figure 15.2   Petunin circle C

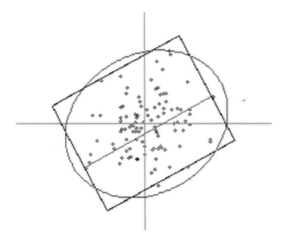

Figure 15.3   Petunin ellipse E

Case d $>$ 2. Construct the convex hull of $X = \{x_1, ..., x_n\}$. Find a diameter of the convex hull and its ends (x_k, y_k) and (x_l, y_l). Using rotation and translation let us align the diameter along to Ox_1'. Project all the points $(x_1', y_1'), (x_2', y_2'), ..., (x_n', y_n')$ to the orthogonal complement of Ox_1'. Using convex hull construction, rotation and translation until the orthogonal complement becomes two-dimensional we obtain a rectangle Π. Perform the algorithm for case $d = 2$. Construct an axis-aligned parallelogram in d dimensional space containing the images $x'_1, ..., x'_n$ of input points. Transform this parallelogram to hypercube. Denote its center by x_0 and find distances $r_1, r_2, ..., r_n$ from it to $x'_1, ..., x'_n$. Let $R = \max(r_1, r_2, ..., r_n)$. Construct a hypersphere with the center x_0 and radius R. Performing inverse transformations we obtain an ellipsoid in the input space.

The Petunin ellipsoid have some wonderful features. First, it uniquely arranges the points according their statistical depth because by construction at the surface of Petunin ellipsoid only one point lays. Second, the probability that random points from the same distribution lays within Petunin ellipse is equal $\frac{n-1}{n+1}$. Thus, we have not only ellipse containing a given set of n random points with the exact probability, but also may find outliers. Peeling these ellipsoids we may construct an influential set of points in design space.

The main complexity problem of the construction of the Petunin ellipsoid in a space of high dimension is the procedure of finding of two most distant points (ends of the set diameter). As it was noted in [4], the complexity of the construction of a convex hull in $\mathbb{R}^d$ is $o\left(n^{\lfloor \frac{n}{2} \rfloor}\right)$ But there are some effective and fast procedures [2, 5] that are not optimal but extremely smooth the problem.

15.2.2 Petunin Ellipses and Linear Constraints

The above algorithm does not take into account the linear constraints. It constructs an ellipse just containing all random points with given probability. But the design of this algorithm allow easy modification so that we can select some Petunin ellipse with the given confidence level that lays within the polygon formed by the constraints $Ax \leq b$

To do this we use a procedure proposed in [1]. Let us return to the step in case $d = 2$ when we construct the concentric circles. These circles cannot be lying within the polygon. Now, finding the side that intersects with the outer circles (with radius) we may construct the normal vector to this side and shrink the circle such that it become tangent to this side. Repeating this procedure as many times as the linear constraints are violated we obtain a desired ellipse satisfying the linear constrains. Only drawback is that this ellipse may have a lower confidence level since this level depends on the points covered by the ellipse. This procedure may be extended to the multidimensional case easily.

15.2.3 Statistical Properties of Petunin's Ellipses

Let G be a population of random values following an unknown distribution function F and B is such that $P\{x \in B\} = 1 - \alpha$, where is an arbitrary sample element and α is a given significance level. If sample values $x_1, x_2, ..., x_n$ are exchangeable identically distributed random values following absolutely continuous distribution function, then [9],

$$p\left(x_{n+1} \in \left(x_{(i)}, x_{(j)}\right)\right) = \frac{j-1}{n+1}$$

where x_{n+1} is the next sample value from G, and x_i, x_j are ith and jth order statistics.

The following theorems hold [8].

Theorem 15.1. If $x_1, x_2, ..., x_{n+1}$ are exchangeable random values following absolutely continuous joint distribution function such that $P\{x_k = x_{(j)}\} = 0$ if $k \neq m$, then

$$P\{x_k \geq x_1, ..., x_k \geq x_{k-1}, x_k \geq x_{k+1}, ..., x_k \geq x_{n+1}\} = \frac{1}{n+1}.$$

Theorem 15.2. If $x_1, x_2, ..., x_{n+1}$ are exchangeable random values following absolutely continuous joint distribution function such that $P\{x_k = x_{(j)}\} = 0$ if $k \neq m$, and $x_{(1)} \leq x_{(2)} \leq ... \leq x_{(n)}$ the a variational series, then $P\left(x_{n+1} \in \left(x_{(i)}, x_{(j)}\right)\right) = \frac{1}{n+1}$.

Corollary. $p\left(x_{n+1} \in \left(x_{(i)}, x_{(j)}\right)\right) = \frac{j-1}{n+1} \forall 0 \leq i < j \leq n+1$, where $\eta_0 = -\infty \eta_{n+1} = +\infty$.

Theorem 15.3. If $\vec{x}_1, \vec{x}_2, ..., \vec{x}_n$ are exchangeable random vectors from G, E_n is the Petunin ellipsoid, containing $\vec{x}_1, \vec{x}_2, ..., \vec{x}_n$, and $\vec{x}_{n+1} \in G$, then $p(\vec{x}_{n+1} \in E) = \frac{n}{n+1}$.

Corollary 15.1. The significance level of the Petunin ellipsoid is less than 0,05, if n > 19.

Corollary 15.2. If n>19, then the area of the Petunin ellipsoid can be decreased without increasing of the significance level, deleting farthest points $r_{(l)}, \cdot, r_{(n)}$, if n >19.

This algorithm has remarkable features: 1) only one point always lays at the surface of the Petunin ellipsoid; 2) the Petunin ellipsoid contains n point with probability $\frac{n-1}{n+1}$; 3) in practice, to guarantee the significance level which is less than 0.05 a set of points may contain only 20 random samples.

15.2.4 Numerical Experiments

Hypothetically, the more points we have the more confidence level of the Petunin ellipsoid covering all the point, and therefore, higher confidence level of the Petunin ellipsoid satisfying the linear constraints. In order to test this hypothesis, we have carried out several numerical experiments with two-dimensional samples following the uniform and Gaussian distributions and satisfying constraints that are defined by a square, pentagon and hexagon. We suppose that the more circular-like a polygon defining constraint the less points we loss during peeling the Petunin ellipsoid. We 100 times generated N two-dimensional samples, where $N = 40, 50, ..., 100$, constructed the optimal Petunin ellipses covering part of these points and satisfying a system of linear constraints and averaged the number of points that are not covered by the Petunin ellipse satisfying a system of linear constraints (Table 15.1–15.2). The purpose of these experiments where to determine the degree of decreasing of the confidence level of ellipses depending on the kind if constraints (Figures 15.1–15.7).

Table 15.1   The number of uniformly distributed points dropped out and the significance level of the Petunin ellipse satisfying square constraints

Initial Number	Number of Dropped out Points	Initial Significance Level	Result Significance Level
40	19	0.976	0.955
50	7	0.980	0.977
60	9	0.984	0.981
70	19	0.986	0.981
80	24	0.988	0.982
90	50	0.989	0.976
100	66	0.990	0.971

Table 15.2   The number of Gaussian distributed points dropped out and the significance level of the Petunin ellipse satisfying square constraints

Initial Number	Number of Dropped out Points	Initial Significance Level	Result Significance Level
40	7	0.976	0.971
50	14	0.980	0.973
60	12	0.984	0.983
70	13	0.986	0.985
80	8	0.988	0.986
90	8	0.989	0.988
100	9	0.990	0.989

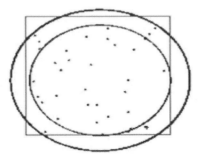

Figure 15.4 Uniformly distributed points with square constraints

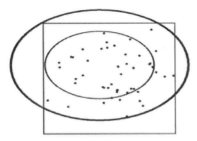

Figure 15.5 Gaussian distributed points with square constraints

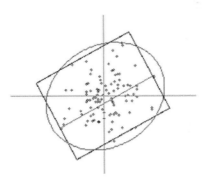

Figure 15.6 Uniformly distributed point with pentagonal constraints

As we see in the uniform case, when the sample size is greater than 40 the decreasing of the confidence level of the Petunin ellipses after peeling is varied in the region (0.03; 0.13). When the sample size is large (90 or 100) we

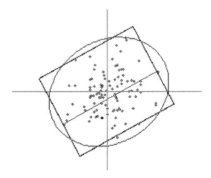

Figure 15.7 Uniformly distributed point with hexagonal constraints

have clear gift in the number of points (55%–66%) without significant loss of the confidence level of the ellipse.

As we see in the Gaussian case, when the sample size is greater than 50 the decreasing of the confidence level of the Petunin ellipses after peeling is varied in the region (0.01; 0.02). However, we had not achieved a significant decreasing of the number of representative points. Thus, we may suppose that this effect is connected with the kind of distribution.

Figure 15.8 illustrates changing of area of Petunin ellipses after dropping one point per step. As far as the Petunin ellipses covering points are concentric the area is monotonically decreasing. Absence of abrupt jumps in these curves justify that the sample is homogeneous and fill the region densely.

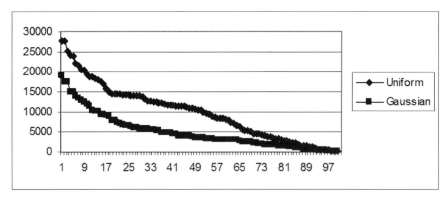

Figure 15.8 Area of Petunin ellipses covering 100 uniformly and Gaussian distributed points after dropping one point per step

15.2.5 Prediction Sets

Generalization of training sample and forecast future values is the primary problem of machine learning. The main tools for solving this problem are "predictive sets" that are also named "tolerance regions" and "minimum volume sets". Consider the general problem of constructing conformal prediction set: given i.i.d. points $x_1, x_2, ..., x_n \in \mathbb{R}^d$ find a prediction set $E(x_1, x_2, ..., x_n) \subset \mathbb{R}^d$ such that $p(x_{n+1} \in E) \geq 1 - \alpha$, where $0 < \alpha < 1$ is a given significance level, thus $p(x_{n+1} \in E)$ is the confidence level of the predictive set.

15.2.6 Hill's Assumption $A_{(n)}$

Let $x_1, x_2, ..., x_n$ be a sample drawn by the simple sampling from a population following absolutely continuous distribution F. Arrange it in the increasing order and construct the variance series $x_{(1)} \leq, x_{(2)} \leq ...x_n$, where $x_{(i)}$ is ith order statistics. The order statistics $x_{(1)}, x_{(2)}, ..., x_{(n)}$ in contrast to initial sample values are dependent and have different distributions. The distribution $F_k(u)$ of the kth order statistics $x_{(k)}$ is $F_k(u) = \sum_{i=k}^{n} C_n^i [F(u)]^i [1 - F(u)]^{n-i}$, where $\pi e F(u) = p(x \leq u)$. The Hill's assumptions A_n [9] states that if x_{n+1} drawn from the population following distribution then

$$p\left(x_{n+1} \in \left(x_{(i)}, x_{(j)}\right)\right) = \frac{j-1}{n+1}, j > i.$$

The Hill's assumption was proved in papers of Yu.I. Petunin et al. [10, 11] for i.i.d. and exchangeable random variables, respectively.

Let us consider the proof provided in [21]. If random variables and are independent, then

$$p\left(\xi < \eta\right) = \int_{-\infty}^{\infty} F_\xi(u) dF_\eta(u),$$

where $F_\xi(u)$ and $F_\eta(u)$ are the distribution functions of ξ and η, respectively. The probability density of ith order statistics is:

$$f_k(u) = \sum_{i=k}^{n} C_n^i [F(u)]^i [1 - F(u)]^{n-i} = \sum_{i=k}^{n} G_i(u).$$

Hence,

$$F'_k(u) = \sum_{i=k}^{n} G'_i(u)$$

$$G'_k(u) = C_n^k[k(F(u))^{k-1}(1 - F(u))^{n-k}f(u) -$$
$$- (F(u))^k(n - k)(1 - F(u))^{n-k-1}f(u)].$$

$$G'_{k+1}(u) = \left[C_n^{k+1}(F(u))^{k+1}(1 - F(u))^{n-k-1}\right]'$$
$$= C_n^{k+1}[(k + 1)(F(u))^k(1 - F(u))^{n-k-1}f(u) -$$
$$- (F(u))^{k+1}(1 - F(u))^{n-k-2}f(u)].$$

The second term annihilates with the first term:

$$-C_n^k(n - k) + C_n^{k1}(k + 1) = \frac{n!(n - k)}{(n - k - 1)!k!} + \frac{n!(k + 1)}{(n - k - 1)!(k + 1)!}$$

$$= -\frac{n!}{(n - k - 1)!k!} + \frac{n!}{(n - k - 1)!k!} = 0.$$

The last term of previous sum is zero

$$(F(u))^n(1 - F(u))^0(n - n)f(u) = 0.$$

Thus,

$$f_k(u) = nC_{n-1}^{k-1}[F(u)]^{k-1}[1 - F(u)]^{n-k}f(u).$$

Let us find $p(x < x^{(i)})$ and $p(x < x^{(j)})$. Using the above equations, we have

$$p(x < x^{(i)}) = \int_{-\infty}^{\infty} F(u)dF_i(u) = nC_{n-1}^{m+1}\int_{-\infty}^{\infty}[F(u)]^i[1 - F(u)]^{n-k}dF(u)$$

$$= nC_{n-1}^{i-1}\int_0^1 v^i(1 - v)^{n-1}dv$$

.

It is well-known that,

$$\int_0^1 x^{p-1}(1 - x)^{q-1}dx = \frac{\Gamma(p)\Gamma(q)}{\Gamma(p + q - 1)} = \frac{(p - 1)!(q - 1)!}{(p + q - 1)!}$$

Thus,

$$\int_0^1 x^{i+1-1}(1-x)^{n-i+1-1}dx = B(i+1, n-i+1)$$

$$= \frac{\Gamma(i+1)\Gamma(n-i+1)}{\Gamma(i+1+n+1-i)} = \frac{\Gamma(i+1)\Gamma(n-i+1)}{\Gamma(n+2)} = \frac{i!(n-i)!}{(n+1)!}$$

$$p(x_{n+1} < x_i) = n_{n-1}^{i-1}\frac{j!(n-i)!}{(n+1)!} =$$

$$= n\frac{(n-1)!j!(n-j)!}{(n-1-j+1)!(j-1)!(n+1)!} =$$

$$= \frac{n(n-1)!(j-1)j!(n-j)!}{(n-j)!(j-1)!n(n+1)(n-1)!} = \frac{j}{n+1}$$

Therefore,

$$p\left(x_{n+1} \in \left(x_{(i)}, x_{(j)}\right)\right) = p(x_{n+1} < x_i) = \frac{j}{n+1} - \frac{i}{n+1} = \frac{j-i}{n+1}$$

Thus, if a random variable x does not depend on $x_1, x_2, ..., x_n$ and it is drawn by sample sampling from the same population following the distribution $F(u)$, then

$$p\left(x \in \left(x_{(1)}, x_{(n)}\right)\right) = \frac{n-1}{n+1}$$

Remark 15.1. The confidence level of the tolerance interval $J = \left(x_{(1)}, x_{(n)}\right)$ is $\frac{n-1}{n+1}$, thus for $n \geq 39$ the confidence level of this interval is less than 0,05.

15.2.7 Testing Statistical Properties for Petunin Ellipses

To test the proposed algorithm we used two-dimensional samples following the uniform distribution in $[0,100] \times [0,100]$ and Gaussian distribution $N(\mu,)$, where $\mu = (50,50)$ and $= \text{diag}(10, 10)$. We computed the Petunin ellipses covered 40 points distributed in and constructed the curve of area of every concentric ellipse.

At Figure 15.9 we see that six (35–40) outer points are outliers because the area of the corresponding ellipses significantly decreases. Also, we see natural clustering of the almost equiprobable points. The borders of the clusters are the points that are not outliers but the corresponding ellipse area

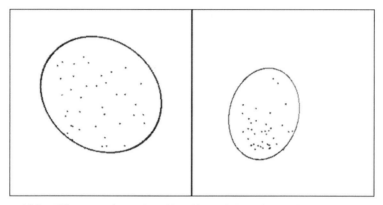

Figure 15.9   Ellipse covering point with uniform (left) and Gaussian (right) distribution

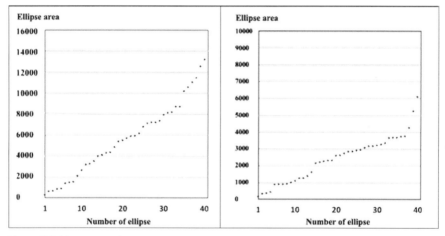

Figure 15.10   Areas of the ellipses covering point with uniform (left) and Gaussian (right) distribution

significantly decreases. We detected the following clusters of points: 1–5, 6–8, 9–16, 17–23, 24–28, 29–34, 35–38, and 39–40.

At Figure 15.10 we see that three (38–40) outer points are outliers because the area of the corresponding ellipses abruptly decreases. More flat curve comparing the curve for the uniform distribution is a consequence of more compact arranging of points with the Gaussian distribution. The number of clusters of equiprobable points following the Gaussian distribution is less than for the uniformly distributed points. We detected the following clusters of points: 1–4, 5–14, 15–19, 17–32, and 33–37.

The Petunin ellipses accurately reflect the statistical properties of distributions. More dispersed distribution (uniform) demonstrate the greater number of clusters of equiprobable points and greater area of ellipses, but more compact distribution (Gaussian) produce the less number of clusters and less area of ellipses.

Also, we tested whether the Petunin ellipses satisfy the condition on the coverage probability $P\left(x_{n+1} \in E\right) = \frac{n-1}{n+1}$. We generated 40 points from the uniform distribution in $[0,100] \times [0,100]$ and constructed the corresponding covering Petunin ellipse E. Then, we generated next 100 points from the same distribution and found the relative frequency of the event $P\left(x_{n+1} \in E\right)$ We repeated these trials 20 times and averaged the relative frequencies. To estimate how the average relative frequency is close to the expected probability $\frac{n-1}{n+1}$ we used the Wilson confidence interval and the $2s$–rule. The results are provided in Table 15.1. The expected probability is covered both by the Wilson interval and by the $2s$–interval, therefore, the experiment confirms the properties of the Petunin ellipse.

Table 15.3 The results of testing confidence level of the Petunin ellipse for uniform distribution

Trial	Observable Relative Frequency	Descriptive Statistics	
1	0.98	Expected probability	0.9048
2	0.95	Mean	0.9616
3	0.96	Standard Error	0.0073
4	0.92	Median	0.9700
5	0.99	Mode	0.9800
6	0.99	Standard Deviation	0.0318
7	0.94	Sample Variance	0.0010
8	0.95	Kurtosis	0.7694
9	0.98	Skewness	−0.9680
10	1	Range	0.1200
11	1	Maximum	1.0000
12	0.98	Minimum	0.8800
13	0.97	Geometric Mean	0.9610
14	0.88	Harmonic Mean	0.9605
15	0.95		
16	0.98		
17	0.99		
18	0.98		
19	0.94		

(Continued)

Table 15.3 Continued

Trial	Observable Relative Frequency	Descriptive Statistics		
20	0.92			
Expected probability	0.9048			
Wilson interval	0.7248	0.8880	1.0000	
2s-interval	0.8988	0.9625	1.0000	

15.3 Conclusion

The proposed algorithm of constructing optimal Petunin ellipsoid in design space is effective and statistically robust. It allows decreasing of the point in design space without loss of confidence level of the ellipsoids. This algorithm demonstrates robustness to outliers, high confidence level with modest number of covering points and useful possibility to arrange the point on the probability to cover them by the ellipsoid.

The proposed algorithm of constructing conformal prediction sets using Petunin ellipsoid is effective and has high performance. It has theoretically precise confidence level and allows unambiguous multivariate points arranging, computing statistical depth every points, finding median of the set, and detecting outliers. The experiments confirmed the theoretical properties of the Petunin ellipses.

The Petunin ellipsoid is useful tool for arrange multivariate data and finding anomalies using statistical depth. Thus, the future research directions within the domain of the topic are developing computationally effective algorithms for construction of high-dimensional conformal prediction sets using genetic and other algorithms and inventing new multivariate statistical tests for homogeneity of multivariate samples in high-dimensional spaces.

References

[1] A. Bedrintsev, V. Chepyzhov, "Description of the design space by extremal ellipsoids in data representation problems", J. Commun. Technol. El., vol. 61, pp. 688–694, 2016.

[2] S. Har-Peled, "A practical approach for computing the diameter of a point set", In: Symposium on Computational Geometry (SOCG'2001), pp. 177–186, 2001.

[3] S. Lyashko, D. Klyushin, V. Alexeenko, "Multivariate ranking using elliptical peeling", Cybern. Syst. Anal., 49, pp. 511–516, 2013.

[4] S. Lyashko., D. Klyushin., V. Semenov, M. Prysiazhna, M. Shlykov, "Nonparametric Ellipsoidal Approximation of Compact Sets of Random Points", Chapter in: Optimization and Applications in Control and Data Sciences (ed. B.Goldengorin), Springer, Optimization and Its Applications, vol. 115, pp. 327–340, 2016.

[5] G. Malandain, J.-D. Boissonnat, "Computing the Diameter of a Point Set. International Journal of Computational Geometry and Applications", IJCGA, vol. 12, pp. 197–208, 2002.

[6] Yu. Petunin, B. Rublev, "Pattern recognition with the help of quadratic discriminant function", J. Math. Sci., vol. 97, pp. 3959–3967, 1999.

[7] S.I. Lyashko, B.V. Rublev, "Minimal Ellipsoids and Maximal Simplexes in 3D Euclidean Space", Cybernetics and Systems Analysis, vol. 39, pp. 831–834, 2003.

[8] S.I. Lyashko, D.A. Klyushin, V. V. Alexeyenko, "Multivariate ranking using elliptical peeling", Cybernetics and Systems Analysis, vol. 49, pp. 511–516, 2013.

[9] B.M. Hill, "Posterior distribution of percentiles: Bayes" theorem for sampling from a population", J. Am. Stat. Assoc., vol. 63, pp. 677–691, 1968.

[10] I. Madreimov, Yu.I. Petunin, "Characterization of the uniform distribution using order statistics", Teor. Veroyatn. Mat. Statist., vol. 27, pp. 96–102, 1982.

[11] R.I. Andrushkiw, D.A. Klyushin, Yu. I. Petunin, V. N. Lysyuk. "Construction of the bulk of general population in the case of exchangeable sample values", Proc. Intern. Conf. of Mathemat. and Engineer. Techniq. in Medicine and Biolog. Sci. (METMBS'03), Las Vegas, Nevada, USA, pp. 486–489, 2003.

[12] V. Barnett, "The ordering of multivariate data", Journal of the Royal Statistical Society. Series A (General), vol. 139, pp. 318–355, 1976.

[13] J.W. Tukey, "Mathematics and the picturing of data", Proc. of the International Congress of Mathematician, Montreal, Canada, pp. 523–531, 1975.

[14] H. Oja, "Descriptive statistics for multivariate distributions", Statistics and Probability Letters, vol. 1, pp. 327–332, 1983.

[15] G. Koshevoy, K. Mosler, K. "Zonoid trimming for multivariate distributions", Annals of Statistics, vol. 25, pp. 1998–2017, 1997.

[16] R.J. Liu, "On a notion of data depth based on random simplices", Annals of Statistics, vol. 18, pp. 405–414, 1990.

[17] Y. Zuo, R. Serfling, "General notions of statistical depth function", Annals of Statistics, vol. 28, pp. 461–482, 2000.

[18] I. Cascos, "Depth function as based of a number of observation of a random vector", Working Paper 07–29, Statistic and Econometric Series 07, vol. 2, pp. 1 28, 2007.

[19] T.I. Lange, P.F. Mozharovsky, "Determination of the depth for multivariate data sets", Inductive simulation of complex systems, vol. 2, pp. 101–119, 2010.

[20] S.S. Wilks, "Statistical Prediction With Special Reference to the Problem of Tolerance Limits", Annals of Mathematical Statistics, vol. 13, pp. 400–409, 1942.

[21] A. Wald, "An extension of Wilks method for setting tolerance limits", The Annals of Mathematical Statistics, vol. 14, pp. 45–55, 1943.

[22] J. Tukey, "Nonparametric estimation. II. Statistical equivalent blocks and multivarate tolerance regions", The Annals of Mathematical Statistics, vol. 18, pp. 529–539, 1947.

[23] D.A.S. Fraser, I. Guttman, "Tolerance regions", The Annals of Mathematical Statistics, vol. 27, pp. 162–179, 1956.

[24] I. Guttman, "Statistical Tolerance Regions: Classical and Bayesian Griffin", Hartigan, J., editor. London.; Clustering Algorithms John Wiley; New York: 1970.

[25] J. Aichison, I.R. Dunsmore, "Statistical Prediction Analysis" Cambridge University Press, 1975.

[26] I.G. Bairamov, Yu. I. "Invariant confidence intervals for the main mass of values from the distribution in a population", J Math Sci, vol. 66, pp. 2534–2538, 1993.

[27] G. Shafer, V.A. Vovk, "A Tutorial on Conformal Prediction", Journal of Machine Learning Research, vol. 9, pp. 371–421, 2008.

[28] J. Lei, J. Robins, L. Wasserman, "Distribution Free Prediction Sets", Journal of the American Statistical Association, vol. 108, pp. 278–287, 2013.

[29] E. Ndiaye, I. Takeuchi, "Computing Full Conformal Prediction Set with Approximate Homotopy", NeurIPS Proceedings, pp. 1384–1393, 2019.

[30] V.M. Kuntsevich et al. (Eds). Control Systems: Theory and Applications. Series in Automation, Control and Robotics, River Publishers, Gistrup, Delft, 2018.

[31] Y.P. Kondratenko, V.M. Kuntsevich, A.A. Chikrii, V.F. Gubarev (Eds). Advanced Control Systems: Theory and Applications. Series in Automation, Control and Robotics, River Publishers, Gistrup, 2021.

[32] Y.P. Kondratenko, A.A. Chikrii, V.F. Gubarev, J. Kacprzyk (Eds). Advanced Control Techniques in Complex Engineering Systems: Theory and Applications. Dedicated to Professor Vsevolod M. Kuntsevich. Studies in Systems, Decision and Control, Vol. 203. Cham: Springer Nature Switzerland AG, 2019.

[33] V.M. Kuntsevich, "Estimation of Impact of Bounded Perturbations on Nonlinear Descrete Systems", In: V.M. Kuntsevich, et al. (Eds). Control Systems: Theory and Applications. Series in Automation, Control and Robotics, River Publishers, Gistrup, Delft, 2018, pp. 3–15.

[34] J. Kacprzyk, et al. A Status Quo Biased Multistage Decision Model for Regional Agricultural Socioeconomic Planning Under Fuzzy Information. In: Y.P. Kondratenko, A.A. Chikrii, V.F. Gubarev, J. Kacprzyk, (Eds) Advanced Control Techniques in Complex Engineering Systems: Theory and Applications. Dedicated to Prof. V.M.Kuntsevich. Studies in Systems, Decision and Control, Vol. 203. Cham: Springer Nature Switzerland AG, 2019, pp. 201–226.

16

On Real-Time Calculation of the Rejected Takeoff Distance

V.I. Vyshenskyy, A.A. Belousov, and V.V. Kuleshyn

V.M. Glushkov Institute of Cybernetics of National Academy of Sciences
of Ukraine, Kyiv, Ukraine
E-mail: vyshenskyy@ukr.net; belousov@nas.gov.ua;
v.v.kuleshin@gmail.com

Abstract

A rejected takeoff is the situation when it is decided to cancel the takeoff of an airplane. When a crucial obstacle to the takeoff occurs (e.g. engine failure), the crew compares the current velocity with the "decision-making velocity". Based on this velocity one can estimate whether the aircraft can safely stop within the available distance of the interrupted flight. These values are taken from the Aircraft Flight Manual before the takeoff. But the real rejected takeoff distances can differ from the pre-calculated values. The goal of our investigation is to obtain real-time estimations of the distance to full stop. The proposed real-time algorithm of the calculation of the distance of the interrupted flight uses the step-wise interpolation of the thrust. Using this interpolation, the Riccati equation of the aircraft motion has explicit solutions. Besides using known time intervals of interpolation, the time to reach a given speed can be explicitly calculated. To check and study the proposed algorithm for estimating the rejected takeoff distance in real time, a computer simulation system was created. To verify the adequacy of the algorithm, the emulation results were compared with the values obtained from the real aircraft nomograms. The algorithm takes into account different aircraft and environment parameters (thrust, velocity, temperature etc.) before the moment of decision-making. As a result, the decision about the rejected takeoff will be more reliable and safe.

Keywords: rejected takeoff, braking distance, real-time algorithm, the Riccati equation.

16.1 Introduction

A rejected takeoff is the situation when it is decided to cancel the takeoff of an airplane. Analysis of accidents and serious incidents in civil aviation shows that an aircraft exiting the runway and terminal safety strip does not happen often when the takeoff is cancelled, but it is extremely dangerous. The reasons for these accidents and incidents may be the following: too late actions of the crew to stop takeoff; inconsistency of the conditions for the interrupted takeoff with the calculated ones. In the first case, the pilot or crew reacts to the appearance of hazard factors with a time delay greater than the allowable one. In the second, the conditions for a rejected takeoff do not correspond to those that were used in the preliminary calculations. In both cases, the aircraft does not have enough time to stop within the available distance of an interrupted takeoff (ASDA, Accelerate-Stop Distance Available), since the real distance of an interrupted takeoff (ASD, Accelerate-Stop Distance) exceeds ASDA.

When a crucial obstacle to the takeoff occurs (e.g. engine failure), the crew has just several seconds to analyse the situation and to make a "go–no go" decision–to continue a takeoff, or to abort it. During this time, the aircraft speed increases from v_{EF}(engine failure velocity) to $v_{EF} + \Delta v$. The crew compares current velocity to the "decision-making velocity" V_1. Based on V_1 velocity, it is possible to estimate if the airplane still can safely stop within the available distance of the airport. If the current speed is less than the "decision-making velocity", the crew makes a decision to interrupt the takeoff; otherwise, it is necessary to continue the takeoff, avoiding overrun. It is assumed that if the crew decides to terminate the takeoff at a speed less than or equal to V_1, the aircraft will stop within the ASDA. The velocity V_1 and ASDA distance are calculated in advance using the "Aircraft Flight Manual" [1]. However, the real rejected takeoff distances (the length of the braking section until the aircraft stops completely) can differ from the pre-calculated values used in flight manuals and nomograms due to different hazard factors [2, 3]. Among these factors, we should consider possible differences between the actual characteristics of the engines, the take-off weight of the aircraft and the conditions of the airfield (especially friction during takeoff and braking) from the characteristics used in calculating the airplane flight curves and nomograms. Therefore, a more precise real-time calculation of the rejected takeoff dista2nce performed by an automatic control system of the aircraft can help the crew to make accurate and safer "go–no go" decision.

This chapter continues some researches of previous fundamental and applied works [4–13]. These publications were devoted to the researches in complex control systems, conflict-controlled processes, dynamic games of approach, aircraft control during takeoff. The goal of our investigation is to obtain real-time estimations of the distance to full stop. The proposed algorithms use step interpolation of thrust, and also of other aircraft parameters. With such interpolation, the Riccati equation of the plane's motion has explicit solutions. In addition to using interpolation time intervals, the time to reach a given speed can be explicitly calculated, because some parameters change abruptly upon reaching a certain speed. Mathematical modelling of the rejected takeoff distance calculation makes it possible to evaluate the accuracy and speed of the algorithm, as well as its relevance to the real data of nomograms and airplane flight manual. Simulations can also investigate the influence of various hazard factors on rejected takeoff parameters.

16.2 Dynamic Model of Aircraft Movement along the Runway

The aircraft motion on a runway can be described by differential equations [14]

$$m\frac{dv}{dt} = P\cos\alpha_P - Q - mg\sin\theta + f(Y - mg)\cos\theta \tag{16.1}$$

$$\frac{dl}{dt} = v, \tag{16.2}$$

$$v(0) = 0, \; l(0) = 0, \tag{16.3}$$

where t–time, v–velocity, l–distance on the airport runway, m–mass, P–thrust of all aircraft engines, α_P–aircraft engine angle of attack, $Q = c_x\frac{\rho v^2}{2}S$–aircraft drag force, $Y = c_y\frac{\rho v^2}{2}S$–aircraft lift, c_x–aircraft drag coefficient, c_y–aircraft lift coefficient, S–relevant surface area, ρ–air density, g–acceleration of gravity, θ–runway angle, f–friction coefficient.

Equation (16.1) is a Riccati equation

$$\frac{dv}{dt} = -\Lambda v^2 + G, \tag{16.4}$$

$$\Lambda = \frac{\rho S}{2m} \cdot (c_x - f \cdot c_y), \tag{16.5}$$

$$G = \frac{\cos\alpha_P}{m}P - g(\sin\theta - f\cos\theta), \tag{16.6}$$

where in (16.4) both Λ and G depend on aircraft and environment parameters (aircraft mass, friction coefficient, air density, lift and drag coefficients, etc.), but only G depends on the thrust.

16.3 Analytical Solution of the Aircraft Motion Equation

Consider the function $x(t) = e^{\Lambda l(t)}$. For this function, we have

$$\frac{de^{\Lambda l}}{dt} = \Lambda e^{\Lambda l}\frac{dl}{dt} = \Lambda e^{\Lambda l}v,$$

$$\frac{d^2 e^{\Lambda l}}{dt^2} = \Lambda\cdot\left[\frac{de^{\Lambda l}}{dt}v + e^{\Lambda l}\frac{dv}{dt}\right] = \Lambda\cdot\left[\Lambda e^{\Lambda l}v^2 + e^{\Lambda l}\left(-\Lambda v^2 + G\right)\right] = \Lambda G e^{\Lambda l}.$$

1. If $\Lambda G < 0$, then $\ddot{x} = -\omega^2 x$, where $\omega = \sqrt{-\Lambda G}$. This is the classical equation of small oscillations of a pendulum [15]. General solution of this equation is

$$x(t) = C_1 \sin \omega t + C_2 \cos \omega t, \tag{16.7}$$

   with constants C_1, C_2 that can be obtained from the initial conditions.
2. If $\Lambda G > 0$, then $\ddot{x} = \gamma^2 x$, where $\gamma = \sqrt{\Lambda G}$. This is the equation of the "inverted pendulum" [15] with general solution

$$x(t) = C_1 \operatorname{sh} \gamma t + C_2 \operatorname{ch}\gamma t, \tag{16.8}$$

   $\operatorname{sh} t = \frac{e^t - e^{-t}}{2}$, $\operatorname{ch} t = \frac{e^t + e^{-t}}{2}$ - hyperbolic sine and cosine.
3. If $\Lambda G = 0$, we obtain a trivial equation with general solution that describes the motion by inertia

$$x(t) = C_1 t + C_2. \tag{16.9}$$

16.4 Engine Thrust Modelling During Switching off and Reverse

Let us analyse the actions of pilots and the change in thrust of the engines during rejected takeoff. Before the cause of the aborted takeoff (for example, engine failure) is detected, the airplane travels the distance l_0 and accelerates to the speed v_{EF}. Then the pilot spends some time δ identifying the cause and making a decision. If the current airplane speed is less than the decision-making velocity V_1, the process of the rejected takeoff usually starts. The thrust of the engines at the time $t_0 + \delta$ is practically equal to the thrust P_0 at the time t_0. At this time, the engine control levers are moved to the "Small Gas" position. As a result, the thrust of the engines is reduced to almost zero in 1.5–2 seconds. Then the breaking systems are activated, including reversal thrust. When the reverse is turned on, the negative thrust of the engines changes from 0 to the full reverse thrust U. This process requires about 7–11 seconds. It is assumed that the transient processes end when they reach 0.95 of the steady-state values [16].

Complex mechanical-mathematical and thermodynamic relations describe the transients of modern turbojet engines. However, from the point of view of control theory, the integral effect of engine thrust on an aircraft during the transient process is described with good accuracy by linear differential equations [16, 17].

The equation for decreasing the takeoff thrust of the engine to zero can be written as follows [17]

$$\frac{dP}{dt} = -\frac{P}{T_1},$$

$$P(t) = P_0 \cdot e^{-t/T_1}, \tag{16.10}$$

where the value T_1 is called the engine shutdown time constant and is an important characteristic of a particular type of turbojet engine. Function (16.10) satisfies the relation

$$P(3T_1) = P_0 \cdot e^{-3} \approx 0,05P_0, \tag{16.11}$$

$$\int_0^{3T_1} P(t)\, dt = P_0 \int_0^{3T_1} e^{-t/T_1}\, dt = P_0 \cdot T_1 \left(1 - e^{-3}\right) \approx 0,95 \cdot$$

$$P_0 \cdot T_1 = 0,95 \cdot \int_0^{T_1} P_0\, dt.$$

We can substitute the thrust function (16.10) in the Riccati Equation (16.4) and numerically find the distance travelled by the aircraft during the transition process. However, given the relation (16.11), instead of the exponential thrust function (16.10) it is possible to use the step-wise constant function

$$P(t) = \begin{cases} P_0, & 0 \le t < T_1, \\ 0, & T_1 \le t \le 3T_1, \end{cases}$$

that fairly well approximates the impact of thrust (16.10).

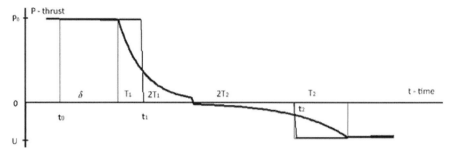

Figure 16.1   Dependence of engine thrust on time

Similarly, we can consider the process of the engine reversal from zero to the nominal reverse thrust, which is described by the function

$$P(t) = U \cdot e^{t/T_2 - 3}, \quad 0 \le t \le 3T_2, \tag{16.12}$$

where T_2 is the reverse time constant.

Thus, the processes (16.12) of switching off the thrust of the engine and the activating of reverse mode are proposed to be emulated by a step function

$$P(t) = \begin{cases} P_0, & t_0 \le t < t_1 = t_0 + \delta + T_1; \\ 0, & t_1 \le t < t_2 = t_1 + 2T_1 + 2T_2; \\ U, & t_2 \le t. \end{cases} \tag{16.13}$$

Dependence of engine thrust on time described by (16.10) and (16.12), as well as their emulated values described by (16.13), are shown on Figure 16.1.

Note that the actual shutdown of the takeoff thrust starts at the time $t_0 + \delta$, and the start of reverse mode at the time $t_0 + \delta + 3T_1$.

16.5 Rejected Takeoff Distance Calculation

Denote $l_i = l(t_i)$, $v_i = v(t_i)$, $\Delta_i = t_i - t_{i-1}$, $\Delta l_i = l_i - l_{i-1}$, $i = 1, 2, 3$.

Determine the distance travelled by the aircraft during time interval $[t_0, t_1]$ with constant thrust P_0. In this case, the value $G > 0$, and the general solution has the form (16.8)

$$e^{\Lambda l(t)} = C_1 sh\gamma t + C_2 ch\gamma t, \quad \gamma = \sqrt{\Lambda G},$$

its derivative

$$\frac{de^{\Lambda l(t)}}{dt} = \Lambda e^{\Lambda l(t)} v(t) = C_1 \gamma ch\gamma t + C_2 \gamma sh\gamma t.$$

We can find the constants C_1 and C_2 from the initial conditions $l(0) = 0$ and $v(0) = v_0$:

$$e^{\Lambda \cdot 0} = C_1 sh(0) + C_2 ch(0) = C_2,$$

$$\Lambda e^{\Lambda \cdot 0} v_0 = C_1 \gamma ch(0) + C_2 \gamma sh(0) = C_1 \gamma,$$

so we have

$$C_1 = \frac{\Lambda v_0}{\gamma}, C_2 = 1.$$

We get the relation between the distance Δl_1 and the speed v_1 over time $\Delta_1 = \delta + T_1$

$$e^{\Lambda \Delta l_1} = \frac{\Lambda v_0}{\gamma} sh(\gamma \Delta_1) + ch(\gamma \Delta_1),$$

$$\Lambda e^{\Lambda \Delta l_1} v_1 = \Lambda v_0 ch(\gamma \Delta_1) + \gamma sh(\gamma \Delta_1),$$

thus

$$\Delta l_1 = \frac{1}{\Lambda} \ln \left[\frac{\Lambda v_0}{\gamma} sh(\gamma \Delta_1) + ch(\gamma \Delta_1) \right], \tag{16.14}$$

$$v_1 = e^{-\Lambda \Delta l_1} \left[v_0 ch(\gamma \Delta_1) + \frac{\gamma}{\Lambda} sh(\gamma \Delta_1) \right]. \tag{16.15}$$

During time interval $[t_1, t_2]$ (with zero thrust) and $[t_2, t_3]$ (with reverse thrust) we have $\Lambda > 0$, $G < 0$, and general solution is (16.7)

$$e^{\Lambda l(t)} = C_1 \sin \omega t + C_2 \cos \omega t, \tag{16.16}$$

where $\omega = \sqrt{-\Lambda G}$. Its derivative

$$\Lambda e^{\Lambda l(t)} v(t) = \omega C_1 \cos \omega t - \omega C_2 \sin \omega t. \tag{16.17}$$

The constants C_1, C_2 can be found from initial conditions $l(0) = 0$ and $v(0) = v_2$, that is $e^{\Lambda \cdot 0} = C_2$, $\Lambda e^{\Lambda \cdot 0} v_2 = \omega C_1$, so

$$C_1 = \frac{\Lambda v_2}{\omega}, \quad C_2 = 1.$$

Then, substituting the condition of stop $v = 0$ into (16.17), we obtain

$$tg\omega \Delta_3 = \frac{\Lambda v_2}{\omega},$$

and the time to full stop

$$\Delta_3 = \frac{1}{\omega} arctg \left(\frac{\Lambda v_2}{\omega} \right).$$

We can find the distance Δl_3 from (16.16)

$$e^{\Lambda \Delta l_3} = \frac{\Lambda v_2}{\omega} \sin \omega \, \Delta_3 + \cos \omega \, \Delta_3 = \frac{1}{\cos \omega \, \Delta_3} = \sqrt{1 + tg^2(\omega \, \Delta_3)}$$

$$= \sqrt{1 + \frac{\Lambda^2 v_2^2}{\omega^2}} = \sqrt{1 - \frac{\Lambda v_2^2}{G}}$$

so

$$\Delta l_3 = \frac{1}{2\Lambda} \ln \left[1 - \frac{\Lambda v_2^2}{G} \right]. \tag{16.18}$$

Similarly, we can find the distance Δl_2 during time interval $[t_1, t_2]$.

Thus, taking into account all three stages of braking, the total distance travelled by the aircraft during an emergency stop will be $l_0 + \Delta l_1 + \Delta l_2 + \Delta l_3$. If this distance is within the length of the runway, then we assume that a safe interrupted takeoff is possible. The total time spent on the interrupted takeoff will be $t_0 + \delta + 3T_1 + 2T_2 + \Delta_3$.

16.6 Interpolation of Aircraft Parameters using Constant Values

Section 16.5 proposes an algorithm for estimating the distance of interrupted takeoff in real time, which uses stepwise thrust interpolation. However, the real process of acceleration and interrupted takeoff is much more complicated. Engine thrust during acceleration is not constant. For example, An–124 aircraft at the beginning of the takeoff uses only 0.7 of the nominal thrust during first 3 seconds. So we assume that $P(t) = 0.7P_0$, $0 < t < t^*$, $P(t^*) = P_0$. Engine thrust depends on speed, $P = P(v)$ – it decreases when the speed increases. After deciding to stop the takeoff, in addition to switching to reverse thrust, the crew increases the aerodynamic drag and reduces the lift of the aircraft, using interceptors. As a result, aerodynamic coefficients c_x, c_y in (16.1) are changed. Wheel brakes are also used and the friction coefficient f changes. The reverse thrust decreases gradually with decreasing speed, and when the speed of the aircraft reaches a certain value, the reverse thrust decreases abruptly to the value of the so-called small gas reverse. Thus, when emulating the motion of the aircraft using time intervals with constant values of the parameters, we can assume that some values change at fixed known moments of time, and at these time intervals $\Delta_i = t_{i+1} - t_i$ we need to estimate $l_{i+1} = l(t_{i+1})$, $v_{i+1} = v(t_{i+1})$ using known t_i, t_{i+1}, $l(t_i)$, $v(t_i)$. We assume that engine failure occurs at some speed v_{EF}, and after that, the rejecting takeoff begins, with known time parameters δ, T_1, T_2 etc. We do not know the exact time t_{EF} of the engine failure, this time should be calculated, but we know that $P(t_{EF}) = P(v_{EF})$. As a result, during the time interval $[t^*, \ t_{EF}]$ we can interpolate the thrust with constant value $\frac{1}{2}[P_0 + P(v_{EF})]$. For more accuracy, we can split $[t^*, \ t_{EF}]$ into two or more parts, e.g. $[t^*, \ \tau]$, $[\tau, \ t_{EF}]$, where $v(\tau) = \frac{1}{2}[v(t^*) + v_{EF}]$, and we can use two constant values $\frac{1}{2}[P_0 + P(v(\tau))]$ and $\frac{1}{2}[P(v(\tau)) + P(v_{EF})]$. Similarly, we assume that we know the time when the coefficients c_x, c_y change their values – it takes 1–2 second after start of the reverse thrust. The time of small gas reverse is not known, and it must be calculated during emulation.

16.7 Estimation of Speed, Path and Time using Fixed Parameters of the Aircraft

Equation (16.4) with constant Λ, G and initial conditions

$$v(t_0) = v_0, \ l(0) = 0, \tag{16.19}$$

allows to obtain both explicit dependences of path and velocity on time, $l(t)$, $v(t)$, and also analytically define the dependency of time on velocity $t(v)$. We will use explicit solutions (16.7) and (16.8). The case $\Lambda G = 0$ does not occur in practice during aircraft movement.

Consider the function $x(t) = e^{\Lambda l(t)}$. For this function we have

$$\frac{dx}{dt} = \Lambda e^{\Lambda l} \frac{dl}{dt} = \Lambda x v, \tag{16.20}$$

$$\frac{d^2 x}{dt^2} = \Lambda \cdot \left[\frac{de^{\Lambda l}}{dt} v + e^{\Lambda l} \frac{dv}{dt} \right] = \Lambda \cdot \left[\Lambda e^{\Lambda l} v^2 + e^{\Lambda l} \left(-\Lambda v^2 + G \right) \right] = \Lambda G x.$$

Let $\Lambda G > 0$, and denote $\gamma = \sqrt{\Lambda G}$. Then

$$x(t) = C_1 sh\gamma t + C_2 ch\gamma t. \tag{16.21}$$

Taking into account the initial conditions, we can obtain explicit dependence of time on speed [10]

$$t = \frac{1}{2\gamma} \ln \left(\frac{1+z}{1-z} \right), \ z = \frac{\Lambda \gamma (v_0 - v)}{\Lambda^2 v_0 v - \gamma^2}. \tag{16.22}$$

We can also obtain explicit formulas to calculate $l(t), \quad v(t)$

$$l(t) = \frac{1}{\Lambda} \ln \left(\frac{\Lambda v_0}{\gamma} sh(\gamma t) + ch(\gamma t) \right),\tag{16.23}$$

$$v(t) = e^{-\Lambda l(t)} \left[v_0 ch(\gamma t) + \frac{\gamma}{\Lambda} sh(\gamma t) \right].\tag{16.24}$$

For the case $\Lambda G < 0$ we denote $\omega = \sqrt{-\Lambda G}$, and the solution is

$$x(t) = C_1 \sin \omega t + C_2 \cos \omega t.\tag{16.25}$$

Taking into account the initial conditions (16.19), we can obtain explicit dependency of time t from velocity v

$$t = -\frac{1}{\omega} arctg \frac{\Lambda \omega (v - v_0)}{\Lambda^2 v_0 v + \omega^2}.\tag{16.26}$$

We can also calculate $l(t), \quad v(t)$:

$$l(t) = \frac{1}{\Lambda} \ln \left(\frac{\Lambda v_0}{\omega} \sin \omega t + \cos \omega t \right),\tag{16.27}$$

$$v(t) = e^{-\Lambda l} \left(v_0 \cos \omega t - \frac{\omega}{\Lambda} \sin \omega t \right).\tag{16.28}$$

Thus, at each interval of interpolation of the aircraft parameters with fixed values, we can analytically express the dependences of distance, time and speed under given initial conditions. If time is known, we can use (16.23), (16.24) or (16.27), (16.28), depending on the ΛG sign. If the time is unknown, but we know the speed at the end of the time interval, then the time can found from (16.22) or (16.26).

16.8 Computer Simulation

To verify and study the proposed algorithms for estimating the distance of the interrupted takeoff in real time, a computer simulation system was created. With this system, we can compare the rejected takeoff distance obtained using the proposed algorithms with the distance calculated from full motion model (16.1)–(16.3) using the numerical solution of the differential equations. The system allows varying numerous aircraft parameters (more than 50) and initial states. The calculations were performed based on the real data of the An-124 aircraft. In addition to the above-mentioned parameters, the temperature, slope of the runway, air density, fuel burn-up rate, number of running engines, thrust dependence on speed, etc. are also taken into account. You can set different initial conditions (mass of the aircraft, the speed of decision-making, after which there is braking, etc.). You can change the time parameters that affect the process of starting and braking (for example, the reaction time of the crew, the time characteristics of the transition from full thrust to reverse, time intervals of change). You can also vary the speed of transition from full reverse thrust to small gas reverse thrust. With the help of the system, both the entire path from start to stop (takeoff acceleration and braking) and the process of only takeoff interruption were investigated (the speed of decision-making is set and the distance to the moment of stop is calculated). Different values of engine failure velocity v_{EF} were investigated, including the values near the decision-making velocity V_1. Actually, for real-time calculations it is necessary to use the part of the system that emulates only takeoff interruption. For such calculations, it is possible to use only 8 time intervals with approximation of parameters of system by the fixed values. It took 10 time intervals to simulate both takeoff and braking, with sufficient accuracy of emulation. To increase the accuracy of the emulation of the acceleration part of the rejected takeoff, we can split the time interval just before the reaching the v_{EF} into several parts (up to 4); in his case we will use at most 13 time intervals.

The Java programming language was chosen to develop the system. We used a Java Development Kit version 1.8 (JDK 1.8). Apache Maven version 3.6.3 was used as a project management tool. The user interface (GUI) was developed using Java Swing.

The main libraries used to develop the system:
Ini4J (Java API for processing configuration files in .ini format for Windows)–configuration and initial values of system parameters.

Apache Commons Mathematics Library (library of light, independent components of mathematics and statistics)-numerical integration of differential equations describing the motion of the aircraft.

JChart2D (an easy-to-use component written in Java for displaying two-dimensional tracks in a coordinate system)–visualization of changes in parameters, speed and coordinates of the aircraft in real time.

Junit, hamcrest, powermock-api-mockito–libraries for the program testing.

A general view of the user interface is shown in Figure 16.2. Using the sliders, you can change up to 50 parameters; their number, names, as well as the initial

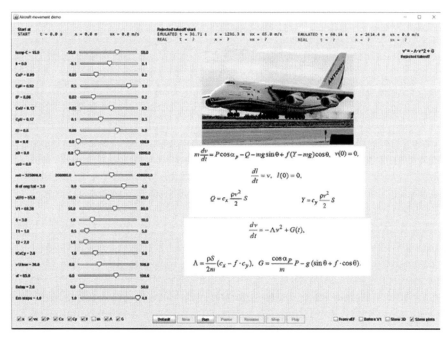

Figure 16.2   User interface

values of these parameters and the ranges of their changes are specified in the configuration file config.ini. In Figure 16.2, about 30 parameters are available for changing by sliders, and the remaining 20 are set directly in config.ini.

After changing necessary parameters, we press "Run" button, and the system performs emulation and equation integration. The results will be shown as plots, see Figure 16.3. Using checkboxes (left bottom), we can select the parameters to display. The simulation results are displayed in real time. For each parameter displayed on the screen, such as speed v, two values are displayed simultaneously in real time. The bold line (Real) shows the real value of the parameter obtained from the complete model (16.1)– (16.3) by numerical integration; thin line (Emulated) represents piecewise linear approximation of the parameter using the proposed algorithm. The upper left corner of the screen shows the initial distance, velocity and time. Top central part shows the distance, velocity and time of the interrupted takeoff. REAL values are obtained by numerical integration of the complete model, EMULATED values are obtained using the described algorithm using two or more (up to 5) piecewise constant approximations, as selected by

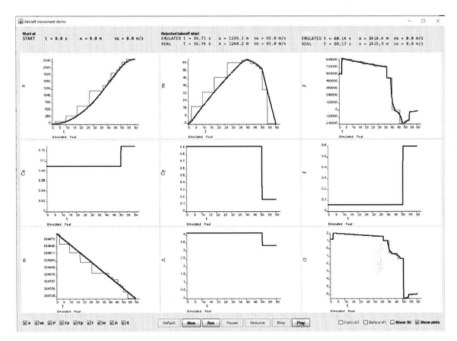

Figure 16.3   Emulation and integration results

corresponding slider. The upper right corner of the screen shows the values of the total distance and the time of the interrupted takeoff, at the moment of full stop with $v = 0$. REAL values are obtained by numerical integration of the complete model, EMULATED values are obtained using the described algorithm using piecewise constant approximations.

16.9 Emulation Results

Figure 16.4 shows the distance, velocity, and thrust from the moment of engine failure. Such calculations, from the moment of danger, should be used for calculations in real takeoff. The emulation shows that we can use only 8 time steps while aircraft parameters are changed significantly. The thick line (Real) shows the aircraft distance and velocity on the runway. These distance and velocity are calculated using the numerical solution of the differential equations with the thrust shown on the figure, and with other parameters that depend on time and velocity (lift, drag, and friction coefficients, mass, etc.). These parameters are based on values for An–124 aircraft. The thin

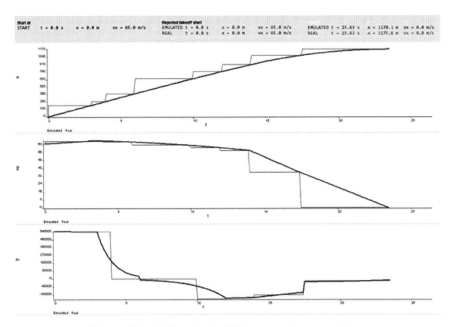

Figure 16.4   Rejected takeoff distance, velocity, and thrust

line (Emulated) shows the constant emulated values for the distance, velocity, and thrust that are calculated using constant estimations of the aircraft parameters.

Table 16.1 shows the typical values of the distance to the stop after reaching the decision-making speed, calculated by this algorithm, in comparison with the results of numerical integration of the complete model. From Table 16.1 we can see that the calculation of the interrupted takeoff distance using the proposed algorithm corresponds to the results of the integration of the motion Equations (16.1)–(16.3). Thus, the piecewise

Table 16.1   Comparison of emulation and integration

Mass, ton	250	276	300	325	350	376
Decision-making velocity, m/s	65.6	66.5	68	68.4	69.5	70.7
Engine failure velocity, m/s	58.4	60.17	62.4	63.3	64.9	66.6
Distance to full stop, m – emulation using algorithm	1060	1089	1128	1144	1174	1206
Distance to full stop, m – numerical integration	1057	1085	1125	1141	1172	1204

Table 16.2   Comparison of emulation and nomograms

Mass, ton	250	275	300	325	350	375
Decision-making velocity, m/s	65.6	66.6	67.5	68.4	69.4	70.7
Rejected takeoff distance, m, nomogram	2008	2164	2319	2504	2689	2875
Rejected takeoff distance, m – emulation using algorithm	1962	2127	2305	2499	2710	2982

constant interpolation of engine thrust and other parameters of the aircraft is legitimate.

16.10 Model Verification

Confirmation of the model adequacy was performed by comparing the results of mathematical modeling of the motion of the An-124 aircraft during interrupted takeoff with the results obtained from the nomograms from the An–124 flight manual. An interrupted takeoff simulation was performed for several values of decision-making speed V_1 and aircraft takeoff mass m. It was taken into account that the values of time intervals used in the model are not strictly defined in the regulations, and therefore allow some variation over the commonly used values.

Table 16.2 shows the results of comparing the lengths of the interrupted takeoff, calculated using mathematical modeling, and nomograms, with the same values of the model parameters (nominal thrust, drag coefficients, lift, friction, etc.).

As we can see from Table 16.2, the simulation allows to obtain results quite close to the data from the flight nomograms. Thus, the model we used is adequate and acceptable for studying the influence of hazard factors on the characteristics of interrupted takeoff.

16.11 Conclusion

The proposed algorithm for calculating the rejected takeoff distance uses step interpolation of aircraft thrust and other parameters. This interpolation allows to obtain explicit solutions of the Riccati equation that describes the aircraft motion. The calculations by explicit formulae on a small number of time intervals can be performed very quickly, in real time. In any moment of the takeoff, we can obtain an estimation of the distance to the full stop. Algorithm takes into account aircraft parameter just in the moment of emergency. With this algorithm, the accuracy of the rejected takeoff distance estimation is better comparing to the pre-calculated values. Computer simulation shows high accuracy of the algorithm. Its simplicity and high speed make it suitable for using in on-board automatic control systems to provide recommendations to the crew at the time of an emergency. So the decision about the takeoff interrupt will be more accurate and safe.

References

[1] Airplane AN-124-100. "Manual for flight" [in Russian], Book 1, 1993.

[2] NLR Air Transport Safety Institute, Report no. NLR-TP-2010-177, April 2010, https://reports.nlr.nl/xmlui/bitstream/handle/10921/15 8/TP-2010-177.pdf.

[3] "Reducing the risk of runway excursion". RUNWAY SAFETY INITIA-TIVE. Flight Safety Foundation, May 2009, https://flightsafety.org/files /RERR/fsf-runway-excursions-report.pdf.

[4] Kondratenko, Y.P., Kuntsevich, V.M., Chikrii, A.A., Gubarev, V.F. (Eds). Advanced Control Systems: Theory and Applications. Series in Automation, Control and Robotics, River Publishers, Gistrup, 2021.

[5] Kondratenko, Y.P., Chikrii, A.A., Gubarev, V.F., Kacprzyk, J. (Eds). Advanced Control Techniques in Complex Engineering Systems: Theory and Applications. Dedicated to Professor Vsevolod M. Kuntsevich. Studies in Systems, Decision and Control, Vol. 203. Cham: Springer Nature Switzerland AG, 2019.

[6] A.A. Chikrii. "Control of Moving Objects in Condition of Conflict". In: Kuntsevich, V.M. et al. (Eds). Control Systems: Theory and Applications. Series in Automation, Control and Robotics, River Publishers, Gistrup, Delft, 2018, pp. 17–42.

[7] A.A. Chikrii, et al. Method of Resolving Functions in the Theory of Conflict-Controlled Processes. In: Kondratenko, Y.P., Chikrii, A.A., Gubarev, V.F., Kacprzyk, J. (Eds) Advanced Control Techniques in Complex Engineering Systems: Theory and Applications. Dedicated to Prof. V.M.Kuntsevich. Studies in Systems, Decision and Control, Vol. 203. Cham: Springer Nature Switzerland AG, 2019, pp. 3–33. DOI: https://doi.org/10.1007/978-3-030-21927-7_1

[8] A.A. Belousov, V.V. Kuleshyn, "Game approach to control of running start of aircraft on its take off", J. of Automation and Information Sciences, 2012, Volume 44, Issue 8, pp. 78–84.

[9] A.A. Belousov, V.V. Kuleshyn, V.I. Vyshenskiy, "Real–time algorithm for calculation of the distance of the interrupted take–off", J. of Automation and Information Sciences, 2020, Volume 52, Issue 4, pp. 38–46

[10] V.I. Vyshenskiy, V.V. Kuleshyn., A.A. Belousov , "Computer Simulation of Accelerate-Stop Distance Calculation", J. of Automation and Information Sciences, 2020, Volume 52, Issue 11, pp. 72–80

[11] A.A. Chikrii, S.D. Eidelman, "Control game problems for quasi-linear systems with Riemann-Liouville fractional derivatives", Cybernetics and Systems Analysis, 2001, vol. 36, No 6, pp. 836–864.

[12] A.A. Chikrii, G.Ts. Chikrii, "Matrix resolving functions in dynamic games of approach", Cybernetics and Systems Analysis, 2014, vol. 50, No 2, pp. 201–217.

[13] A.A. Chikrii, I.I Matichin, K.A. Chikrii, "Conflict controlled processes with discontinuous trajectories", Kibernetika i Sistemnyi Analiz, 2004, vol. 40, No 6, pp. 15–29 (in Russian)

[14] N.M. Lysenko, "Flight dynamics" [in Russian], VVIA im. prof. N.E. Zhukovskogo, Moscow, 1967.

[15] V.I. Arnold, "Ordinary differential equations", Universitext, Springer–Verlag, Berlin, 2006, 334 pp.

[16] Popov E.P., "Automation regulation and control" [in Russian]. Nauka, Moscow, 1966. –386 pp.

[17] Shtoda A.V., Morozov F.N., Shiukov A.G., "Systems of control and regulation of aviation engines" [in Russian], VVIA im. prof. N.E. Zhukovskogo, Moscow, 1977.

17

Automated Control Problem for Dynamic Processes Applied to Cryptocurrency in Financial Markets

Viktor Romanenko, Yurii Miliavskyi, and Heorhii Kantsedal

Institute of Applied System Analysis of National Technical University
of Ukraine "Igor Sikorsky Kyiv Polytechnic Institute",
37 Peremohy av., Kyiv, 03056, Ukraine
E-mail: romanenko.viktorroman@gmail.com; yuriy.milyavsky@gmail.com;
g.kantsedal@protonmail.com

Abstract

In this chapter we develop a cognitive map (CM) of the use of cryptocurrency in the financial market and describe dynamic model of CM impulse processes as a difference equation system (Roberts equations) based on that CM. We chose an external control vector for the CM impulse process provided by means of CM nodes varying. A closed-loop CM impulse process control system was implemented. This system includes multidimensional discrete controller, based on automated control theory methods, that generates selected control vector and directly affects respective CM nodes by varying their coordinates. We solved three problems of a discrete controller design for automated dynamic processes control applied to cryptocurrency in financial markets. The first problem is unstable cryptocurrency rate stabilization based on modal CM impulse process control. The second problem is constrained external and internal disturbances suppression during CM impulse processes control based on invariant ellipsoid method. The third problem is minimizing generalized variance of CM nodes coordinates and controls for stabilizing coordinates at given levels. In this chapter, we developed a system for identifying CM weighting coefficients based on a recurrent least-squares method. We did some performance research for each of the designed discrete controllers.

Keywords: Cognitive map, linear matrix inequalities, cryptocurrency, state controller, invariant ellipsoid method, modal control.

17.1 Introduction

To study the dynamic processes applied to cryptocurrency the chapter uses cognitive modelling, which is one of the most relevant areas of scientific and practical research of complex systems of different nature now [1–4]. It is based on the notion of a cognitive map (CM), which is a weighted directed graph [5]. The nodes (nodes) are coordinates (factors) of the complex system and weighted edges (arcs) of the graph describe causal links between CM nodes. When disturbances affect CM nodes, we can observe impulse transitional process. Equation (17.1) describes the dynamics of that process [6]:

$$\Delta y_i(k+1) = \sum_{j=1}^{n} a_{ij} \Delta y_j(k), \qquad (17.1)$$

where $\Delta y_i(k) = y_i(k) - y_i(k-1), i = 1, 2, ..., n$, a_{ij}- weight of an edge connecting the jth node and the ith one. Equation (17.1) describes the free motion of the ith node of CM without external control impact. We can write this equation in vector-matrix form (17.2):

$$\Delta \bar{Y}(k+1) = A\Delta \bar{Y}(k), \qquad (17.2)$$

where A is a weighted adjacency matrix of CM.

In order to implement CM impulse process control based on modern control theory [7, 8] it is necessary to be able to physically change some coordinates of CM nodes as control actions. Then we can describe the forced motion of the CM impulse process under external control as (17.3):

$$\Delta \bar{Y}(k+1) = A\Delta \bar{Y}(k) + B\Delta \bar{U}(k), \qquad (17.3)$$

where $\Delta \bar{U}(k) = \bar{U}(k) - \bar{U}(k-1)$– vector of controls increments with size $m \leq n$. The operator fills the control matrix B and usually uses ones and zeros.

In [9] stabilization of CM impulse processes in complex systems for $m \leq n$ based on modal control described in [10] is suggested. We investigated options of modal control with one and with several controls and carried out the simulation of a real modal control system of the CM impulse process in the commercial bank.

17.1.1 Problem Statement

The first problem is to develop a dynamic control model of the CM impulse process for the application of cryptocurrency in the financial market. The second problem is the development and research of a modal control algorithm to stabilize an unstable cryptocurrency rate. The third problem is to develop a closed-loop CM impulse process control system to suppress constrained external and internal influences. The control system is based on the invariant ellipsoids method. In the fourth problem it is necessary to implement a discrete-time controller for stabilizing the CM nodes coordinates at specified levels based on the criterion of minimal generalized variance of the nodes coordinates and the control vector under the influence of perturbations. The fifth problem is to implement adaptive control of the CM impulse process with unknown or varying coefficients of the adjacency matrix A. Adaptive control includes estimating the elements of matrix A during the transient impulse process and using the estimated values of A to generate the control vector $\Delta \overline{U}(k)$.

17.2 Cognitive Mapping of Cryptocurrency Usage in Financial Markets

Cryptocurrency (bitcoin in particular) occupies an intermediate position between conventional currency and securities. A lot of scientific research has been devoted to the exploration of this phenomenon [11–14]. Bitcoin has users who informally create a kind of a "gross domestic product" and there are companies that accept payments in this cryptocurrency. We can estimate the number of users by the number of transactions between the so-called "wallets" and transfers between them (this is publicly available information, according to which you can see transfers made in bitcoin or other cryptocurrency, but "wallets" do not have any information about people). People create crypto exchanges in order to create contracts to purchase bitcoins and receive credits. Exchange has a certain number of users and a certain amount of money comes in every day (trading volume). Several exchanges can function at the same time (and actually, we have a lot of them now). Each exchange is itself a common company. Users entrust their money to it and actually support its development. Exchanges form the price of bitcoin. During formation of the initial price of bitcoin (during mining), it's worth in the range from 2000 to 3000 dollars, but expectations of bitcoin becoming a conventional payment system allow it to maintain its price at 9000-10000 dollars. But any financial news might drastically decrease the price.

There are various perturbations to the processes of using cryptocurrency, aimed at reducing the level of confidence in the use of cryptocurrency. Therefore, there are the following risks that can arise in crypto operations.

1. Risk of losing users, which leads to a decrease in the price of bitcoin;
2. Risk associated with inappropriate general expectations of many users at the same time, which traders create by manipulating the trades on the exchanges or breaking crypto-related news;
3. The risk of a sharp change in the price of cryptocurrency as a result of fraud on exchanges like high-frequency trading that usually lies in the high speed of buying monetary assets before the large operation of an investor and then selling it to the investor, before the information about the purchase reaches all exchanges;
4. The risk associated with the lack of guarantee of the safety of capital invested in the purchase of cryptocurrency, which leads to a certain hysteria of users during trading on exchanges;

Figure 17.1 shows the CM, which is developed on the basis of causal relationships.

We introduce the following notations of the CM nodes:

1. rate of cryptocurrency (bitcoin value),
2. trading volume of cryptocurrency,
3. capitalization volume,
4. number of cryptocurrency users,
5. investment volume (interest in bitcoin from institutional investors),
6. volume of speculation in cryptocurrency,
7. supply of cryptocurrency,
8. variance of cryptocurrency rate,
9. demand for cryptocurrency,
10. indirect profit,
11. level of confidence in cryptocurrency,
12. integrated level of risk when using cryptocurrency.

We can measure or compute nodes 1-8. However, we cannot measure nodes 9-12 or it's hard to measure them.

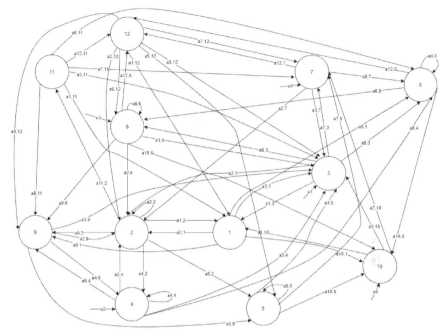

Figure 17.1 CM cryptocurrency usage in financial markets

As the control actions at the expense of varying the CM node coordinate resources can be used:

- Varying of cryptocurrency trading volume ($\Delta u_1(k)$);
- Varying of cryptocurrency capitalization volume ($\Delta u_2(k)$);
- Varying of investments volume ($\Delta u_3(k)$)
- Varying of speculation volume ($\Delta u_4(k)$)
- Varying the offer of cryptocurrency ($\Delta u_5(k)$).

The adjacency matrix A has the following form:

$$
A = \begin{pmatrix}
1.3 & 0.25 & 0.15 & 0.4 & 0.6 & 0.3 & 0 & 0 & 0.2 & 0 & 0.4 & -0.5 \\
0 & 0 & 0.15 & 0.2 & 0.1 & 0.7 & 0.7 & -0.4 & 0.8 & 0 & 0.7 & -0.5 \\
0.5 & 0 & 0.9 & 0 & 0.4 & 0 & 0 & 0 & 0 & 0 & 0.6 & 0 \\
0.65 & 0 & 0 & 0.85 & 0 & 0 & 0 & 0 & 0 & 0 & 0.7 & -0.7 \\
0 & 0.4 & 0 & 0 & 0.9 & 0 & 0 & 0 & 0 & 0.5 & 0 & -0.4 \\
0 & 0.7 & 0 & 0.4 & 0 & 0 & 0.75 & -0.2 & 0 & 0 & 0 & -0.5 \\
0 & 0 & 0 & 0.4 & 0.4 & 0 & 0 & 0 & 0.5 & 0.3 & 0 & 0 \\
-0.3 & 0 & 0 & 0 & 0 & 0 & 0 & 0 & 0 & 0 & 0 & 0 \\
0.05 & 0.8 & 0 & 0 & 0 & 0.5 & 0.7 & -0.3 & 0 & 0 & 0 & -0.4 \\
0 & 0.7 & 0.1 & 0 & 0 & 0.5 & 0 & -0.5 & 0.1 & 0.9 & 0.05 & 0 \\
0.5 & 0 & 0.5 & 0 & 0.3 & 0 & 0 & -0.3 & 0.4 & 0 & 0 & -0.5 \\
0 & 0 & 0 & 0 & -0.4 & 0.6 & 0 & 0.4 & 0 & -0.2 & 0 & 0
\end{pmatrix}.
$$

We use the calculation of the maximum sample variance [15] to determine the variance of the cryptocurrency rate (node 8). The following algorithm could define the variance on the interval NT_0:

$$\mathrm{var}\left\{y_1\left(kT_0\right)\middle| y_1\left((k-1)\,T_0\right),...,y_1\left((k-N)\,T_0\right)\right\} =$$
$$= \frac{1}{N}\sum_{i=1}^{N}\left\{y_i\left((k-i)\,T_0\right)-\frac{1}{N}\sum_{j=1}^{N}y_j\left((k-j)\,T_0\right)\right\}^2 \qquad (17.4)$$

We chose the sampling time interval NT_0 experimentally. We calculate the maximum sample conditional variance using the sliding duration $N_{\min}T_0 \leq NT_0 \leq N_{\max}T_0$. At each value of N in the specified range with a sampling rate of one sampling period N_0 we calculate the sampling conditional variance and its maximum value is set as:

$$V_{8\max} = \sup N_{\min}T_0 \leq NT_0 \leq N_{\max}T$$
$$\left[\mathrm{var}\left\{y_1\left(kT_0\right)\middle| y_1\left((k-1)\,T_0\right),...,y_1\left((k-N)\,T_0\right)\right\}\right]$$

Let us consider the nature of the main perturbations that affect the CM nodes:

1. Informational influences. Since the cryptocurrency environment is quite open and a large number of cryptocurrency users have no professional education, the information effects have a significant impact on the fluctuations of the cryptocurrency rate due to the rapid increase in supply or demand, depending on positive or negative news. A vivid example of informational impact was Elon Musk's recommendation to use the messenger "Signal". In this case, the company's shares rose by 1300% in one hour. This kind of perturbation can affect the nodes 3,4,5,7,9,11 of the CM. (At the time of writing this article, Elon Musk made an announcement about Tesla's $1.5bln acquisition of bitcoin, but it did not affect the rate as rapidly as in the first example - only a 40% increase.)

2. Fluctuations in the global economy create uncertainty for institutional investors (such as large banks), forcing them to look for new areas for investment. At the same time, the entry of a significant number of stock market players leads to a sharp increase in the bitcoin rate. Their behavior in the market is similar to long-term investing and affects nodes 3, 5, 6 CM, which leads to the increase in demand for cryptocurrencies (node 9). The opposite effect is also possible. If the economy stabilizes, some investors will move to the stock market, leading to a sharp increase in the supply of cryptocurrency (node 7).

3. Fluctuations in the price of energy, which is one of the factors of instability in the world economy. This way the decrease in the price

of energy attracts new users to the process of mining and a significant number of people start to have cryptocurrency. This leads to an increase in the amount of speculations (node 6) and, naturally, to the growth of the cryptocurrency rate (node 1).

4. Legislative outrage. For example, the introduction of cryptocurrency as a unit of payment in Denmark has fuelled interest in cryptocurrency and dramatically increased the number of users (node 4) and the growth of speculation (node 6).
5. The mismatch between the level of confidence (node 11) and the volume of investment (node 5), which is caused by the contradiction between large and small investors. This mismatch can cause significant fluctuations in the exchange rate of cryptocurrency. Because small investors use borrowed money, they will be in danger. This leads to a decrease in the number of cryptocurrency users (node 4).

17.3 Design of Unstable Cryptocurrency Rate Stabilization System

To stabilize an unstable impulse process in CM, we use the method of modal control. We design the state controller:

$$\Delta \overline{U}(k) = -K \Delta \overline{Y}(k). \tag{17.5}$$

Here the size m of the control vector is smaller than the size n of the vector of CM nodes coordinates $\overline{Y}(k)$. This version of the modal control can be found in [10]. If the pair (A, B) in (17.3) is controllable, the following algorithm determines the feedback matrix K.

1. We set the desired spectrum of modes $\lambda_1, \ldots, \lambda_n$ of the closed-loop system $\Delta \overline{Y}(k+1) = (A - BK)\Delta \overline{Y}(k)$, where all $\lambda_j, j = 1, \ldots, n$ are different and modulo less than one. Moreover, there are no eigenvalues of matrix A among λ_j.
2. We introduce $\overline{R}_j, j = 1, \ldots, n$ – eigenvectors of the state matrix of the closed system (A-BK), for which this relation is satisfied:

$$(A - BK)\overline{R}_j \Delta(k) = \lambda_j \overline{R}_j \tag{17.6}$$

We can write equation in the following form:

$$(A - \lambda_j I)\overline{R}_j = BK\overline{R}_j = B\overline{P}_j, \tag{17.7}$$

where vector-columns $\overline{P_j} = K\overline{R_j}$ have dimension m, same as control vector $\Delta\overline{U}$.

3. We set an arbitrary matrix P of dimension $m \times n$ so that it has full rank and has no zero columns, $\overline{P_j} = (\overline{P_1 P_2 \ldots P_n})$.

4. We calculate the following vectors from equation (17.7):

$$\overline{R_j} = (A - \lambda_j I)^{-1}BK\overline{R_j}, j = 1, \ldots, n, \qquad (17.8)$$

After that, we form a matrix $R = (\overline{R_1 R_2 \ldots R_n})$ of dimension $n \times n$, which is non-singular.

5. We calculate the feedback matrix of the modal regulator (17.5) of dimension $m \times n$:

$$K = PR^{-1}, \qquad (17.9)$$

which, by construction, provides the desired set of modes λ_j of the closed system. This approach is possible if $(A - \lambda_j I)^{-1}$ exists, so λ_j is not an eigenvalue of matrix A. The choice of the matrix P affects the nature of the controls $\Delta\overline{U}(k)$, but the spectrum of the closed system remains invariant with respect to P.

We define constraints $\Delta u_{i_{\min}}(k) \leq \Delta u_i(k) \leq \Delta u_{i_{\max}}(k)$ for the changes of $\Delta u_i(k)$ increments in CM. The constraints are defined based on the existing resource constraints of the CM nodes, to which the synthesized $\Delta u_i(k)$ is fed. If you do not take this into account, in a real system, the control actions will be cut off at the level of constraints, so the control system will not be linear. If information about the maximum possible increments $\Delta y_i(k)$ of the CM node coordinates is available, the constraints on the value of $\Delta u_i(k)$ in the controller (17.5) can be satisfied by reducing the values of the elements of the rows of the feedback matrix K. This can be achieved experimentally by initial selection of vectors $\overline{P_j}$ to calculate eigenvectors $\overline{R_j}$ in (17.8).

17.3.1 Experimental Study of the Stabilization System of Unstable Cryptocurrency Rate

The overall It is easy to see that the system under consideration is unstable, its modes are equal to 2.352; 1.011; $0.95 \pm$ j0.2921; 0, .5277 $\pm$ j0.4123; $-0.6412 \pm$ j0.1454; $-0.0753 \pm$ j0.3622; $-0.0208 \pm$ j0.0849 (the absolutes of the first two eigenvalues are greater than one). We create the matrix B depending on the number of control actions. During modelling we set desired spectrum of eigenvalues λ_j of the closed-loop control system state matrix $A - BK$ as following: -0.6; -0.5; -0.4; -0.3; -0.2; -0.1; 0.1; 0.2; 0.3; 0.4; 0.5; 0.6, with absolutes that are less than one.

Consider the case when control actions $\Delta u_1(k)$, $\Delta u_2(k)$, $\Delta u_3(k)$, $\Delta u_4(k)$, and $\Delta u_5(k)$ are applied to the following nodes of CM respectively: 2 - trading volume, 3 - capitalization volume, 5 - investment volume, 6 - speculation volume, 7 - supply of cryptocurrency. In this case, the matrix B has the following form:

$$B^{\mathrm{T}} = \begin{pmatrix} 0 & 1 & 0 & 0 & 0 & 0 & 0 & 0 & 0 & 0 & 0 & 0 \\ 0 & 0 & 1 & 0 & 0 & 0 & 0 & 0 & 0 & 0 & 0 & 0 \\ 0 & 0 & 0 & 0 & 1 & 0 & 0 & 0 & 0 & 0 & 0 & 0 \\ 0 & 0 & 0 & 0 & 0 & 1 & 0 & 0 & 0 & 0 & 0 & 0 \\ 0 & 0 & 0 & 0 & 0 & 0 & 1 & 0 & 0 & 0 & 0 & 0 \end{pmatrix}.$$

We form the matrix P for the design of the modal controller as follows:

$$P = \begin{pmatrix} 1 & 1 & 1 & 1 & 1 & 1 & 1 & 1 & 1 & 1 & 1 & 1 \\ 1 & 0 & 0 & 1 & 1 & 0 & 0 & 1 & 0 & 1 & 1 & 0 \\ 0 & 0 & 1 & 0 & 0 & 1 & 0 & 0 & 0 & 0 & 1 & 1 \\ 0 & 1 & 1 & 1 & 1 & 1 & 0 & 1 & 0 & 1 & 1 & 1 \\ 1 & 1 & 0 & 0 & 0 & 0 & 0 & 1 & 1 & 1 & 0 & 0 \end{pmatrix}.$$

Based on methods (17.6), (17.7), (17.8), we define the feedback matrix of the modal controller (17.9).

Figure 17.2 shows the results of modelling the CM system of equations under five control actions, given that initial negative impulse "-1" affects node 11 "confidence level". The graphs show transients of the CM node coordinates y_i, $i = 1, \ldots, n$, coordinate increments $\Delta y_i(k)$, and controls increments $\Delta u_1(k)$, $\Delta u_2(k)$, $\Delta u_3(k)$, $\Delta u_4(k)$, and $\Delta u_5(k)$.

Based on the analysis of the graphs, we can state the following patterns of changes in the CM node coordinates and their increments during the transition process:

- The coordinates: y_7 (supply of cryptocurrency) and y_{10} (indirect profit) increased;
- The following coordinates decreased: y_1 (cryptocurrency rate), y_2 (trading volume) and y_4 (number of users), y_8 (rate variance), y_{12} (risk level).

The node coordinate increments $\Delta y_i(k)$, $i = 1, \ldots, 12$ during the stabilization transient converge to zero, i.e. we suppress all perturbations. The increments of all control actions $\Delta u_i(k)$ converge to zero after the coordinates of nodes y_i are stabilized.

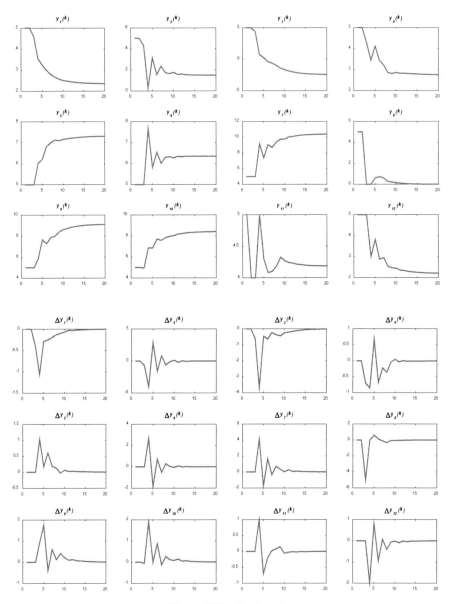

Figure 17.2 Continued

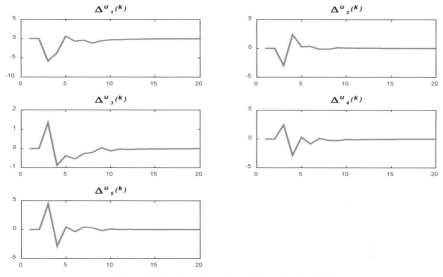

Figure 17.2 Results of modelling for the CM

17.4 The Problem of Constrained Internal and External Disturbances Suppression in Control of the Cryptocurrency CM Impulse Process

In [16, 17] theoretical positions on the suppression of arbitrary constrained external disturbances in terms of invariant ellipsoids based on the design of a static state feedback, which minimizes the size of the invariant ellipsoid of the dynamical system, are presented. In this case, we implemented a robust control, the analysis and synthesis problem of which is reduced to equivalent conditions in the form of linear matrix inequalities (LMI), solved numerically on the basis of semi definite programming.

In [18] we solve the problem of suppression of constrained external disturbances based on the invariant ellipsoids approach in the implementation of a closed-loop control system of impulse processes in CM of complex systems. The general model of the dynamics of impulse processes (17.2) is decomposed into two mutually related systems of difference equations:

$$\Delta \overline{X}(k+1) = A_1 \Delta \overline{X}(k) + D \Delta \overline{Z}(k); \qquad (17.10)$$

$$\Delta \overline{Z}(k+1) = C \Delta \overline{Z}(k) + \Psi \Delta \overline{X}(k). \qquad (17.11)$$

Here $\overline{X}$ is the vector of measurable coordinates of CM nodes; $\overline{Z}$ is the vector of unmeasurable or hardly measurable coordinates. The matrices A_1, D,

C, Ψ are compiled from the coefficients of the matrix of the initial model (17.2) of the CM impulse process. The matrices D, Ψ show the relationships between the first (17.10) and the second (17.11) parts of the initial CM (17.2). The values of non-measurable coordinates $\Delta\overline{Z}(k)$ are taken into account as external constrained perturbations with unknown probability characteristics in the first system of equations (17.10) of the CM model compiled for impulse processes with measurable coordinates $\overline{X}$.

We synthesized a control vector to suppress constrained perturbations $\Delta\overline{Z}(k)$ by implementing static state feedback:

$$\Delta\overline{U}(k) = -K_p\Delta\overline{X}(k), \tag{17.12}$$

which acts directly on the measured coordinates of the nodes $\overline{X}$ of the first control system according to the state equation:

$$\Delta\overline{X}(k+1) = A_1\Delta\overline{X}(k) + B\Delta\overline{U}(k) + D\Delta\overline{Z}(k) \tag{17.13}$$

The control is performed by changing the resources of the CM nodes, which are influenced by the vector $\Delta\overline{U}(k)$.

In this chapter, the change in the weight coefficients $\Delta A_1(k)$ with respect to the known basic values of the estimated matrix $\hat{A}_1(k)$ is proposed to be considered as the internal perturbations in the CM impulse process model (17.10). For this purpose, we propose to consider the initial model (17.10) in the form:

$$\Delta\overline{X}(k+1) = A_1\Delta\overline{X}(k) + \Delta A_1\Delta\overline{X}(k) + D\Delta\overline{Z}(k), \tag{17.14}$$

where $\Delta A_1 = A_1 - A_{1\text{var}}(k)$ is the change in the adjacency matrix of CM (17.10) during the sampling period, where $A_{1\text{var}}(k)$ is the real unknown value of the matrix A_1, which changes as the complex system evolves.

Let us denote the increment of internal perturbations in (17.14) as $\Delta A_1(k)\Delta\overline{y}(k) = \Delta\overline{w}(k)$. Then the equation of the uncontrolled impulse process (17.14) will take the form:

$$\Delta\overline{X}(k+1) = A_1\Delta\overline{X}(k) + (\ I_1 \quad D\ )\begin{bmatrix} \Delta\overline{w}(k) \\ \Delta\overline{Z}(k) \end{bmatrix}, \tag{17.15}$$

in which the vectors and matrices have the following dimensions: $\dim \Delta\overline{X} = n$; $\dim \Delta\overline{Z} = p$; $\dim \Delta\overline{w} = n$; $A_1(n \times n)$; $D(n \times p)$, I is a unit matrix of dimension $n \times n$. We assume that the internal and external perturbations are jointly constrained by the norm l_∞, so that:

$$
\left\| \begin{bmatrix} \Delta \overline{w}(k) \\ \Delta \overline{Z}(k) \end{bmatrix} \right\|_\infty = \sup \left\{ \begin{bmatrix} \Delta \overline{w}^T(k) & \Delta \overline{Z}^T(k) \end{bmatrix} \begin{bmatrix} \Delta \overline{w}(k) \\ \Delta \overline{Z}(k) \end{bmatrix} \right\}^{1/2} \leq 1
$$

$$(17.16)$$

We propose invariant ellipsoids on state variables to describe the characteristic of the effect of perturbations of the type (17.16) on the trajectory of a dynamic discrete system (17.15), as in [16, 17]:

$$
\varepsilon_{\Delta \overline{X}} = \left\{ \Delta \overline{X}(k) \in R^n : \Delta \overline{X}^T P^{-1} \Delta \overline{X} \leq 1 \right\}, P > 0, \qquad (17.17)
$$

if from $\Delta \overline{X}(0) \in \varepsilon_{\Delta \overline{X}}$ the condition $\Delta \overline{X}(k) \in \varepsilon_{\Delta \overline{X}}$ follows for all discrete moments of time $k = 1, 2, 3, \ldots$. Then the matrix P is called the matrix of the ellipsoid $\varepsilon_{\Delta \overline{X}}$.

Let us prove the invariance condition of the ellipsoid (17.17) under perturbations (17.16). For this purpose, using the method [17], we introduce the quadratic Lyapunov function $V(\overline{X}(k)) = \Delta \overline{X}^T Q \Delta \overline{X}$ at $Q > 0$, constructed on the solutions of the system (17.15). To ensure that the trajectories $\Delta \overline{X}(k)$ of the system (17.15) do not go beyond the boundary of the ellipsoid $\varepsilon_{\Delta \overline{X}} = \left\{ \Delta \overline{X}(k) \in R^n : V(\overline{X}(k)) \leq 1 \right\}$, it is required that $V(\Delta \overline{X}(k)) \leq 1$ at $V(\overline{X}(k)) \leq 1$, so we have:

$$
\Delta \overline{X}^T(k+1) Q \Delta \overline{X}(k+1) =
$$

$$
= \left[\Delta \overline{X}^T A_1{}^T + \begin{bmatrix} \Delta \overline{w}^T(k) & \Delta \overline{Z}^T(k) \end{bmatrix} \begin{bmatrix} I_1 \\ D^T \end{bmatrix} \right] Q
$$

$$
\left[A_1 \Delta \overline{X}(k) + \begin{pmatrix} I_1 & D \end{pmatrix} \begin{bmatrix} \Delta \overline{w}(k) \\ \Delta \overline{Z}(k) \end{bmatrix} \right] \leq 1
$$

By multiplication, we get:

$$
\Delta \overline{X}^T(k+1) Q \Delta \overline{X}(k+1) =
$$

$$
= \begin{bmatrix} \Delta \overline{X}^T & \begin{bmatrix} \Delta \overline{w}^T(k) & \Delta \overline{Z}^T(k) \end{bmatrix} \end{bmatrix}
$$

$$
\begin{bmatrix} A_1^T Q A_1 & A_1^T Q \begin{pmatrix} I_1 & D^T \end{pmatrix} \\ \begin{bmatrix} I_1 \\ D^T \end{bmatrix} Q A_1 & \begin{bmatrix} I_1 \\ D^T \end{bmatrix} Q \begin{pmatrix} I_1 & D^T \end{pmatrix} \end{bmatrix} \begin{bmatrix} \Delta \overline{X}(k) \\ \begin{bmatrix} \Delta \overline{w}(k) \\ \Delta \overline{Z}(k) \end{bmatrix} \end{bmatrix}
$$

Let us apply the S-procedures [17, 19]. Let $S = \begin{bmatrix} \Delta \overline{X}(k) \\ \begin{bmatrix} \Delta \overline{w}(k) \\ \Delta \overline{Z}(k) \end{bmatrix} \end{bmatrix}$. Then the

quadratic forms we can represent as follows:

$$f_0(\overline{S}) = \overline{S}^T M_0 \overline{S} =$$
$$= \begin{bmatrix} \Delta \overline{X}^T & \begin{bmatrix} \Delta \overline{w}^T(k) & \Delta \overline{Z}^T(k) \end{bmatrix} \end{bmatrix}$$

$$\begin{bmatrix} A_1{}^T Q A_1 & A_1{}^T Q \begin{pmatrix} I_1 & D^T \end{pmatrix} \\ \begin{bmatrix} I_1 \\ D^T \end{bmatrix} Q A_1 & \begin{bmatrix} I_1 \\ D^T \end{bmatrix} Q \begin{pmatrix} I_1 & D^T \end{pmatrix} \end{bmatrix} \begin{bmatrix} \Delta \overline{X}(k) \\ \begin{bmatrix} \Delta \overline{w}(k) \\ \Delta \overline{Z}(k) \end{bmatrix} \end{bmatrix} \leq 1$$

;

$$f_1(\overline{S}) = \overline{S}^T M_1 \overline{S} = \begin{bmatrix} \Delta \overline{X}^T & \begin{bmatrix} \Delta \overline{w}^T(k) & \Delta \overline{Z}^T(k) \end{bmatrix} \end{bmatrix}$$

$$\begin{bmatrix} Q & 0 \\ 0 & 0 \end{bmatrix} \begin{bmatrix} \Delta \overline{X}(k) \\ \begin{bmatrix} \Delta \overline{w}(k) \\ \Delta \overline{Z}(k) \end{bmatrix} \end{bmatrix} \leq 1;$$

$$f_2(\overline{S}) = \overline{S}^T M_2 \overline{S} = \begin{bmatrix} \Delta \overline{X}^T & \begin{bmatrix} \Delta \overline{w}^T(k) & \Delta \overline{Z}^T(k) \end{bmatrix} \end{bmatrix}$$

$$\begin{bmatrix} Q & 0 \\ 0 & I_2 \end{bmatrix} \begin{bmatrix} \Delta \overline{X}(k) \\ \begin{bmatrix} \Delta \overline{w}(k) \\ \Delta \overline{Z}(k) \end{bmatrix} \end{bmatrix} \leq 1,$$

where I_2 is the unit matrix of dimension $(n + p) \times (n + p)$. According to the statement of S-procedure [17] we have $M_0 \leq \sum_{i=1}^{2} \tau_i m_i$, that is:

$$\begin{bmatrix} A_1{}^T Q A_1 & A_1{}^T Q \begin{pmatrix} I_1 & D^T \end{pmatrix} \\ \begin{bmatrix} I_1 \\ D^T \end{bmatrix} Q A_1 & \begin{bmatrix} I_1 \\ D^T \end{bmatrix} Q \begin{pmatrix} I_1 & D^T \end{pmatrix} \end{bmatrix} \leq \tau_1 \begin{bmatrix} Q & 0 \\ 0 & 0 \end{bmatrix} + \tau_2 \begin{bmatrix} 0 & 0 \\ 0 & I_2 \end{bmatrix}$$

or

$$\begin{bmatrix} A_1{}^T Q A_1 & A_1{}^T Q \begin{pmatrix} I_1 & D^T \end{pmatrix} \\ \begin{bmatrix} I_1 \\ D^T \end{bmatrix} Q A_1 & \begin{bmatrix} I_1 \\ D^T \end{bmatrix} Q \begin{pmatrix} I_1 & D^T \end{pmatrix} \end{bmatrix} \leq 0 \qquad (17.18)$$

Using Schur's formula, inequality (17.18) will take the following form:

$$A_1{}^T Q A_1 - \tau_1 Q \leq A_1{}^T Q \begin{pmatrix} I_1 & Q^T \end{pmatrix} \begin{bmatrix} \begin{bmatrix} I_1 \\ D^T \end{bmatrix} Q \begin{pmatrix} I_1 & Q^T \end{pmatrix} - \tau_2 I_2 \end{bmatrix}^{-1}$$

$$\begin{bmatrix} I_1 \\ D^T \end{bmatrix} Q A_1$$

After performing elementary transformations, we convert this inequality to the following form:

$$\tau_1 Q \geq A_1^T$$

$$\left[Q - Q \begin{pmatrix} I_1 & Q^T \end{pmatrix} \left[\begin{bmatrix} I_1 \\ D^T \end{bmatrix} Q \begin{pmatrix} I_1 & Q^T \end{pmatrix} - \tau_2 I_2 \right]^{-1} \begin{bmatrix} I_1 \\ D^T \end{bmatrix} Q \right] A_1$$

If $\tau_2 = 1 - \tau_1$ then we get:

$$\tau_1 Q \geq A_1^T$$

$$\left[Q + Q(I_1 \ Q^T) \left[(1 - \tau_1) I_2 - \begin{bmatrix} I_1 \\ D^T \end{bmatrix} Q \begin{pmatrix} I_1 & D \end{pmatrix} \right]^{-1} \begin{bmatrix} I_1 \\ D^T \end{bmatrix} Q \right] A_1 \tag{17.19}$$

According to the lemma of matrix inversion [20], we will have:

$$Q + Q \begin{pmatrix} I_1 & Q^T \end{pmatrix} \left[(1 - \tau_1) I_2 - \begin{bmatrix} I_1 \\ D^T \end{bmatrix} Q \begin{pmatrix} I_1 & D \end{pmatrix} \right]^{-1} \begin{bmatrix} I_1 \\ D^T \end{bmatrix} Q$$

$$== \left[Q^{-1} - (1 - \tau_1)^{-1} \begin{pmatrix} I_1 & D \end{pmatrix} \begin{bmatrix} I_1 \\ D^T \end{bmatrix} \right]^{-1}$$

Then expression (17.19) we can write as follows:

$$\tau_1 Q = A_1^T \left[Q^{-1} - (1 - \tau_1)^{-1} \begin{pmatrix} I_1 & D \end{pmatrix} \begin{bmatrix} I_1 \\ D^T \end{bmatrix} \right]^{-1} A_1$$

Let us perform an elementary transformation in condition $P = Q^{-1}$:

$$\tau_1 P^{-1} \geq \left[A_1^{-1} \left[Q^{-1} - (1 - \tau_1)^{-1} \begin{pmatrix} I_1 & D \end{pmatrix} \begin{bmatrix} I_1 \\ D^T \end{bmatrix} \right] (A_1^T)^{-1} \right]^{-1}$$

After inverting the left and right parts, we get:

$$\frac{P}{\tau_1} \geq A_1^{-1} \left[P - (1 - \tau_1)^{-1} \begin{pmatrix} I_1 & D \end{pmatrix} \begin{bmatrix} I_1 \\ D^T \end{bmatrix} \right] (A_1^T)^{-1}$$

Multiply from the left by A_1, and then from the right by A_1^T, and renominate $\tau_1 = \alpha$. Then the linear matrix inequality will take the final form:

$$\frac{1}{\alpha} A_1 P A_1^T - P + \frac{I_1 + DD^T}{(1 - \alpha)} \leq 0 \tag{17.20}$$

17.4.1 Algorithm for the State Controller Design for the CM Impulse Process

The equation of state of the controlled impulse process of CM (17.15) under additional internal perturbation takes the form:

$$\Delta \overline{X}(k+1) = A_1 \Delta \overline{X}(k) + B \Delta \overline{U}(k) + \begin{pmatrix} I_1 & D \end{pmatrix} \begin{bmatrix} \Delta \overline{w}(k) \\ \Delta \overline{Z}(k) \end{bmatrix} \quad (17.21)$$

When the state controller (17.12) is applied, the equation of the closed-loop CM impulse process control system is written as follows:

$$\Delta \overline{X}(k+1) = (A_1 - BK_p)\Delta \overline{X}(k) + \begin{pmatrix} I_1 & D \end{pmatrix} \begin{bmatrix} \Delta \overline{w}(k) \\ \Delta \overline{Z}(k) \end{bmatrix} \quad (17.22)$$

It is assumed that the pair $\begin{pmatrix} A_1 & B \end{pmatrix}$, in the model (17.21) is controllable. Then the LMI (17.20) for the closed system takes the form:

$$\frac{1}{\alpha}(A_1 - BK_p)P(A_1 - BK_p)^T - P + \frac{I_1 + DD^T}{(1-\alpha)} \leq 0 \quad (17.23)$$

We consider the minimization of the trace of the ellipsoid matrix (17.17) as the optimality criterion for the synthesis of the regulator (17.12) in this chapter:

$$tr P(\alpha) \to \min, \alpha^* \leq \alpha < 1, \quad (17.24)$$

This ensures minimization of the size of the invariant ellipsoid (17.17) with the largest suppression of perturbations $\begin{bmatrix} \Delta \overline{w}(k) \\ \Delta \overline{Z}(k) \end{bmatrix}$, which are constrained only by the maximum range (17.16). After multiplying the factors in the inequality (17.23), we obtain:

$$\frac{1}{\alpha}(A_1 P A_1^T - BK_p P A_1^T - A_1 P K_p^T B^T + BK_p P K_p^T B^T)$$
$$- P + \frac{I_1 + DD^T}{(1-\alpha)} \leq 0 \quad (17.25)$$

Inequality (17.25) is nonlinear with respect to P and K_p, which need to be optimized. In [17] a linearization by replacement $L = K_p P$ and introduction of an additional constraint is proposed:

$$\begin{bmatrix} R & L \\ L^T & P \end{bmatrix} \geq 0, \quad (17.26)$$

where $P = R^T$. This inequality is equivalent to $R \geq LP^{-1}L^T = K_p P K_p{}^T$ according to the Schur's formula at $P > 0$. Then to fulfil inequality (17.25) it is sufficient that:

$$\frac{1}{\alpha}(A_1 P A_1{}^T - BLA_1{}^T - A_1 L^T B^T + BRB^T) - P + \frac{I_1 + DD^T}{(1-\alpha)} \leq 0 \quad (17.27)$$

Minimization of criterion (17.24) under constraints (17.26), (17.27) is performed on variables P, L, R by semidefinite programming method by using Matlab-based SeDuMi Toolbox.

Then the matrix $\hat{K}_p$ of the optimal state regulator (17.12) we define as:

$$\hat{K}_p = \hat{L}\hat{P}^{-1} \quad (17.28)$$

with the estimated values of $\hat{\alpha}, \hat{P}, \hat{L}, \hat{R}$, providing minimization of criterion (17.24) under constraints (17.26), (17.27).

17.4.2 Experimental Study of the System of Constrained Internal and External Disturbances Suppression in the Cryptocurrency CM Impulse Process Control

Consider a stable version of the cognitive map of cryptocurrency shown in Figure 17.1. Nodes coordinates 1, 2, 3, 4, 5, 6, 7, 8 are used as measurable and computable coordinates $\bar{X}$ in model (10). The group of coordinates $\bar{Z}$ that are not measured or difficult to measure includes demand for cryptocurrency (node 9), indirect profit (node 10), confidence level (node 11) and integral risk level (node 12). After decomposing model (17.3) with updated coefficients into (17.10) and (17.11), the matrices A_1, D are the following:

$$A_1 = \begin{pmatrix} 0.3 & 0.1 & 0.15 & 0.2 & 0.2 & 0.3 & 0 & 0 \\ 0 & 0 & 0.1 & 0.1 & 0.1 & 0.35 & 0.35 & 0 \\ 0.25 & 0 & 0.5 & 0 & 0.1 & 0 & 0 & 0 \\ 0.4 & 0 & 0 & 0.3 & 0 & 0 & 0 & 0 \\ 0 & 0.4 & 0 & 0 & 0.4 & 0 & 0 & 0 \\ 0 & 0.5 & 0 & 0.2 & 0 & 0 & 0.6 & -0.2 \\ 0 & 0 & 0 & 0.35 & 0.3 & 0 & 0 & 0 \\ -0.3 & 0 & 0 & 0 & 0 & 0 & 0 & 0 \end{pmatrix},$$

$$D = \begin{pmatrix} 0.2 & 0 & 0.3 & -0.3 \\ -0.3 & 0 & 0.3 & -0.2 \\ 0 & 0 & 0.5 & 0 \\ 0 & 0 & 0.2 & -0.2 \\ 0 & 0.4 & 0 & -0.3 \\ 0 & 0 & 0 & -0.4 \\ 0.5 & 0.2 & 0 & 0 \\ 0 & 0 & 0 & 0 \end{pmatrix}.$$

The control matrix B has the form given in the example of paragraph 3 (cut to 8 rows).

We consider the step impacts with the amplitude of 1, acting at the initial moment of time on the two unmeasured nodes 10 and 11 in the direction of reducing the indirect profit (node 10) and confidence level (node 11) for modelling the dynamics of the closed-loop control system of the impulse process of this CM cryptocurrency using the proposed method, as external disturbances. The internal perturbations in the simulation are formed as follows: the value of non-zero matrix coefficients vary at each sampling period by the formula $A_{1_{var}}(k) = A_1 \xi(k)$, where $\xi(k)$ is a normally distributed random variable (Gaussian white noise). Only the values of A_1 are used for control, while $A_{1_{var}}$ remains unknown. The initial levels of all coordinates of the CM nodes are assumed zero for convenience.

The charts of the transients of the CM node coordinates, their increments and controls are shown in Figure 17.3.

17.5 The Problem of Stabilizing the Coordinates of Cryptocurrency CM Nodes at Given Levels Based on Varying of Edge Weights and Nodes Coordinates

Provided that $\lim_{k \to \infty} \Delta X_i(k) = 0$, in order to stabilize the coordinates of measured CM nodes X_i at specified levels G_i at each sampling period it is necessary to form the control actions, which according to the synthesized control law affect directly the CM nodes X_i in the closed-loop control system.

However, in practice, difficulties arise in controlling impulse processes in CM due to small resources for varying the coordinates of CM nodes when implementing control actions. Usually if the CM dimension is large, there are not enough nodes available for variation, which leads to a significant difference between the dimension n of the vector of output-controlled coordinates X_i and the dimension m of the vector of control actions u_i. Therefore,

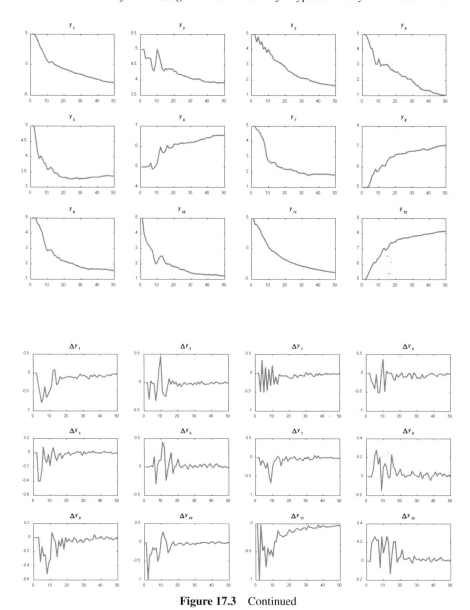

Figure 17.3 Continued

we decided [21] to apply also another approach, specifically, to generate an external control vector by simultaneously varying the resources of the CM node coordinates and the weight coefficients of the adjacency matrix in the

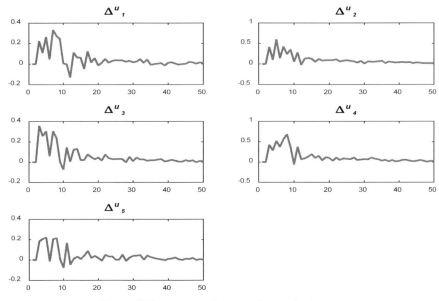

Figure 17.3   CM coordinates and controls changes

closed-loop control system. In this case, the general mathematical model of the controlled impulse CM process (17.13) in vector-matrix form in full nodes coordinates is represented in the form:

$$\overline{X}(k+1) = (I + A_1 - A_1 q^{-1})\overline{X}(k) + \begin{bmatrix} B & L(k) \end{bmatrix} \begin{bmatrix} \Delta \overline{U}(k) \\ \Delta \overline{a}(k) \end{bmatrix} + \xi(k),$$

$$(17.29)$$

where the matrix $L(k)$ is composed of the measured CM coordinates $X_\mu(k)$, which through edges with varying weight coefficients $\Delta a_{i\mu_l}(k)$ will be used as the control action; q^{-1} - backshift operator for one sampling period; $\xi(k) = D\Delta \overline{z}(k)$ - vector of uncontrollable perturbations, according to (13).

Let us formulate the rules for the formation of the vectors of increments $\Delta \overline{U}(k)$ and $\Delta \overline{a}(k)$, as well as the matrix B, $L(k)$ in the model (17.29).

1. Only one control action can influence on each node of CM $X_i(k), i = 1, \ldots, n$ by varying the resources of this node Δu_i or by varying the weighting coefficient $\Delta a_{i\mu_l}$. Therefore, we can guarantee the autonomy of the control system.

2. The vector of increments of weight coefficients $\Delta \overline{a}(k)$ of dimension p at $p < n$ contains only nonzero elements $\Delta a_{i\mu_l} \neq 0$, where μ is the number of the CM node that affects the ith node X_i through the

edge with the coefficient $a_{i\mu}$. $l = 1, 2, \ldots, p$ are the ordinal numbers of this coefficient in the vector $\Delta\bar{a}(k)$. If the weight coefficient $a_{i\mu}$ for a CM edge entering a node X_i cannot be varied, then the increment $\Delta\bar{a}_{i\mu_l}(k) = 0$ and in the vector $\Delta\bar{a}(k)$ are not considered.

3. The matrix B in (17.29) is created by a designer of the CM impulse process control system. This matrix is designed to organize scaling and switching of the synthesized control action $\Delta\bar{u}_i$, $i = 1, 2, \ldots, m$. The dimension of the matrix B is equal to $n \times m$, where n is the dimension of the vector $\overline{X}$, and m is the dimension of the vector $\Delta\bar{u}(k)$. As a rule, the elements of B are ones and zeros. Elements $b_{i\mu} = 1$ if the ith control action affects the μth node of the matrix. Thus in each row of the matrix B only one element can be equal to one, and the remaining elements will be equal to zero.

4. The matrix $L(k)$ with dimension $n \times p$ contains only one element in each i row equal to the measured coordinate $X_\mu(k)$, which is placed in the lth row, where l is the ordinal number of the increment $\Delta a_{i\mu_l}(k)$ in the vector $\Delta\bar{a}(k)$, and μ is the number of the CM node, through which the increment $\Delta a_{i\mu_l}(k)$ controls the CM node X_i.

5. All elements in the ith row of the matrix $L(k)$ are equal to zero if the ith node X_i does not include a CM edge with a varying weight coefficient, i.e. $\Delta a_{i\mu} = 0$.

Thus, the total number of controls in a CM can be equal to $m + p \leq n$.

17.5.1 Design of a Discrete Controller

We implement the synthesis of the combined vector of control $\begin{bmatrix} \Delta\overline{U}(k) \\ \Delta\bar{a}(k) \end{bmatrix}$ based on minimization of the following quadratic criterion of optimality:

$$J(k+1) = E\left\{ \left[\overline{X}(k+1) - \overline{G}\right]^T \left[\overline{X}(k+1) - \overline{G}\right] + \begin{bmatrix} \Delta\overline{U}(k) \\ \Delta\bar{a}(k) \end{bmatrix}^T \right.$$

$$\left. \begin{bmatrix} R_1 & 0 \\ 0 & R_2 \end{bmatrix} \begin{bmatrix} \Delta\overline{U}(k) \\ \Delta\bar{a}(k) \end{bmatrix} \right\},$$

where $\overline{G}$ is the reference vector to stabilize the coordinates of the CM nodes at the given levels; R_1, R_2 are the weight diagonal positive definite matrices; E is the expectation operator. Taking into account Equation (17.29), let us perform minimization of this criterion with respect to the combined control vector.

$$\frac{\partial J(k+1)}{\partial \begin{bmatrix} \Delta \overline{U}(k) \\ \Delta \overline{a}(k) \end{bmatrix}} == 2 \begin{bmatrix} B^T \\ L^T(k) \end{bmatrix}$$

$$\left\{ \left[I + A_1 - A_1 q^{-1} \right] \overline{X}(k) + \begin{bmatrix} B & L(k) \end{bmatrix} \begin{bmatrix} \Delta \overline{U}(k) \\ \Delta \overline{a}(k) \end{bmatrix} + \xi(k) - \overline{G} \right\} +$$

$$+ 2 \begin{bmatrix} R_1 & 0 \\ 0 & R_2 \end{bmatrix} \begin{bmatrix} \Delta \overline{U}(k) \\ \Delta \overline{a}(k) \end{bmatrix} = 0$$

From where we get the law of the combined optimal control:

$$\begin{bmatrix} \Delta \overline{U}(k) \\ \Delta \overline{a}(k) \end{bmatrix} = \begin{bmatrix} B^T \\ L^T(k) \end{bmatrix} \begin{bmatrix} B & L(k) \end{bmatrix} +$$

$$+ \begin{bmatrix} R_1 & 0 \\ 0 & R_2 \end{bmatrix}^{-1} \begin{bmatrix} B^T \\ L^T(k) \end{bmatrix} \left\{ \left[I + A_1 - A_1 q^{-1} \right] \overline{X}(k) + \xi(k) - \overline{G} \right\}$$

(17.30)

Based on equations (17.29), (17.30), the equation of dynamics of the closed system of the CM impulse process equation can be derived.

$$\overline{X}(k+1) = \left\{ I - \begin{bmatrix} B & L(k) \end{bmatrix} \left[\begin{bmatrix} B^T B & B^T L(k) \\ L^T(k)B & L^T(k)L(k) \end{bmatrix} \right. \right.$$

$$\left. \left. + \begin{bmatrix} R_1 & 0 \\ 0 & R_2 \end{bmatrix} \right]^{-1} \begin{bmatrix} B^T \\ L^T(k) \end{bmatrix} \right\} \times \left[I + A_1 - A_1 q^{-1} \right] \overline{X}(k) +$$

$$+ [B \ L(k)] \left[\begin{bmatrix} B^T B & B^T L(k) \\ L^T(k)B & L^T(k)L(k) \end{bmatrix} + \begin{bmatrix} R_1 & 0 \\ 0 & R_2 \end{bmatrix} \right]^{-1} \overline{G} +$$

$$+ \left\{ I - \begin{bmatrix} B & L(k) \end{bmatrix} \left[\begin{bmatrix} B^T B & B^T L(k) \\ L^T(k)B & L^T(k)L(k) \end{bmatrix} \right. \right.$$

$$\left. \left. + \begin{bmatrix} R_1 & 0 \\ 0 & R_2 \end{bmatrix} \right]^{-1} \begin{bmatrix} B^T \\ L^T(k) \end{bmatrix} \right\} \xi(k)$$

(17.31)

The stability of the closed-loop control system (17.31) is determined by the eigenvalues varying in time of the nonlinear matrix expression:

$$I - [B \ L(k)] \left[\begin{bmatrix} B^T B & B^T L(k) \\ L^T(k)B & L^T(k)L(k) \end{bmatrix} + \begin{bmatrix} R_1 & 0 \\ 0 & R_2 \end{bmatrix} \right]^{-1} \begin{bmatrix} B^T \\ L^T(k) \end{bmatrix}$$

(17.32)

In [21] the proof is given: the matrix $\begin{bmatrix} B^T B & B^T L(k) \\ L^T(k)B & L^T(k)L(k) \end{bmatrix} +$ $\begin{bmatrix} R_1 & 0 \\ 0 & R_2 \end{bmatrix}$ in expression (17.32) is diagonal and positively determined, since on the basis of the rule of formation of matrices B and $L(k)$ it turns out that $B^T L(k) = 0$, $L^T(k)B = 0$. As a result, it is shown in [21] that the eigenvalues of the diagonal matrix (17.32) $I -$ $\begin{bmatrix} B(B^T B + R_1)^{-1}B^T & 0 \\ 0 & L(k)(L^T(k)L(k) + R_2)^{-1}L^T(k) \end{bmatrix}$ will be modulo less than unity. As a result, the closed-loop system (17.31) will be stable.

17.5.2 Experimental Studies of the Stabilization of Cryptocurrency CM Nodes Coordinates at Given Levels

We use the coordinates of nodes 1, 2, 3, 4, 5, 6, 7, 8 of the CM shown in Figure 17.1 as measurable coordinates $\overline{X}$ in the model (17.29). We include nodes 9, 10, 11, and 12 in the group of non-measurable or hard-to-measure coordinates $\overline{Z}$. The values of the matrices A_1, D are given above. The control vector by varying the node coordinate resources is equal to $\Delta \overline{u}(k) = \begin{bmatrix} \Delta u_1(k) & \Delta u_2(k) & \Delta u_3(k) & \Delta u_4(k) \end{bmatrix}^T$, which are fed respectively to the following nodes of CM: 2 - trading volume; 5 - investment volume; 6 - speculation volume and 7 - cryptocurrency supply. Then the control matrix B will be formed by the system designer as follows:

$$B^{\mathrm{T}} = \begin{pmatrix} 0 & 1 & 0 & 0 & 0 & 0 & 0 & 0 \\ 0 & 0 & 0 & 0 & 1 & 0 & 0 & 0 \\ 0 & 0 & 0 & 0 & 0 & 1 & 0 & 0 \\ 0 & 0 & 0 & 0 & 0 & 0 & 1 & 0 \end{pmatrix}.$$

As a result, the product of matrices $B^T B$ will be the unit matrix 4×4. The second control vector based on the increments of the weight coefficients $\Delta \overline{a}$ will have the form $\Delta \overline{a} = \begin{bmatrix} \Delta a_{15} & \Delta a_{35} \end{bmatrix}^T$. Then the control matrix $L(k)$ of dimensionality 8×2 is composed as follows:

$$L^{\mathrm{T}}(k) = \begin{pmatrix} x_5(k) & 0 & 0 & 0 & 0 & 0 & 0 & 0 \\ 0 & 0 & x_5(k) & 0 & 0 & 0 & 0 & 0 \end{pmatrix}.$$

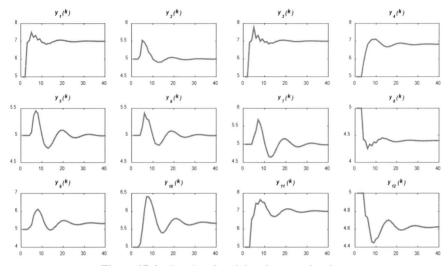

Figure 17.4   Results of applying the control actions

As a result the product of matrices $L^T(k)L(k)$ will be a diagonal matrix 2×2:

$$L^T(k)L(k) = \begin{bmatrix} x_5{}^2(k) & 0 \\ 0 & x_5{}^2(k) \end{bmatrix}$$

Let the initial values of CM node coordinates, which are measured on a 10-point scale, be at the average level of 5. Suppose that it is necessary to transfer the coordinates of the nodes x_1 (cryptocurrency rate), x_3 (capitalization volume) to a higher level 7. Applying the control law (17.30), we obtain the dynamics of the impulse process control presented in Figure 17.4.

17.6 Design of a System for Identifying CM Weighting Coefficients Based on Recurrent Least Squares Method

The model of the controlled impulse process (17.3) of the input-output CM type can be represented as:

$$(I - A_1 q^{-1})\Delta \overline{Y}(k) = Bq^{-1}\Delta \overline{U}(k) \qquad (17.33)$$

Weighting coefficients of the incidence adjacency matrix A are usually determined by applying expert evaluations. In the process of evolving of the cryptocurrency application system, these coefficients may change over time, depending on changes in the influence of one of the CM nodes, i.e. both control and evaluation of the adjacency matrix A parameters (coefficients) must be performed simultaneously.

Let us describe the control (17.33) coordinate-wise:

$$\Delta y_i(k) = \sum_{j=1}^{n} a_{ij}\Delta y_j(k-1) + b_i\Delta u_i(k-1) + \xi_i(k) \qquad (17.34)$$

It is assumed that the perturbations $\xi_i(k)$, caused by inaccurate knowledge of the model coefficients, are white noise. It should be taken into account that the structure of the matrix A is known and some of the coefficients a_{ij} are obviously equal to zero (in those cases when there are no connections between the corresponding CM nodes). Also, when the dimensionality of the control vector $m < n$, then in (17.34) some control actions $\Delta u_i(k-1)$ will be equal to zero.

Let the nonzero elements in the ith row of the matrix A be the coefficients $a_{ij_1} \ldots a_{ijP_1}$. Denote $\bar{X}_i^T(k) = [\Delta Y_{j1}(k-1), \ldots, \Delta Y_{jP_i}(k-1)]$, $\overline{\Theta}_i = [a_{ij_1} \ldots a_{ijP_1}]^T$. Then expression (17.34) can be written as follows:

$$\Delta y_i(k) - b_i\Delta u_i(k-1) = \bar{X}_i^T(k)\overline{\Theta}_i + \xi_i(k) \qquad (17.35)$$

The current estimate of the vector $\overline{\Theta}_i$ is denoted by $\hat{\overline{\Theta}}_i(k)$. To estimate the weight coefficients of the matrix A we apply the recurrent least squares method [22–25]:

$$\hat{\overline{\Theta}}_i(k) = \hat{\overline{\Theta}}_i(k-1) + K_i(k)\left(\Delta y_i(k) - b_i\Delta u_i(k-1) - \bar{X}_i^T(k)\hat{\overline{\Theta}}_i(k-1)\right)$$

$$K_i(k) = \frac{1}{1 + \bar{X}_i^T(k)P_i(k-1)\overline{X}_i(k)}P_i(k-1)\overline{X}_i(k)\bar{X}_i^T(k)P_i(k-1)$$

$$(17.36)$$

$$P_i(k) = P_i(k-1) - \frac{1}{1 + \bar{X}_i^T(k)P_i(k-1)\overline{X}_i(k)}$$
$$P_i(k-1)\overline{X}_i(k)\bar{X}_i^T(k)P_i(k-1).$$

The recurrent procedure (17.36) should be performed for each node of CM $\Delta y_i(k), i = 1, 2, \ldots, n$ at each sampling step. We use the obtained estimates $\hat{\overline{\Theta}}_i(k)$ as the values of the adjacency matrix A coefficients at the current step of the control algorithm. For parametric identification of the adjacency matrix A, we can also apply non-recurrent identification methods, outlined in [8, 26].

17.7 Conclusion

This chapter considers the cryptocurrency CM and solves three problems of its application in the financial markets. The first problem covers the case of instability when the cryptocurrency rate is unstable and decision-makers want to stabilize it. For this case control actions are designed based on the modal controller, which help to stabilize the system and assign desired values to the closed-loop system poles. The second problem considers the cryptocurrency system affected by unmeasurable internal and external disturbances, and the only thing we know about these disturbances is that they are constrained. For this case controls are designed based on the modified invariant ellipsoids method which ensures that the system's "trajectory" is limited inside of the minimal size ellipsoid. The third problem discussed in the chapter relates to the case when it is required to set some of the CM nodes at the new levels. This problem is solved based on minimization of the generalized variance criterion. Additional feature of this approach is that we may not only change some resources associated with CM nodes but also change how one node affects another one, i.e. to vary CM edges weights. As a result of solving all three problems decision-makers obtain a sequence of control actions which they can apply to the real cryptocurrency market. All the results were mathematically simulated and the effectiveness of the suggested approaches was proven.

References

[1] E. Papageorgiou, J. Salmeron, "A review of fuzzy cognitive maps research during the last decade", *IEEE transactions on fuzzy systems*, vol. 21, no. 1, 2013, pp. 66–79.

[2] G. Gorelova, N. Pankratova (Eds.). Innovational development of socio-economic systems based on methodologies of foresight and cognitive modeling, Kyiv, Naukova dumka, 2015, 464 p. (In Russian)

[3] C. Stylios, P.P. Groumpos, "Modeling complex systems using fuzzy cognitive maps", *IEEE Transactions on Systems, Man, and Cybernetics-Part A: Systems and Humans*, 34((1), 155–162, 2004.

[4] J. Aguilar, "A survey about fuzzy cognitive maps papers", International Journal of Computational Cognition, 3, N 2, P. 27–33, 2005.

[5] R. Axelrod, "The structure of decision: Cognitive maps of political elites", Princeton University Press, 1976.

[6] F. Roberts, "Discrete mathematical models with applications to social biological and environmental problems", Englewood Cliffs: Prentice-Hall, 1976.

[7] V.M. Kuntsevich, et al (Eds). Control Systems: Theory and Applications. Series in Automation, Control and Robotics, River Publishers, Gistrup, Delft, 2018.

[8] Y.P. Kondratenko, V.M. Kuntsevich, A.A. Chikrii, V.F. Gubarev, (Eds). Advanced Control Systems: Theory and Applications. Series in Automation, Control and Robotics, River Publishers, Gistrup, 2021.

[9] V.D. Romanenko, Yu.L. Milyavsky. "Stabilization of impulse processes in cognitive maps of complex systems on the basis of modal state regulators", *Cybernetics and Computer Engineering*, vol. 179, pp. 43–55, 2015. (In Russian)

[10] Methods of Classical and Modern Automatic Control Theory: book in 3 Volumes. – T.2: Synthesis of Controllers and Optimization Theory of Automatic Control Systems / Ed. by N.D. Egupov. Moscow: Bauman Moscow State Technical University, 2000, 736 p. (In Russian)

[11] Z. Dvulit, Kh. Peredalo, R. Tylipska, R. Terno, R. Stubel, "Cryptocurrency: state and trends of development", Economy and state, 2019, N. 1, P. 10–14. (In Ukrainian)

[12] R. Orastean, S. Marginean, R. Sava, "Bitcoin in the scientific literature – A bibliometric study", *Studies in Business and Economics*, 14(3), p. 160–174, 2019.

[13] A.F. Bariviera, M.J. Basgall, W. Hasperue, M. Naiouf, "Some stylized facts of the bitcoin market", Physica a-Statistical Mechanics and Its Applications, 484, p. 82–90, 2017.

[14] D. Ron, A. Shamir, "Quantitative analysis of the full bitcoin transaction graph", International Conference on Financial Cryptography and Data Security, pp. 6–24, 2013.

[15] V.D. Romanenko, Yu.L. Milyavsky. "Forecasting of maximal conditional dispersions for multidimensional processes with mulirate discretization on the basis of adaptive GARCH models", *System Research and Information Technologies*, 2009, No. 4, pp. 92–108. (In Russian)

[16] B. Polyak, R. Shcherbakov. Robust Stability and Control. Moscow, Nauka, 2002, 303 p. (In Russian)

[17] S. Nasin. B. Polyak, M. Topunov. "Suppression of constrained external perturbations using the method of invariant ellipsoids", *Automatics and telemechanics*, 2007, No. 3, pp. 106–125.

[18] V.D. Romanenko, Yu.L. Milyavsky. " Control automation of impulse processes in cognitive maps with constrained disturbance suppression based on invariant ellipsoids method", *System Research and Informational Technologies*, 2017, no. 2, pp. 29–39. (In Russian)

[19] B.T. Polyak. "Convexity of quadratic transformations and its use in control and optimization", *Journal of Optimization Theory and Applications*, 1998, vol. 99, pp. 553–583.

[20] W.W. Hager. "Updating the inverse of a matrix", *SIAM Review*, 1989, vol. 31(2), pp. 221–239.

[21] V.D. Romanenko, Yu.L. Milyavsky. "Control automation method in cognitive maps based on the synthesis of increments of weighting coefficients and nodes coordinates", *System Research and Information Technologies*, 2019, no. 3, pp. 89–99. (In Russian)

[22] R. Isermann. Digital Control Systems: Springer-Verlag, Berlin, Heidelberg, 1981.

[23] K.J. Astrom, B. Wittenmark. Computer controlled systems, Prentice-Hall, Inc., Englewood Cliffs, 1984.

[24] L. Ljung, "Systems identification. Theory for the user", Prentice-Hall, 1986.

[25] D.W. Clarke, P.J. Gawthrop. "Self tuning control", *Proceedings of the IEE*, vol. 126, 1979, pp. 633–640.

[26] V. Gubarev. Modeling and Identification of Complex Systems, Kyiv, Naukova Dumka, 2019, 247p. (In Ukrainian)

Index

About the Editors

Yuriy P. Kondratenko, Doctor of Science (habil.), Professor, Honour Inventor of Ukraine (2008), Corr. Academician of Royal European Academy of Doctors – Barcelona 1914 (2000), Head of Intelligent Information Systems Department at Petro Mohyla Black Sea National University, Ukraine. Professor Kondratenko received a Ph.D. (1983) and a D.Sci. (1994) in Computer and Control Systems from Odessa National Polytechnic University. He has received several international grants and scholarships for conducting research at the Institute of Automation of Chongqing University, P.R. China (1988–1989), Ruhr-University Bochum, Germany (2000, 2010), Nazareth College and Cleveland State University, USA (2003). In 2015 he received a Fulbright grant for conducting research during 9 months at the Department of Electrical Engineering and Computer Science, Cleveland State University, USA.

He is an author of more than 700 publications, including 140 patents and 14 books published by Springer, River Publishers, etc. He is a member of the Scientific Committee of the National Council of Ukraine on Development of Science and Technology, National Committee of Ukrainian Association on Automatic Control, as well as visiting lecturer at the universities in Rochester, Cleveland, Kassel, Vladivostok and Warsaw. He is a member of Editorial Boards of such journals as Journal of Automation and Information Sciences, International Journal of Computing, Eastern European Journal of Enterprise Technologies, International Research and Review: Journal of Phi Beta Delta, Quantitative Methods in Economics and others. His research interests include intelligent decision support systems, automation, sensors and control systems, fuzzy logic, soft computing, modeling and simulation, and robotics.

Vsevolod M. Kuntsevich, Ph.D. (1959), D.Sci. (1965), Professor (1967), Academician of the National Academy of Sciences of Ukraine (1992), Honorary Director of Space Research Institute of NASU, Kyiv, Ukraine. Honored Figure of Science and Technology of Ukraine (1999), laureate of the State Prize of the Ukrainian SSR (1978, 1991) and Ukraine (2000) in the field of science and technology, S. Lebedev Award (1987), V. Glushkov Award

(1995), V. Mikhalevich Award (2003). Professor Kuntsevich graduated from Kyiv Polytechnic Institute (1952) and has worked at the Institute of Electrical Engineering (1958–1963), Institute of Cybernetics (1963–1996), and Space Research Institute (from 1996). He is part of the editorial staff of several journals including Problemy Upravleniya I Informatiki and Cybernetics and Systems Analysis. He is the founder of the national school in the field of discrete control systems and he made a significant contribution to the development of the modern theory of adaptive and robust control under uncertainty. He is the author of 8 books and over 250 articles and during 25 years was a Chairman of the National Committee of the Ukrainian Association on Automatic Control (1993-2018).

Arkadii A. Chikrii, Ph.D. (1972), D.Sci. (1979), Professor (1989), Academician of the National Academy of Sciences of Ukraine (2018), Head of the department "Optimization of Controlled Processes" of the Institute of Cybernetics, NAS of Ukraine. Laureate of the State Prize of Ukraine in the field of science and technology (1999), Glushkov Award (2003), Laureate of the State Prize of Ukraine in the field of education (2018), Mikhalevich Award (2021). After graduation from Ivan Franko Lviv University (1968), Professor Chikrii has been working at the Institute of Cybernetics, Professor of Taras Shevchenko Kyiv National University, Professor of Ihor Sikorski National Technical University, and Professor of Yurij Fedkovich Chernivtsi National University. He is the editor-in-chief of the journal Problemy Upravleniya I Informatiki (since 2020). He is a specialist in the field of applied nonlinear analysis, theory of extremal problems, mathematical theory of control, theory of dynamic games, theory of search for moving objects, and computer technologies for analysis of conflict situations. He is a disciple and follower of L.S. Pontryagin, N.N. Krasovskii and B.N. Pshenichnyi and is the author of 6 books and over 550 articles. He is a Chairman of the National Committee of the Ukrainian Association on Automatic Control (since 2019).

Vyacheslav F. Gubarev, Doctor of Science (1992), Professor, Corresponding Member of the National Academy of Science of Ukraine (2006), Professor of Mathematics Methods of System Analysis Department, Kyiv National Technical University, Ukraine, Head of Control Department, Space Research Institute, National Academy of Science, Ukraine. Professor Gubarev received a Ph.D. (1971) and a D.Sci. (1992) in System Analysis and Automatic Control from the Institute of Cybernetics of the National Academy of Science, Ukraine. He has taken part in several international grants with the Russia Academy of Science, Moscow University, and others. His research interests include mathematical modeling of complex systems, automatic control, estimation and identification, ill-posed mathematical problems, dynamic and control under uncertainty, and spacecraft control systems. He is part of the editorial staff of several journals including Problemy Upravleniya I Informatiki and Cybernetics and Systems Analysis. He is vicechairman of the Ukrainian Association of Automatic Control, which is an NMO of IFAC.